Lecture Notes in Mathematics

Edited by A. Dold and B. Eckmann

415

Ordinary and Partial Differential Equations

Proceedings of the Conference
held at Dundee, Scotland, 26–29 March, 1974

Edited by
B. D. Sleeman and I. M. Michael

Springer-Verlag
Berlin · Heidelberg · New York 1974

Dr. B. D. Sleeman
Dr. I. M. Michael
Dept. of Mathematics
University of Dundee
Dundee DD1 4HN/Scotland

Library of Congress Cataloging in Publication Data

Conference on Ordinary and Partial Differential Equations, Dundee, Scot., 1974.
Ordinary and partial differential equations.

(Lecture notes in mathematics, v. 415)
1. Differential equations--Congresses. 2. Differential equations, Partial--Congresses. I. Everitt, William Norrie, ed. II. Sleeman, B. D., ed. III. Michael, Ian MacRae, 1940- ed. IV. Title. V. Series: Lecture notes in mathematics (Berlin), v. 415
QA3.L28 no. 415 [QA371] 510'.8s [515'.35] 74-18467

AMS Subject Classifications (1970): 34A10, 34A40, 34B15, 34B20, 34B25, 34C10, 34C25, 34C35, 34D15, 34D99, 34J99, 35A05, 35A15, 35A20, 35A40, 35A99, 35B20, 35B35, 35B45, 35D10, 35G10, 35G15, 35J10, 35J60, 35K10, 35K55, 35L05, 35N10, 35P15, 45D05, 46E20, 47A10, 47A40, 47A55, 47E05, 47F05, 81A09, 81A10.

ISBN 3-540-06959-3 Springer-Verlag Berlin · Heidelberg · New York
ISBN 0-387-06959-3 Springer-Verlag New York · Heidelberg · Berlin

Offsetdruck: Julius Beltz, Hemsbach/Bergstr.

PREFACE

These Proceedings form a record of the lectures at the Conference on Ordinary and Partial Differential Equations held at the University of Dundee, Scotland during the four days 26 to 29 March 1974.

The Conference was attended by 150 mathematicians from the following countries: Belgium, Berlin, BRD (Germany), Canada, Czechoslovakia, DDR (Germany), Finland, Ireland, The Netherlands, Norway, South Africa, Sweden, Switzerland, the United Kingdom and the United States of America.

Both invited and contributed lectures formed an important part of the work of the Conference and a list of all these lectures will be found in the contents of these Proceedings.

The Conference was organised by the following Committee: W. N. Everitt (Chairman); B. D. Sleeman and I. M. Michael (Organising Secretaries).

I take this opportunity to record that at the 1972 Dundee Conference (see Lecture Notes in Mathematics (Springer-Verlag) Volume 280) the Organising Committee named as Honorary Presidents of the Conference

Professor Dr Wolfgang Haack (Berlin)

Professor Einar Hille (United States of America).

At the 1974 Dundee Conference the Organising Committee named as Honorary Presidents

Academician Otakar Borůvka (Czechoslovakia)

Professor Dr Ernst Mohr (Berlin).

On behalf of the Committee I thank all mathematicians who took part in the work of the Conference, especially those who travelled long distances to be in Dundee in March 1974. The Committee thanks: the University of Dundee for generously providing facilities which made it possible to hold the Conference in Dundee; the Warden and Staff of Belmont Hall for accommodation; and colleagues in the Department of Mathematics.

In particular the Committee thanks, as in 1972, Mrs Norah Thompson, Secretary in the Department of Mathematics, for her sustained efforts over several weeks which made possible the preparation of the papers for the Conference and several of the contributions of these Proceedings.

Finally I thank my senior colleagues Dr B. D. Sleeman and Dr I. M. Michael for their efforts and cheerfulness over several months which culminated in a successful and most enjoyable Conference.

W. N. Everitt

Dundee, Scotland

CONTENTS

Invited lectures

Contributed lectures

Lectures which do not appear here

J. Ball:	Continuity conditions for nonlinear semigroups
F. H. Cooper:	Existence of solutions to a class of variational inequalities
J. de Graaf:	Evolution equations with lateral boundary conditions
E. Hille:	Finiteness of the order of transcendental meromophic solutions of non-linear ordinary differential equations
H. Kalf:	The quantum mechanical Virial theorem and the absence of positive energy bound states of Schrödinger operators
B. Karlsson:	A theorem about the deficiency of pairs of differential expressions
V. Krishna Kumar:	A strong limit-2 classification for a fourth-order formally-symmetric differential expression
R. Martini:	On differential operators degenerating at the boundary
E. Mohr:	Über die Rayleighsche Vermutung betr den tiefsten Gründton einer platte von fester Fläche
F. Musa:	A class of matrices in variational methods
J. Persson:	A generalization of Caratheodory's existence theory for ordinary differential equations
A. J. B. Potter:	A generalized degree theory
C. G. Simader :	Some remarks on Schrödinger operators with singular potentials
R. A Smith:	Elliptic ball stability criterion
J. Walter:	Proof of Peano's existence theorem without using the notion of the definite integral
J. R. L. Webb:	On the Dirichlet problem for strongly nonlinear second order equations in arbitrary domains

LIST OF SPEAKERS

Invited Speakers

C. Andreian Cazacu:	Institut de Mathematique, Calea Grivitei 21, Bucarest 12, Romania.
J. V. Baxley:	Department of Mathematics, Wake Forest University, Winston-Salem, North Carolina 27109, USA.
N. W. Bazley:	Mathematical Institute, University of Cologne, 5 - Cologne/Lindenthal 41, Weyertal 86, West Germany.
J. W. Bebernes:	Department of Mathematics, University of Colorado, Boulder, Colorado 80302, USA.
H. E. Benzinger:	Department of Mathematics, University of Illinois, 273 Altgeld Hall, Urbana, Illinois 61801, USA.
O. Borůvka:	The Mathematical Institute, Janackovo nam 2a, 66295 Brno, Czechoslovakia.
B. L. J. Braaksma:	Mathematisch Instituut, Universiteit Groningen, Postbus 800, Groningen, Netherlands.
P. J. Browne:	Department of Mathematics, University of Calgary, Calgary, Alberta T2N 1N4, Canada.
L. Collatz:	Institut für Angewandte Mathematik, Universität Hamburg, 2, Hamburg 13, Roghenbaumchaussee 41, West Germany.
H. S. V. de Snoo	Mathematisch Instituut, Universiteit Groningen, Postbus 800, Groningen, The Netherlands.
A. Devinatz:	Department of Mathematics, Northwestern University, Evanston, Illinois 60201, USA.
M. S. P. Eastham:	Department of Mathematics, Chelsea College, Manresa Road, London SW3.

D. E. Edmunds: Mathematics Division, University of Sussex, Falmer, Brighton BN1 9QH, Sussex.

M.R. Essén: Department of Mathematics, Royal Institute of Technology, 100 44 Stockholm 70, Sweden.

M. Faierman: Department of Mathematics, Loyola College, 7141 Sherbrooke Street West, Montreal, Quebec, Canada.

I. Fenyő: H - 1125 Budapest, Istenhegyi UT 48/A , Hungary.

A. Friedman: Department of Mathematics, Northwestern University, Evanston, Illinois 60201, USA.

M. Giertz: Department of Mathematics, Royal Institute of Technology, 100 44 Stockholm 70, Sweden.

P. Habets: Institut de Mathematique Pure et Appliquée, Université Catholique de Louvain, B-1348 Louvain-La-Neuve, Belgium.

W. A. Harris: Department of Mathematics, University of Southern California, Los Angeles, USA.

P. Hess: Mathematics Institute, University of Zürich, Freiestrasse 36, 8032 Zürich, Switzerland.

E. Hille: Department of Mathematics, University of California, San Diego, La Jolla, California 92037, USA.

D. B. Hinton: Department of Mathematics, University of Tennessee, Knoxville, Tennessee 37916, USA.

I. W. Knowles: Department of Mathematics, University of the Witwatersrand, Johannesburg, South Africa.

K. Kreith: Department of Mathematics, University of California, Davis, California 95616, USA.

J. B. McLeod: The Mathematical Institute, 24-29 St Giles, Oxford.

E. Mohr: Technische Universität Berlin West, 1 Berlin 12, Strasse des 17 Juni 35, Germany.

H.-D. Niessen: Department of Mathematics, University of Essen, Gesamthochschule, 43 Essen, Unionstrasse 2, Germany.

Å. V. C. Pleijel: Mathematiska Institutionen, Sysslomansgatan 8, 75223 Uppsala, Sweden.

K. Schmitt: Mathematisches Institut I, Universität Karlsruhe, 75 Karlsruhe, West Germany.

G. R. Sell: School of Mathematics, University of Minnesota, Minneapolis, Minnesota 55455, USA.

I. Stakgold: Department of Mathematics, Northwestern University, Evanston, Illinois 60201, USA.

F. Stummel: Johann Wolfgang Goethe-Universität, Mathematisches Seminar, 6 Frankfurt-am-Main, Robert-Mayer-Strasse 10, West Germany.

J. Walter: Institute of Mathematics, Technical University, 51 Aachen, Germany.

J. Weidmann: Fachbereich Mathematik der Universität Frankfurt, 6 Frankfurt am Main, Robert Mayer-Strasse 10, Germany.

J. S. W. Wong: Department of Mathematics, University of Iowa, Iowa City, USA.

A. Zettl: Department of Mathematical Sciences, Northern Illinois University, Dekalb, Illinois 60115, USA.

Other Speakers (contributed lectures)

F. M. Arscott: Department of Applied Mathematics, University of Manitoba, Winnipeg, Manitoba, Canada.

J. M. Ball: Department of Mathematics, Heriot-Watt University, Riccarton, Midlothian.

M. F. Barnsley: Department of Mathematics, University of Bradford, Bradford, Yorkshire.

B. M. Brown: Department of Pure Mathematics, University College, Cardiff.

E. J. Brändas: Quantum Chemistry Group, University of Uppsala, S-751 20 Uppsala 1, Sweden.

J. Carr: The Mathematical Institute, 24-29 St Giles, Oxford.

P. J. Caudrey: Department of Mathematics, The University of Manchester, PO Box 88 Sackville Street, Manchester.

F. H. Cooper: Department of Mathematics, University of Glasgow, Glasgow.

J. de Graaf: Department of Mathematics, University of Groningen, Groningen, The Netherlands.

J. W. de Roever: Mathematisch Centrum, Amsterdam, The Netherlands.

J. Dyson: Brasenose College, Oxford.

W. D. Evans: Department of Pure Mathematics, University College, Cardiff.

W. N. Everitt: Department of Mathematics, The University, Dundee DD1 4HN.

D. L. Farnsworth: Division of Science and Mathematics, Eisenhower College, Sencca Falls, New York 13148, USA.

H. I. Freedman: School of Mathematics, University of Minnesota, 127 Vincent Hall, Minneapolis, Minnesota 55455, USA.

B. H. Gilding: School of Mathematical and Physical Sciences, University of Sussex, Falmer, Brighton.

W. S. Hall: Department of Mathematics, University of Pittsburgh, 615 Schenley, Pittsburgh, PA 15213, USA.

S. G. Halvorsen: Inst. Num. Mat. University of Trondheim, N.T.H., 7034 Trondheim, Norway.

M. Hehenberger: Quantum Chemistry Group, Uppsala University, Box 518, 751 20 Uppsala 1, Sweden.

G. Jank: Lehrakanzel und Institut für Mathematik, Technische Hochschule in Graz, A-8010 Graz, Kopernikusgasse 24, Austria.

H. Kalf: Institut of Mathematics, Technical University, 51 Aachen, Germany.

P. A. Källström: Department of Mathematics, University of Dundee, Dundee, Scotland.

B. A. Karlsson: Department of Mathematics, University of Uppsala, Sysslomansgatan 8, 752 23 Uppsala, Sweden.

V. Krishna Kumar: Department of Mathematics, University of Dundee Dundee, Scotland.

N. G. Lloyd: St John's College, Cambridge.

S. O. Londen: Institute of Mathematics, Helsinki University of Technology, Otaniemi, Finland.

R. Martini: Afdeling Wiskunde T.H. Delft, Julianalaan 132, Delft, The Netherlands.

F. A. Musa: Department of Computational Science, Liverpool University, Liverpool, England.

K. L. Nickel: Institut fur Praktische Mathematik, Universität, D(75) Karlsruhe, Germany.

L. A. Peletier: Mathematics Division, University of Sussex, Falmer Brighton.

J. Persson: The Auroral Observatory, University of Tromsø, Postboks 953, N-9001, Tromsø, Norway.

A. T. Plant: Fluid Mechanics Research Institute, University of Essex, Wivenhoe Park, Colchester.

A J B Potter: Mathematics Department, King's College, University of Aberdeen, Aberdeen, Scotland.

G. F. Roach: Department of Mathematics, University of Strathclyde, Glasgow, Scotland.

P. D. Robinson: School of Mathematics, Bradford University, Bradford, Yorks.

C. G. Simader: Mathematisches Institut der Universität Munchen, D 8 München 2, Theresienstr. 39, Germany.

B. D. Sleeman: Department of Mathematics, The University, Dundee, Scotland.

R. A. Smith: Department of Mathematics, University of Durham, Durham.

J. F. Toland: Fluid Mechanics Research Institute, University of Essex, Wivenhoe Park, Colchester.

U. Trottenberg: Mathematisches Institut der Universität Köln, Weyertal 86-90, D 50000-Köln 41, Germany.

P. Volkmann: Mathematisches Institut I, Der Universität (TH), 75 Karlsruhe 1, Germany.

L. J. Walpole: School of Mathematics and Plysics, University of East Anglia, Norwich.

J. R. L. Webb: Department of Mathematics, The University of Glasgow, Glasgow, Scotland.

A. D. Wood: Department of Mathematics, Cranfield Institute of Technology, Cranfield, Bedford.

Partial differential equations related to extremal problems for quasiconformal mappings

Cabiria Andreian Cazacu

Under a geometric configuration we mean a bordered or non bordered Riemann surface possibly with distinguished points. An important method in the conformal classification of the geometric configurations associates to them one or more conformal invariants called modules, their coincidence characterizing the conformal equivalence. We shall consider in this paper only modules expressed by Ahlfors' and Beurling's extremal length of some curve families, which -as it is well known - is a conformal invariant.

In the quasiconformal frame these modules are not invariant but their variation is bounded - as in Grötzsch's inequalities - so that extremal problems arise in order to determine in a family of quasiconformal mappings of the configuration those which extremize the modules. Starting from the classical extremal problems of H.Grötzsch, who considered the whole family of Q-quasiconformal mappings, O.Teichmüller and later L.I.Volkoviskii proposed to determine the extremal mapping under certain conditions imposed on M.A.Lavrentiev's characteristics of the quasiconformal mappings. The present paper represents a synthesis of several studies, which we published between 1964-1970, [2] - [9] and in which we gave a method to solve such problems, obtaining the system of partial differential equations verified by the extremal mapping: Our method is based upon a transformation formula under quasiconformal mappings which we established for the extremal length with weight in the Ohtsuka sense [3] . After some preliminaries about quasiconformal mappings and extremal length, we shall present the main results, their application, some examples and remarks among which we mention the generalization of the method to the n-dimensional case [9].

§I. Preliminaries

1. Quasiconformal mappings. There are many equivalent definitions for the notion of two-dimensional quasiconformal mapping which are systematically discussed in the fundamental monograph by O.Lehto and K.I.Virtanen [11]. We recall here only the following one:

A quasiconformal mapping $w=w(z)$ of a domain G in the z-plane onto the domain $G^*= w(G)$ in the w-plane is a homeomorphic generalized solution of a Beltrami equation

$$w_{\bar{z}} = \mu w_z \tag{1}$$

where μ is a measurable *) complex function, defined a.e. in G such that

$$\|\mu\| = \operatorname*{ess\,sup}_{z \in G} |\mu(z)| < 1. \tag{2}$$

This means that $w(z)$ is ACL in G and the derivatives $w_z, w_{\bar{z}}$ (that exist a.e) satisfy (1) a.e. in G.

One proves that w_z and $w_{\bar{z}}$ belong to L^2 in every compact subset of G and that w is differentiable with positive Jacobian $J(z) > 0$ a.e. in G.

We shall call a regular point of w each point $z \in G$ where w is differentiable and $J(z) > 0$. The affine mapping associated to w in a regular point z carries a family of similar ellipses with the center z called the characteristic ellipses of w in z in circles with the center $w(z)$. These ellipses are determined by the so-called M.A.Lavrentiev's characteristics: by $p(z)$ the ratio of the

*) We work with Lebesgue measure and integral and use the following usual abbreviations a.e. for almost everywhere, ACL for absolutely continuous on lines, ℓ. for locally (i.e. on every compact).

major to the minor axis (called also the quotient of dilatation), and, if $p(z) > 1$, by $\theta(z)$ the angle between the x-axis and the direction of the major axis, $\theta \in [0, \pi)$. In a regular point

$$\frac{w_{\bar{z}}}{w_z} = -\frac{p-1}{p+1} e^{i 2\theta} \tag{3}$$

the so called complex dilatation and it coincides with μ a.e. in G.

Our method applies also to the homeomorphic generalized solutions $w=w(z)$ of the equation

$$w_{\bar{z}} = \mu(z) w_z + \mu_1(z) \overline{w_z} \tag{4}$$

where μ and μ_1 are complex measurable functions defined a.e. in G and such that

$$\|\mu\| + \|\mu_1\| < 1.$$

Almost all points of G are regular points for w and in a regular point the affine transformation associated to w carries similar ellipses with the center z and the characteristics $p(z)$ and (if $p(z) > 1$) $\theta(z)$ in similar ellipses with the center w and the characteristics $p_1(z)$ and (if $p_1(z) > 1$) $\theta_1(z)$.

We have [13]

$$(5) \quad \mu = -\frac{p - p^{-1}}{p+p^{-1}+p_1+p_1^{-1}} e^{2\theta i} \text{ and } \mu_1 = \frac{p_1 - p_1^{-1}}{p+p^{-1}+p_1+p_1^{-1}} e^{2\theta_1 i}.$$

These so-called quasiconformal mappings with two pairs of characteristics in the sense of B.V.Šabat (and in particular G.N.Polojii's p-and (p,q)-analytic mappings)are also quasiconformal after a fundamental paper of B.Bojarskii [10].

The method applies also to the M.A.Lavrentiev quasiconformal mappings associated to systems of non-linear partial differential equations

$$F_k(x,y,u,v,u_x,u_y,v_x,v_y)=0, \qquad k=1,2,$$

with two unknown functions u(x,y) and v(x,y), [6].

A quasiconformal mapping f: $\mathcal{R} \to R$ between the Riemann surfaces $\mathcal{R}$ and R is a homeomorphism which can be written by means of local parameters z and w in corresponding points as a quasiconformal mapping w=w(z) defined by a Beltrami equation (1) and (2). Here $\mu(z)dz^{-1}d\bar{z}$ is a differential on $\mathcal{R}$, $|\mu(z)|$ a function defined a.e. on $\mathcal{R}$ and $\|\mu\| = \operatorname{ess\,sup}_{\mathcal{R}} |\mu| < 1$. The characteristic p is also a measurable function defined a.e. on $\mathcal{R}$ but θ depends on the local parameter, so that we shall choose on $\mathcal{R}$ a field of directions τ defined a.e.(given for instance by tangents to the curves $\mathcal{K}$ of a family $\{\mathcal{K}\}$ which covers simply and without gaps $\mathcal{R}$) and consider at the points z with $p(z) > 1$ the angle $\alpha(z)$ between the major axis of the characteristic ellipse at z and $\tau(z)$, $\alpha \in [0, \pi)$. Thus we obtain a measurable function α defined at almost every point of $\mathcal{R}$ at which $p > 1$. (If $\varphi(z)$ is the argument of the direction $\tau(z)$ at a regular point z of f(i.e. of w(z)) then $\theta(z) = \varphi(z) + \alpha(z) \pmod{\pi}$.)

Similarly one defines for Riemann surfaces a quasiconformal mapping $\mathcal{R} \to R$ with two pairs of characteristics [3].

Let us introduce now another function on $\mathcal{R}$ which will play a central role in our method, namely for each quasiconformal mapping f: $\mathcal{R} \to R$ and each curve family $\{\mathcal{K}\}$ on $\mathcal{R}$ we shall define at every regular point z at which $\tau(z)$ exists the function

$$d_{\mathcal{K},f}(z) = \left|\frac{dw(z)}{dz}\right|_{\mathcal{K}}^2 J^{-1}(z) \tag{5}$$

where $\left|\frac{dw}{dz}\right|_{\mathcal{K}}$ means the module of the derivative of w(z) on the direction $\tau(z)$ and J is its Jacobian at z. We have

$$d_{\mathcal{K},f}(z)=p^{-1}(z)\cos^2\alpha(z)+p(z)\sin^2\alpha(z) \qquad (6)$$

such that $p^{-1}\leq d\leq p$ a.e. on $\mathcal{R}$.*)

In the case of a quasiconformal mapping f with two pairs of characteristics p,α with respect to $\{\mathcal{K}\}$ on $\mathcal{R}$ and p_1, α_1 with respect to a curve family $\{K\}$ on R,

$$d_{\mathcal{K},f}(z)=\frac{p^{-1}(z)\cos^2\alpha(z)+p(z)\sin^2\alpha(z)}{p_1^{-1}(z)\cos^2\alpha_1(z)+p_1(z)\sin^2\alpha_1(z)} \qquad (7)$$

and similarly in the case of Lavrentiev 's mappings $d_{\mathcal{K},f}$ can be expressed by means of the corresponding characteristics, [6].

2. Extremal length of a curve family. Let $\{K\}$ be a curve family in a domain G of the Riemann surface R and π a measurable function defined a.e. in G with positive finite values. We call admissible with respect to $\{K\}$ each non-negative function ρ defined in G, measurable and such that $\int_K \rho\, ds \geq 1$ for every ℓ. rectifiable curve K. We denote by $\mathcal{A}(K)$ the class of all admissible functions and by $A_{\pi,\rho}(G)=\iint_G \pi\rho^2 dxdy$. Then the extremal length of $\{K\}$ with the weight π is given by

$$\ell_{\pi}(K)=\left[\inf_{\rho\in\mathcal{A}(K)} A_{\pi,\rho}(G)\right]^{-1} \qquad (8)$$

and $M_{\pi}(K)=\ell_{\pi}^{-1}(K)$ is the module of $\{K\}$ with the weight π. For $\pi=1$ one obtains the usual extremal length or module, which we shall denote by $\ell(K)$ and M(K) respectively.

It is obvious that if π is a constant then $\ell_{\pi}(K)=\pi^{-1}\ell(K)$ and if $0<k_0\leq\pi\leq k_1<+\infty$ a.e. on G, with k_0 and k_1 two constants, then

$$k_1^{-1}\,\ell(K)\leq \ell_{\pi}(K)\leq k_0^{-1}\,\ell(K). \qquad (9)$$

Further if π_1 and π_2 are two weights on G such that $\pi_1\geq\pi_2$ a.e.

*) There are many other problems in which it is advantageous to replace p by d, [8].

on G , then

$$\ell_{\pi_1}(K) \leq \ell_{\pi_2}(K). \qquad (10)$$

3. <u>Modular and quasimodular curve families.</u> We call modular a family of level lines of a harmonic function solution or conjugate of solution of a module problem as presented for instance in L.V.Ahlfors and L.Sario [1] or in L.Sario and M.Nakai [12]. Let us recall for instance the module problem for a compact bordered Riemann surface. We divide the boundary of R into disjoint classes of Jordan curves $\beta_0, \beta_1, \ldots \beta_N$. The module M of β_N with respect to β_0 and β_j (j=1,...,N-1) is given by the Dirichlet integral of the harmonic function U on R which is 0 on β_0, constant on each β_j and on β_N and whose conjugate function V has the periods $\int_{\beta_j} dV=0$ and $\int_{\beta_N} dV = 1$. One shows that the module M is equal to the extremal length of the family of arcs or of finite unions of arcs joining β_0 to β_N or β_0 to β_{ν_1}, β_{ν_1} to $\beta_{\nu_2}, \ldots, \beta_{\nu_s}$ to β_N on R, in particular M= $\ell(L_V)$ where $\{L_V\}$ is the family of level lines of V. Further M is equal to the module of the family of finite unions of Jordan curves, which separates β_0 and β_N on R and in particular M= $\ell^{-1}(C_U)$ where $\{C_U\}$ designates the level lines of U. The families $\{C_U\}$ and $\{L_V\}$ are examples of modular families. They decompose R in a finite number of "rectangles " on which U and V form an orthogonal coordinates system (the modular coordinates). Similarly one can consider the problem of the extremal distance [1] .

A <u>quasimodular curve family</u> will be each image of a modular family under a quasiconformal mapping.

After these preliminaries we can give the transformation formula for the extremal length :

<u>Let</u> f: $\mathcal{R} \longrightarrow R$ <u>be a quasiconformal mapping</u>, $\{\mathcal{K}\}$ <u>a quasimodular family of curves on</u> $\mathcal{R}$ <u>and</u> $\{K\}$ <u>its image on</u> R, $f(\mathcal{K})=K$, π

a weight on $\mathcal{R}$, $\pi \circ f^{-1}$ the corresponding weight on R, and $d=d_{\mathcal{K},f}$ the function (5). Then

$$\ell_{\pi\circ f^{-1}}(K) = \ell_{\pi d^{-1}}(\mathcal{K}). \tag{11}$$

We proved (11) in [3] by means of a representation formula for the extremal length with weight for quasimodular curve families.*) The following proposition is partially a consequence of (11) and (1o), the equality case being proved also in [3] :

If $f_j : \mathcal{R} \to R_j$, j=1,2, are two quasiconformal mappings such that for a quasimodular curve family $\{\mathcal{K}\}$ on $\mathcal{R}$ we have $d_{\mathcal{K},f_1} \leq d_{\mathcal{K},f_2}$ a.e. on $\mathcal{R}$, and if π is a weight on $\mathcal{R}$ such that $0 < k_o \leq \pi \leq k_1 < +\infty$ a.e. on $\mathcal{R}$ (k_o, k_1 - constants), then

$$\ell_{\pi\circ f_1^{-1}}(K_1) \leq \ell_{\pi\circ f_2^{-1}}(K_2), \tag{12}$$

where $\{K_j\} = \{f_j(\mathcal{K})\}$, and equality occurs in (12) iff $d_{\mathcal{K},f_1} = d_{\mathcal{K},f_2}$ a.e. on $\mathcal{R}$.

§II. Formulation of the extremal problems

We shall consider a geometric configuration: a compact bordered or non-bordered Riemann surface $\mathcal{R}$ possibly with distinguished points.**) We choose on $\mathcal{R}$ a quasimodular family of curves $\{K_{\mathcal{R}}\}$ and a weight $\pi_{\mathcal{R}}$ with values in an interval $[k_o(\mathcal{R}), k_1(\mathcal{R})]$ where $k_o(\mathcal{R})$ and $k_1(\mathcal{R})$ are constants as in (9).

---

*) Formula (11) remains true under more general conditions, [8], p. 7o-71.

**) In the case when there are such points the considered mappings carry them into the distinguished points of the image.

Further we shall consider a non-void family $\mathcal{F}_P$ of quasi-conformal mappings $f: \mathcal{R} \to R$ whose characteristics verify some properties P and distinguish two cases :

I. On each $R=f(\mathcal{R})$ for $f \in \mathcal{F}_P$ a quasimodular curve family $\{K_R\}$ and a weight π_R with values in $[k_0(R), k_1(R)]$ are given.

II. The family $\{K_R\} = \{f(K_{\mathcal{R}})\}$ and the weight $\pi_R = \pi_{\mathcal{R}} \circ f^{-1}$.

We suppose that the following condition a) is fulfilled :

a) If the mappings $f_j: \mathcal{R} \to R_j$, j=1,2, belong to $\mathcal{F}_P$ then

$$\ell_{\pi_{R_1}}(K_{R_1}) \leq \ell_{\pi_{R_2} \circ f_2 \circ f_1^{-1}} \left[f_1 \circ f_2^{-1}(K_{R_2}) \right] , \tag{13}$$

the equality sign in (13) implying

$$\{K_{R_1}\} = \{f_1 \circ f_2^{-1}(K_{R_2})\} \quad \text{and} \quad \pi_{R_1} = \pi_{R_2} \circ f_2 \circ f_1^{-1} \quad \text{a.e. on } R_1 . \tag{13'}$$

We shall write

$$\mathcal{M} = \ell_{\pi_{\mathcal{R}}}(K_{\mathcal{R}}) \text{ and } M = \ell_{\pi_R}(K_R) \tag{14}$$

and ask for the extremal mapping $f^* \in \mathcal{F}_P$ which makes M minimum or maximum in the family $\mathcal{F}_P$.

§ III. Main results.

Theorem 1. If $f^*: \mathcal{R} \to R$ makes M minimum and if for $\{\mathcal{K}^*\} = \{f^{*-1}(K_{R^*})\}$ there is a mapping $\tilde{f} \in \mathcal{F}_P$, $\tilde{f}: \mathcal{R} \to \tilde{R}$, such that

$$d_{\mathcal{K}^*,\tilde{f}} = \inf_{f \in \mathcal{F}_P} d_{\mathcal{K}^*,f} \qquad \text{a.e. on } \mathcal{R} \tag{15}$$

then

$$d_{\mathcal{K}^*,f^*} = \inf_{f \in \mathcal{F}_P} d_{\mathcal{K}^*,f} \qquad \text{a.e. on } \mathcal{R}. \tag{16}$$

Moreover $\tilde{f}$ makes M minimum too.

Proof. We can write

$$M^* := \ell_{\pi_{R^*}}(K_{R^*}) \underset{1)}{\leq} \tilde{M} := \ell_{\pi_{\tilde{R}}}(K_{\tilde{R}}) \underset{2)}{\leq} \ell_{\pi_{R^*} \circ f^* \circ \tilde{f}^{-1}}(\tilde{f}(\mathcal{K}^*))$$

$$\leq \ell_{\pi_{R^*}} (f^*(\mathcal{K}^*)) = \ell_{\pi_{R^*}} (K_{R^*}) = M^*$$

3)

Indeed we have 1) since M^* is $\min_{\mathcal{F}_p} M$, 2) after condition a) and 3) after (12) since $d_{\mathcal{K}^*,\tilde{f}} \leq d_{\mathcal{K}^*,f^*}$ a.e. on $\mathcal{R}$. Therefore equality holds in 3), which implies after the equality case in (12) that $d_{\mathcal{K}^*,\tilde{f}} = d_{\mathcal{K}^*,f^*}$ a.e. on $\mathcal{R}$, i.e. that f satisfies (16). One sees that $M^* = \tilde{M}$, i.e. $\tilde{f}$ minimizes M too and from a) one deduces that $\{\mathcal{K}^*\} = \{\tilde{f}^{-1}(K_{\tilde{R}})\}$.

Remark. Both existence hypotheses : the existence of the minimizing mapping f^* and that of $\tilde{f}$ are essential. For instance let $\mathcal{R}$ and R be rectangles and let $\mathcal{F}_p$ consist in only two mappings f_1: $\mathcal{R} \to R$ corresponding to a convenient p non-symmetric with respect to the center of $\mathcal{R}$ and $f_2 = f_1 \circ \varphi$, where φ is the symmetry of R relative to its center. Both mappings realize the minimum of M in $\mathcal{F}_p$ but none fulfills (15) or (16). *Example:* $\mathcal{R} = \mathcal{R}_1 \cup \mathcal{R}_2$, $\mathcal{R}_1 = \{z \mid 0 \leq x \leq 1, 0 \leq y \leq 1\}$ $\mathcal{R}_2 = \{z \mid 1 \leq x \leq 2, 0 \leq y \leq 1\}$, f_1: $w = px + iy$ in $\mathcal{R}_1$, $w = z + p - 1$ in $\mathcal{R}_2$, $p > 1$ a constant.

Theorem 1 admits a complement in the case II, because then $\{\mathcal{K}^*\} = \{K_{\mathcal{R}}\}$ (since $\{K_{\mathcal{R}}\} = \{f^{-1}(K_R)\}$) for each $f \in \mathcal{F}_p$:

Theorem 1' for the case II. *If there exist a mapping $\tilde{f} \in \mathcal{F}_p$ verifying*

$$d_{K_{\mathcal{R}},\tilde{f}} = \inf_{f \in \mathcal{F}_p} d_{K_{\mathcal{R}},f} \quad \text{a.e. on } \mathcal{R} \tag{15'}$$

then 1° this mapping minimizes M and 2° each mapping which minimizes M satisfies (15').

Theorem 2. *Suppose that there exist a mapping*

f^*: $\mathcal{R} \to R^*$, $f^* \in \mathcal{F}_p$, such that for $\{\mathcal{K}^*\} = \{f^{*-1}(K_{R^*})\}$ we have

$$d_{\mathcal{K}^*,f^*} = \sup_{f \in \mathcal{F}_p} d_{\mathcal{K}^*,f} \quad \text{a.e on } \mathcal{R}. \tag{17}$$

Then 1° f^* <u>makes M maximum</u> and 2° $\hat{f} \in \mathcal{F}_p$, $\hat{f}: \mathcal{R} \to \hat{R}$, <u>maximizes</u> M <u>iff</u>

$$d_{\hat{\mathcal{K}},\hat{f}} = \sup_{f \in \mathcal{F}_p} d_{\hat{\mathcal{K}},\hat{f}} \quad \text{a.e. on } \mathcal{R} \tag{18}$$

where $\{\hat{\mathcal{K}}\} = \{\hat{f}^{-1}(K_{\hat{R}})\}$.

<u>Proof</u>. 1°. First, we designate by $\hat{f}$ an arbitrary mapping of $\mathcal{F}_p$. Then we can write

$$\hat{M} := \ell_{\pi_{\hat{R}}}(K_{\hat{R}}) \underset{1)}{\leq} \ell_{\pi_{R^*}^{\circ}} f^* \circ \hat{f}^{-1} \quad (\hat{f}\,(\mathcal{K}^*))$$

$$\underset{2)}{\leq} \ell_{\pi_{R^*}} (f^*(\mathcal{K}^*)) = : M^*,$$

where 1) holds after a) and 2) after (12) and after (17) that means $d_{\mathcal{K}^*,\hat{f}} \leq d_{\mathcal{K}^*,f^*}$ a.e. on $\mathcal{R}$.

2°. If $\hat{M} = M^*$ then equality in 2) implies as before $d_{\mathcal{K}^*,\hat{f}} = d_{\mathcal{K}^*,f^*}$ a.e. on $\mathcal{R}$ and equality in 1) implies $\{\mathcal{K}^*\} = \{\hat{\mathcal{K}}\}$ from where (18) follows immediately. The rest of the second assertion results from 1°.

<u>Consequence</u>(<u>Unicity theorem</u>) <u>Under the hypothesis of Theorem 2</u>, if $d_{\mathcal{K}^*,f}$ <u>determines the characteristics of</u> $f \in \mathcal{F}_p$ <u>with respect</u> to $\{\mathcal{K}^*\}$ <u>and</u> $\{f(\mathcal{K}^*)\}$, <u>then the extremal mapping which maximizes</u> M <u>is unique up to a composition with a conformal mapping.</u>

<u>Proof.</u> 1°. If for a quasiconformal mapping f one knows the characteristics p, α with respect to a quasimodular family $\{\mathcal{K}^*\}$ and p_1, α_1 with respect to $\{f(\mathcal{K}^*)\}$ one obtains by an elementary device, [3] 3.2. 2°, at each regular point of f where $\mathcal{K}^*$ has the tangent direction τ, the characteristics $\tilde{p}, \tilde{\alpha}$ with respect to $\{\mathcal{K}^*\}$ of the ellipses transformed in circles by the affine mapping associated to f or, if necessary, the characteristics

$\tilde{p}_1, \tilde{\alpha}_1$ with respect to $\{f(\mathcal{K}^*)\}$ corresponding to the ellipses which are images of circles by the same affine mapping.

2°. From 1° it follows that f^* and $\hat{f}$ verify the same Beltrami equation, hence $f^* \circ \hat{f}^{-1}$ is conformal.

In a similar manner we proved other results on the extremal mappings, [5], from which we shall indicate here only the following one :

Theorem 3. 1° Suppose that $\mathcal{F}_P$ verifies instead of the condition a) the condition

$\bar{a}$) For each $f \in \mathcal{F}_P$, $f: \mathcal{R} \to R$,

$$\ell_{\pi_{\mathcal{R}}}(K_{\mathcal{R}}) \leq \ell_{\pi_{\mathcal{R}} \circ f^{-1}}[f(K_{\mathcal{R}})].$$

If there exists a mapping $\tilde{f} \in \mathcal{F}_P$, $\tilde{f}: \mathcal{R} \to \tilde{R}$, such that

$$d_{K_{\mathcal{R}}, \tilde{f}} = \sup_{f \in \mathcal{F}_P} d_{K_{\mathcal{R}}, f}$$

then
$$M \leq \ell_{\pi_{\mathcal{R}} \circ \tilde{f}^{-1}}[\tilde{f}(K_{\mathcal{R}})] .$$

Further if $\bar{D} = \operatorname{ess\,sup}_{z \in \mathcal{R}} d_{K_{\mathcal{R}}, \tilde{f}}(z)$ then $M \leq \bar{D}\mathcal{M}$.

§IV. Applications. Examples. Remarks .

1. In conclusion Theorems 1 and 2 show that the problem to extremize M reduces to an analogous problem for $d_{\mathcal{K}^*, f}$. If one can determine from the properties P imposed on the characteristics of the mappings $f \in \mathcal{F}_P$ and from (17) for instance, the characteristics of f^* : p, α with respect to $\{\mathcal{K}^*\}$ and p_1, α_1 with respect to $\{K_{R^*}\}$ as functions of the point on R , applying the device indicated in the proof of Consequence, nr.1°, one obtains the characteristics $\tilde{p}, \tilde{\alpha}$ with respect to $\{\mathcal{K}^*\}$ and 1, or respectively 1 and $\tilde{p}_1$, $\tilde{\alpha}_1$ with respect to $\{K_{R^*}\}$.

In the case I the family $\{K_{R^*}\}$ is known, so that we have the characteristics 1; $\tilde{p}_1$, $\tilde{\Theta}_1$ and from (4) an elliptic equation for f^* of the type

$$w_{\bar{z}} = \mu_1 \overline{w_z} \quad . \qquad (19)$$

Generally this equation is quasi-linear, but if $\{K_{R^*}\}$ is a modular family and if we work with the corresponding modular coordinates, it becomes linear.Its solution f^*(if it exists)makes M maximum and each maximizing function satifies such an equation. In the case II the family $\{K_{\mathcal{R}}\}$ is known and we shall use the characteristics $\tilde{p}$, $\tilde{\theta}$: 1,obtaining for f^* a Beltrami equation (3).

2.We shall remark also that theorems 1 and 2 permit to obtain more complete results in the case I for "modular problems " when it is asked to extremize M under the hypothesis that on every R there exist two orthogonal families $\{L_R\}$ and $\{C_R\}$ such that

$$M = \ell(L_R) = \ell^{-1}(C_R). \qquad (20)$$

Indeed in order to obtain the maximum (the minimum) of M we may apply theorem 2 for $\{K_R\} = \{L_R\}$ ($= \{C_R\}$) and theorem 1 for $\{K_R\} = \{C_R\}$ ($= \{L_R\}$ respectively).

We emphasize that in this case conditions a) or $\bar{a}$) are always fulfilled from the extremal character of $\{L_R\}$ and $\{C_R\}$with res - pect to the extremal length (as in I, 3. above).

3. In [4] and [7] we considered different examples of extremal problems which can be studied with our method.We shall outline only one of them [4] :

Let p be a measurable function, defined a.e. on $\mathcal{R}$ with values in an interval [1, Q] and let $\mathcal{F}_p$ be the family of quasiconformal mappings of $\mathcal{R}$ having the quotient of dilatation equal to p a.e. on $\mathcal{R}$.

One sees immediately that if $\{\mathcal{K}^*\}$ is a quasi-modular family on $\mathcal{R}$ and if p and α designate the characteristics of a mapping $f \in \mathcal{F}_p$ with respect to $\{\mathcal{K}^*\}$ then

$$\inf_{f \in \mathcal{F}_p} d_{\mathcal{K}^*, f} \text{ is realized by } \alpha = 0$$

and

$$\sup_{f \in \mathcal{F}_p} d_{\mathcal{K}^*, f} \quad \text{by } \alpha = \frac{\pi}{2} .$$

Let us suppose that we deal with a modular problem (20) Applying theorems 1 and 2 to both families $\{C_R\}$ and $\{L_R\}$ we deduce that <u>a necessary and sufficient condition in order that</u> f^* <u>minimizes (maximizes)</u> M <u>consists in the fact that</u> f^* <u>verifies in the generalized sense the elliptic linear system</u>

$$u_x = p^{-1} v_y, \; u_y = -p^{-1} v_x \; (\text{or } u_x = p v_y \, , \; u_y = -p v_x). \tag{21}$$

4. The method reduces the considered type of extremal problems to the solution of different elliptic systems of equations (1),(4), (19),(21) etc. on Riemann surfaces, a research field which is actual.

5. The method may be extended directly to the n-dimensional case, since its fundaments (the transformation formula for extremal length (module) as well as the representation formula) remain true for q-dimensional surface families (q=1,..,n-1 ; q=1 corresponding to curve families) [9].

6. In connection with the type of extremal problems discussed in this paper we mention also the important results of R.Kühnau (see References nr [11] - [15] given in [9]).

REFERENCES

1. L.V.Ahlfors and L.Sario, Riemann surfaces, Princeton Univ.Press,1970
2. C.Andreian Cazacu ,Sur l'application de la longueur extrémale. . . Mathematica (Cluj),1964,6 (29),1, 5-10.
3. _ , Sur un problème de L.J.Volkovyski, Revue Roum.Math.Pures Appl.,1965,10, 1,43-63.
4. _ , Problèmes extrémaux des représentations quasi-conformes Ibidem,1965,10,4,409-429.
5. _ , Une propriété caractéristique des représentations quasi-conformes extrémales, Ibidem, 1967, 12, 2,167-179.
6. _ , Sur les applications quasi-conformes de M.A.Lavréntieff Ibidem, 1968, 13,9, 1217-1223.
7. _ , Sur un exemple de L.I.Volkovyski, Bul.Inst.Polit.Iaşi,I 1970, 16 (20),fasc.1-2,21-30.
8. _ , Influence of the orientation of the characteristic ellipses . . . . .Proceedings of the Romanian-Finnish Seminar on Teichmüller spaces . . .Publ.House Academy SRR, 1971, 65-85.
9. _ , Some formulae on the extremal length in n-dimensional case,Ibidem, 87-102.
10. Б. В. Боярский, Обобщение решения система дифференциальных уравнений первого порядка эллиптического типа с разрывными коэффициентами, Мат. сб., 1957, 43 (85), 4, 451-503.
11. O.Lehto and K.I.Virtanen , Quasiconformal Mappings in the Plane,II. ed.,Springer-Verlag,Berlin,Heidelberg,New York,1973.
12. L.Sario and M.Nakai, Classification Theory of Riemann Surfaces, Springer-Verlag,Berlin,Heidelberg,New York,1970.
13. I.N.Vekua,Generalized Analytic Functions, Reading (Mass),1964.

SINGULARLY PERTURBED INITIAL VALUE PROBLEMS

John V. Baxley

1. Introduction. We are concerned here with the singularly perturbed initial value problem

$$\varepsilon y'' + f(x,y,y',\varepsilon) = 0, \quad 0 \le x \le b < \infty \tag{1.1}$$

$$y(0) = \alpha(\varepsilon), \quad y'(0) = \beta(\varepsilon) \tag{1.2}$$

where $\varepsilon > 0$ is a small parameter and "prime" denotes differentiation with respect to x. We shall formulate conditions under which the problem (1.1), (1.2) has a unique solution $y(x,\varepsilon)$ existing on the entire interval $[0,b]$, for ε sufficiently small. We shall also obtain explicit bounds on $y(x,\varepsilon)$ and $y'(x,\varepsilon)$ as $\varepsilon \to 0$. In particular, we find conditions under which $y(x,\varepsilon)$ is uniformly bounded as $\varepsilon \to 0$. The conditions we place on $f(x,y,z,\varepsilon)$ are quite different from the standard uniform Lipschitz condition; other than smoothness, our essential requirements are that $f_3(x,y,z,\varepsilon) \equiv \frac{\partial f}{\partial z}$ be strictly positive and bounded away from zero and that $f_2(x,y,z,\varepsilon) \equiv \frac{\partial f}{\partial y}$ be bounded above <u>only for $z = 0$</u>. The precise conditions are stated below. We shall indicate a couple of applications of our results at the end. More serious applications will hopefully appear elsewhere. The spirit of this work is very close to that of [1] and [2].

2. Main Results. In [7, Chapter 1], Protter and Weinberger use the maximum principle both to establish uniqueness of solutions to initial value problems and to derive bounds for these solutions. In order to apply the maximum principle directly to (1.1), (1.2), we would need to assume that $f_2(x,y,y',\varepsilon) \le 0$. However, Protter and Weinberger show that it is often possible, even if $f_2(x,y,y',\varepsilon) \le 0$ is violated, to make a change of dependent variable of the form $y = wu$, where u is the new variable and w is a certain fixed positive function of x, after which the maximum principle is applicable. Exponential functions are often used by Protter and Weinberger for w and it seems natural in the present situation to try $w = \exp(-p(\varepsilon)x)$, where $p(\varepsilon)$ is a positive function of ε to be chosen later. Familiarity with singular

perturbation problems suggests the choice $p(\varepsilon) = q/\varepsilon$, where $q > 0$, because such a choice corresponds to the usual "stretching transformation" for examining the boundary layer region. However, even though such a choice is often appropriate, we prefer to postpone this decision.

Without further ado, we put $y = wu$, where $w = \exp(-p(\varepsilon)x)$ and $p(\varepsilon) > 0$. Substitution leads to the identity

$$\varepsilon y'' + f(x,y,y',\varepsilon) = \varepsilon w[u'' + H(x,u,u',\varepsilon)] \tag{2.1}$$

where

$$H(x,u,u',\varepsilon) = \varepsilon^{-1}[\varepsilon p^2(\varepsilon)u - 2\varepsilon p(\varepsilon)u' + w^{-1}f(x,wu,(wu)',\varepsilon)]. \tag{2.2}$$

Thus y is a solution of (1.1), (1.2) if and only if u is a solution of

$$u'' + H(x,u,u',\varepsilon) = 0 \tag{2.3}$$

$$u(0) = \alpha(\varepsilon), \quad u'(0) = \beta(\varepsilon) + p(\varepsilon)\,\alpha(\varepsilon)\ . \tag{2.4}$$

Our goal is to use the maximum principle to construct upper and lower bounds of u (and hence of y) and thus to prove global existence of y on the interval $[0,b]$. Although other lines of attack are possible, in this note we will seek to bound y by constant functions. We begin with the assumption

H-1: $f(x,y,z,\varepsilon)$, $f_2(x,y,z,\varepsilon)$, and $f_3(x,y,z,\varepsilon)$ are continuous throughout the region

$$R = \{(x,y,z):\ 0 \le x \le b,\quad y^2 + z^2 < \infty\},$$

It follows from the standard theorems [3] that the problem (1.1), (1.2) has unique local solution.

Put $v = c(\varepsilon)\exp(p(\varepsilon)x)$. Then from (2.1),

$$v'' + H(x,v,v',\varepsilon) = (\varepsilon w)^{-1} f(x,c(\varepsilon),0,\varepsilon). \tag{2.5}$$

We now assume that for each ε sufficiently small

H-2: there exist $c_1(\varepsilon)$, $c_2(\varepsilon)$ such that

$$f(x,c(\varepsilon),0,\varepsilon) \ge 0 \quad \text{for } c(\varepsilon) \ge c_1(\varepsilon),\quad 0 \le x \le b,$$

$$\text{and}\quad f(x,c(\varepsilon),0,\varepsilon) \le 0 \quad \text{for } c(\varepsilon) \le c_2(\varepsilon),\quad 0 \le x \le b.$$

It follows then from (2.5) that

$$v'' + H(x,v,v',\varepsilon) \ge 0,\quad 0 \le x \le b,$$

if $c(\varepsilon) \ge c_1(\varepsilon)$ and

$$v'' + H(x,v,v',\varepsilon) \leq 0, \quad 0 \leq x \leq b$$

if $c(\varepsilon) \leq c_2(\varepsilon)$.

We now seek conditions on f under which the maximum principle may be applied to bound u by functions of the form v in (2.5). Toward this end, let

$$C_1(\varepsilon) = \max\{c_1(\varepsilon), \alpha(\varepsilon), \alpha(\varepsilon) + \beta(\varepsilon)/p(\varepsilon)\} \tag{2.6}$$

$$C_2(\varepsilon) = \min\{c_2(\varepsilon), \alpha(\varepsilon), \alpha(\varepsilon) + \beta(\varepsilon)/p(\varepsilon)\}, \tag{2.7}$$

and further, set

$$v_1 = v_1(x,\varepsilon) = C_1(\varepsilon) \exp(p(\varepsilon)x)$$

$$v_2 = v_2(x,\varepsilon) = C_2(\varepsilon) \exp(p(\varepsilon)x) .$$

Let $I \subset [0,b]$ be the largest subinterval of $[0,b]$ on which the local solution $u(x,\varepsilon)$ of (2.3), (2.4) exists. We now wish to show that

$$v_2(x,\varepsilon) \leq u(x,\varepsilon) \leq v_1(x,\varepsilon)$$

$$v_2'(x,\varepsilon) \leq u'(x,\varepsilon) \leq v_1'(x,\varepsilon),$$

for all $x \in I$.

First, from (2.6), (2.7), we deduce

$$v_2(0,\varepsilon) = C_2(\varepsilon) \leq \alpha(\varepsilon) = u(0,\varepsilon) \leq C_1(\varepsilon) = v_1(0,\varepsilon) \tag{2.8}$$

and

$$v_2'(0,\varepsilon) = p(\varepsilon)\, C_2(\varepsilon) \leq \beta(\varepsilon) + p(\varepsilon)\, \alpha(\varepsilon) = u'(0,\varepsilon) \leq p(\varepsilon)\, C_1(\varepsilon) = v_1'(0,\varepsilon). \tag{2.9}$$

Second, since $v_1'' + H(x,v_1,v_1',\varepsilon) \geq 0$ and $u'' + H(x,u,u',\varepsilon) = 0$, subtraction gives

$$0 \leq (v_1-u)'' + \varepsilon^{-1}[\varepsilon p^2(\varepsilon)(v_1-u) - 2\varepsilon p(\varepsilon)(v_1-u)' + w^{-1}(f(x,C_1(\varepsilon),0,\varepsilon) - f(x,wu,(wu)',\varepsilon))]$$

or, using the one-dimensional mean value theorem on the third variable of f,

$$\begin{aligned}(v_1-u)'' + \varepsilon^{-1}[&(\varepsilon p^2(\varepsilon) - p(\varepsilon) f_3(x,wu,\hat{z},\varepsilon))(v_1-u) \\ &+ (f_3(x,wu,\hat{z},\varepsilon) - 2\varepsilon p(\varepsilon))(v_1-u)' \\ &+ w^{-1}(f(x,wv_1,0,\varepsilon) - f(x,wu,0,\varepsilon))] \geq 0,\end{aligned}$$

where $\hat{z} = \hat{z}(x,\varepsilon)$ is between 0 and $(wu)'(x,\varepsilon)$. Now using the mean value theorem again, this time on the second variable of f, we get

$$(v_1-u)'' + \varepsilon^{-1}[f_3(x,wu,\hat{z},\varepsilon) - 2\varepsilon p(\varepsilon)](v_1-u)'$$

$$+ \varepsilon^{-1}[\varepsilon p^2(\varepsilon) - p(\varepsilon)f_3(x,wu,\hat{z},\varepsilon) + f_2(x,\hat{y},0,\varepsilon)](v_1-u) \geq 0 \qquad (2.10)$$

where $\hat{y} = \hat{y}(x,\varepsilon)$ is between $(wu)(x,\varepsilon)$ and $(wv_1)(x,\varepsilon) = C_1(\varepsilon)$.

In order to apply the maximum principle, we need the coefficient of v_1-u in this last inequality to be nonpositive. Thus, we now assume

H-3: $f_3(x,y,z,\varepsilon) \geq a(\varepsilon) > 0$ throughout the region R and
$f_2(x,y,0,\varepsilon) \leq B(\varepsilon) < \infty$ throughout the region $R_0 = \{(x,y): 0 \leq x \leq b, \ |y| < \infty\}$.

It then follows that

$$\varepsilon p^2(\varepsilon) - p(\varepsilon)\, f_3(x,wu,\hat{z},\varepsilon) + f_2(x,\hat{y},0,\varepsilon)$$

$$\leq \varepsilon p^2(\varepsilon) - p(\varepsilon)\, a(\varepsilon) + B(\varepsilon), \qquad (2.11)$$

and we wish to make this last expression non-positive by appropriate choice of $p(\varepsilon) > 0$. We consider here three choices:

1) If $f_2(x,y,0,\varepsilon) \leq 0$, we may take $B(\varepsilon) \equiv 0$. Then choosing $p(\varepsilon) \equiv 1$, we need $\varepsilon - a(\varepsilon) = a(\varepsilon)\left[\frac{\varepsilon}{a(\varepsilon)} - 1\right] \leq 0$, which certainly happens if $\frac{\varepsilon}{a(\varepsilon)} \to 0$ as $\varepsilon \to 0$.

2) Recalling the remark made earlier about "stretching transformations" in singular perturbation problems, we consider the choice $p(\varepsilon) = \varepsilon^{-1} q(\varepsilon)$, with $q(\varepsilon) > 0$. We need

$$\varepsilon p^2(\varepsilon) - p(\varepsilon)\, a(\varepsilon) + B(\varepsilon) = \varepsilon^{-1} q(\varepsilon)[q(\varepsilon) - a(\varepsilon)] + B(\varepsilon) \leq 0.$$

Assuming $B(\varepsilon) > 0$ in general, we need $q(\varepsilon) < a(\varepsilon)$. Picking $q(\varepsilon) = \frac{1}{2} a(\varepsilon)$, we get

$$\varepsilon p^2(\varepsilon) - p(\varepsilon)\, a(\varepsilon) + B(\varepsilon) = \frac{a^2(\varepsilon)}{4\varepsilon}\left[\frac{4\varepsilon B(\varepsilon)}{a^2(\varepsilon)} - 1\right],$$

which is certainly negative if $\frac{\varepsilon B(\varepsilon)}{a^2(\varepsilon)} \to 0$ as $\varepsilon \to 0$.

3) Another choice which sometimes offers advantages is $p(\varepsilon) = \frac{2B(\varepsilon)}{a(\varepsilon)}$; this choice in many cases may give a bounded $p(\varepsilon)$ as $\varepsilon \to 0$. In this case, we have

$$\varepsilon p^2(\varepsilon) - p(\varepsilon)\, a(\varepsilon) + B(\varepsilon) = B(\varepsilon)\left[\frac{4\varepsilon B(\varepsilon)}{a^2(\varepsilon)} - 1\right],$$

which, as before, is negative if $\frac{\varepsilon B(\varepsilon)}{a^2(\varepsilon)} \to 0$ as $\varepsilon \to 0$.

With these possibilities for $p(\varepsilon)$ well in mind, (2.8) - (2.11) permit us to apply the maximum principle together with the standard theory [3, Chapter 1] on the continuation of solutions to obtain the following result.

Theorem 1. Given the initial value problem (1.1), (1.2), suppose that f satisfies the hypotheses H-1, H-2, and H-3. If, in addition, there exists $p(\varepsilon) > 0$ such that

$$\varepsilon p^2(\varepsilon) - p(\varepsilon)\, a(\varepsilon) + B(\varepsilon) \leq 0$$

for all ε sufficiently small, then the local solution $y(x,\varepsilon)$ of (1.1), (1.2) exists globally on [0,b] (for small ε) and satisfies

$$C_2(\varepsilon) \leq y(x,\varepsilon) \leq C_1(\varepsilon)$$

$$-p(\varepsilon)(C_1(\varepsilon) - C_2(\varepsilon)) \leq y'(x,\varepsilon) \leq p(\varepsilon)(C_1(\varepsilon) - C_2(\varepsilon)).$$

for all $x \in [0,b]$.

The choice $p(\varepsilon) = \frac{a(\varepsilon)}{2\varepsilon}$ leads to

Corollary 1. In addition to H-1, H-2, and H-3, suppose that $c_1(\varepsilon)$, $c_2(\varepsilon)$ of H-2, $\alpha(\varepsilon)$, and $\varepsilon\beta(\varepsilon)/a(\varepsilon)$ are all bounded as $\varepsilon \to 0$ and that $\varepsilon B(\varepsilon)/a^2(\varepsilon) \to 0$ as $\varepsilon \to 0$. Then the local solution $y(x,\varepsilon)$ of (1.1), (1.2) exists globally (for small ε) on [0,b], is uniformly bounded on [0,b] as $\varepsilon \to 0$, and further the derivative $y'(x,\varepsilon)$ is $O(\varepsilon^{-1} a(\varepsilon))$ as $\varepsilon \to 0$.

It is interesting to note that in the previous corollary, the initial value $\beta(\varepsilon)$ of $y'(x,\varepsilon)$ is allowed to be singular as $\varepsilon \to 0$, and this singularity in $\beta(\varepsilon)$ reappears in the conclusion of the theorem in the behavior of $y'(x,\varepsilon)$. It is reasonable to guess that if in fact $\beta(\varepsilon)$ is $O(1)$ as $\varepsilon \to 0$, then $y'(x,\varepsilon)$ should be uniformly bounded as $\varepsilon \to 0$. This strengthened conclusion can indeed be obtained with the choice $p(\varepsilon) = \frac{2B(\varepsilon)}{a(\varepsilon)}$.

Corollary 2. In addition to H-1, H-2, and H-3, suppose that $c_1(\varepsilon)$, $c_2(\varepsilon)$ of H-2 $\alpha(\varepsilon)$, $\beta(\varepsilon)$, and $\frac{B(\varepsilon)}{a(\varepsilon)}$ are all bounded as $\varepsilon \to 0$ and that $\frac{\varepsilon B(\varepsilon)}{a^2(\varepsilon)} \to 0$ as $\varepsilon \to 0$. Then the local solution $y(x,\varepsilon)$ of (1.1), (1.2) exists globally (for small ε) on [0,b] and moreover both $y(x,\varepsilon)$ and $y'(x,\varepsilon)$ are uniformly bounded on [0,b] as $\varepsilon \to 0$.

We remark that the hypothesis H-2 certainly is satisfied if in fact $f_2(x,y,0,\varepsilon) \geq b > 0$ throughout R_0, although this condition is much stronger than necessary.

3. Applications. In certain cases, e.g. in [4], one wishes to make hypotheses on $f(x,y,z,\varepsilon)$ only in the region $R' = \{(x,y,z): 0 \le x \le b,\ y \ge 0,\ z \ge 0$. In this case we obtain the following theorem.

Theorem 2. Suppose that

i) $f(x,y,z,\varepsilon)$, $f_2(x,y,z,\varepsilon)$ and $f_3(x,y,z,\varepsilon)$ are continuous throughout the region R' above,

ii) there exists $c_1(\varepsilon)$ such that $f(x,c(\varepsilon),0,\varepsilon) \ge 0$ for $c(\varepsilon) \ge c_1(\varepsilon)$, $0 \le x \le b$.

iii) $f_3(x,y,z,\varepsilon) \ge a(\varepsilon) > 0$ throughout the region R' and $f_2(x,y,0,\varepsilon) \le 0$ throughout the region $R_0' = \{(x,y): 0 \le x \le b,\ y \ge 0\}$.

Then, if $p(\varepsilon) > 0$ satisfies $\varepsilon p(\varepsilon) - a(\varepsilon) \le 0$ for all ε sufficiently small, the local solution $y(x,\varepsilon)$ of the initial value problem

$$\varepsilon y'' + f(x,y,y',\varepsilon) = 0 \tag{3.1}$$

$$y(0) = \alpha(\varepsilon) > 0, \quad y'(0) = \beta(\varepsilon) > 0 \tag{3.2}$$

exists globally on $[0,b]$ (for small ε) and satisfies

$$\alpha(\varepsilon) \le y(x,\varepsilon) \le C_1(\varepsilon)$$

$$0 < y'(x,\varepsilon) \le p(\varepsilon)C_1(\varepsilon)$$

for all $x \in [0,b]$, where $C_1(\varepsilon) = \max\{c_1(\varepsilon),\ \alpha(\varepsilon),\ \alpha(\varepsilon) + \frac{\beta(\varepsilon)}{p(\varepsilon)}\}$.

Proof. An upper bound for $y(x,\varepsilon)$ is constructed as in section 2. If there exists a value of $x_0 \in (0,b]$ such that $y'(x_0,\varepsilon) = 0$ but $y'(x,\varepsilon) > 0$ for $0 \le x < x_0$, we may use the one-dimensional mean value theorem twice as in section 2 to see that for $0 \le x \le x_0$, $y(x,\varepsilon)$ is a solution of

$$\varepsilon u'' + a(x,\varepsilon)\,u' + b(x,\varepsilon)\,u = -f(x,0,0,\varepsilon)$$

where $a(x,\varepsilon) = f_3(x,y,\hat{z},\varepsilon)$, $b(x,\varepsilon) = f_2(x,\hat{y},0,\varepsilon)$, $\hat{z} = \hat{z}(x,\varepsilon)$ satisfies $0 < \hat{z}(x,\varepsilon) < y'(x,\varepsilon)$, and $\hat{y} = \hat{y}(x,\varepsilon)$ satisfies $0 < \hat{y}(x,\varepsilon) < y(x,\varepsilon)$. But $b(x,\varepsilon) \le 0$, $y'(x_0,\varepsilon) = 0$, and $y(x,\varepsilon)$ has an endpoint maximum on $[0,x_0]$ at x_0, in contradiction of the maximum principle [7, p. 7]. Thus no such $x_0 \in (0,b]$ exists and the continuation theory of solutions gives the desired result.

In particular, it follows from Theorem 2 that the initial value problem of the form

$$\varepsilon u'' + f(x,u,u')\, u' = 0$$

$$u(0) = h > 0, \quad u'(0) = A + ah, \quad A \geq 0,\ a > 0,$$

considered in [4] has a unique global solution on [0,1]. The maximum principle approach taken here seems simpler than the approach in [4].

Theorem 1 applies to certain two-parameter singular perturbation problems. Let $\mu = \mu(\varepsilon)$ satisfy $\varepsilon/\mu^2 \to 0$ as $\varepsilon \to 0$. Then from Corollary 1 there follows immediately

Corollary 3. Consider the initial value problem

$$\varepsilon y'' + \mu g(x,y)\, y' + h(x,y,\varepsilon) = 0 \tag{3.3}$$

$$y(0) = \alpha(\varepsilon), \quad y'(0) = \beta(\varepsilon) \tag{3.4}$$

and suppose that $g(x,y)$, $h(x,y,\varepsilon)$, $h_2(x,y,\varepsilon)$ are all continuous, $g(x,y) \geq a > 0$, $0 < b \leq h_2(x,y,\varepsilon) \leq B < \infty$ on $[0,1] \times (-\infty,\infty)$ for ε sufficiently small, and further that $h(x,0,\varepsilon) = O(1)$ uniformly on [0,1] as $\varepsilon \to 0$, $\alpha(\varepsilon) = O(1)$, $\beta(\varepsilon) = O(\mu/\varepsilon)$ as $\varepsilon \to 0$. Then (3.3), (3.4) has a unique global solution $y(x,\varepsilon)$ on [0,1] for ε sufficiently small, and $y(x,\varepsilon) = O(1)$, $y'(x,\varepsilon) = O(\mu/\varepsilon)$, uniformly on [0,1] as $\varepsilon \to 0$.

In the case that (3.3) is linear and we put $R = (\varepsilon/u^2)y$ in (3.3), (3.4), we get O'Malley's crucial lemma ([5, p. 1148], [6, p. 402]), which he proved using integral equations and used to develop an asymptotic expansion for the solution of such two-parameter linear problems.

REFERENCES

1. J. V. Baxley, On singular perturbation of nonlinear two-point boundary value problems, to appear.

2. J. V. Baxley, Global existence and uniqueness for second order ordinary differential equations, to appear.

3. E. A. Coddington and N. Levinson, Theory of Ordinary Differential Equations, McGraw-Hill, New York, 1955.

4. D. S. Cohen, Singular perturbation of nonlinear two-point boundary value problems, J. Math. Anal. Appl. 43 (1973), 151-160.

5. R. E. O'Malley, Jr., Two-parameter singular perturbation problems for second order equations, J. Math. Mech. 16 (1967), 1143-1164.

6. R. E. O'Malley, Jr., Topics in singular perturbations, Advances in Math. 2 (1968), 365-470.

7. M. H. Protter and H. F. Weinberger, Maximum Principles in Differential Equations, Prentice-Hall, Englewood Cliffs, N. J., 1967.

EXISTENCE AND BOUNDS FOR THE LOWEST CRITICAL ENERGY OF THE HARTREE OPERATOR

N.W. Bazley *

INTRODUCTION

The purpose of this paper is to announce some new results of R. Seydel, J. Weyer and the author concerning the existence and estimation of critical values of nonlinear operators. To illustrate the applicability of our results we formulate them in terms of the Hartree operator for helium and remark that they are valid for a much wider class of operators. The last section of the article also contains a brief exposition of recent work of J. Weyer on cyclically monotone operators. Weyer's results show that upper bounds to the critical energy can be obtained by the techniques of convex programming. Some of these results appear in unpublished form in [13] and [19], while other parts have been published in a physical setting in [2].

The Hartree equations have long been used by physicists to approximate complicated atomic and molecular systems. They replace the linear Schrödinger equation by a coupled system of nonlinear equations, which can be approximately solved by iterative techniques. Hartree equations are simpler than the given Schrödinger equation in the sense that each equation involves the space variables of one particle only. On the other hand they form a system of nonlinear differential-integral equations for which the mathematical properties are not well understood. In order to make a first investigation of such operators we consider the simplest case of one equation; namely, the Hartree operator for helium under the assumption that both electrons are in identical states.

* Research sponsored in part by the Institute for Fluid Dynamics of the University of Maryland, College Park, Maryland.

The first existence result for this problem goes back to Reeken [10], who proved the existence of an isolated and pointwise positive eigenfunction for the radial equation. Part of his analysis extends the theory in [3] to unbounded regions. More recently several interesting articles on the mathematical properties of the Hartree operator have appeared [5,20,15,9,16]. These results are primarily concerned with the properties of the eigenfunctions in terms of the eigenvalue parameter.

One problem of physical interest is the existence and estimation of critical energies of the Hartree operator. For simplicity we will consider only the minimal energy. Since the operator is nonlinear this energy will be different from the eigenvalue, although, by the principle of Euler-Lagrange, minimizing vectors will satisfy the eigenvalue equation. Our approach is that of upper and lower bounds. Upper bounds are obtained by a nonlinear analogue of the Rayleigh-Ritz procedure. Our method for lower bounds makes essential use of the energy scalar product and the triangle inequality; it reduces the original nonlinear problem to that of finding lower bounds for the eigenvalues of an associated linear eigenvalue problem containing a trial vector. The method of A. Weinstein [18] is applicable to this last problem. We then show by a special choice of our trial vector that Reeken's solution actually minimizes the energy. Part of our analysis was suggested by the chapter on variational methods in Stakgold's book [14] and further details can be found in [13].

In the last section we discuss recent results of J. Weyer [19]. As applied to the Hartree operator, they relate the convexity of the potential to the boundedness and self-adjointness of the Fréchet derivative. This convexity provides an alternative method for proving the existence of a minimizing solution. Also it allows application of the techniques of convex programming for the calculation of the upper bound.

THE EIGENVALUE PROBLEM

Consider the real Hilbert space $\mathfrak{H} = L^2(\mathbb{R}^3)$. In atomic units we define the Hartree operator by

$$A(u) = -\frac{1}{2}\nabla^2 u - \frac{2u}{|x|} + u(x)\int_{\mathbb{R}^3} \frac{u^2(y)}{|x-y|}\,dy\ , \tag{1}$$

where x and y denote vectors in $\mathbb{R}^3$. Each solution to the eigenvalue problem $A(u) = \lambda u$ is required to satisfy the normalization condition

$$\|u\|^2 = \int_{\mathbb{R}^3} u^2(y)\,dy = 1\ . \tag{2}$$

Thus the Hartree operator has the decomposition

$$A(u) = Bu + C(u)\ , \tag{3}$$

where

$$Bu = -\frac{1}{2}\nabla^2 u - \frac{2u}{|x|} \tag{4}$$

and

$$C(u) = u(x)\int_{\mathbb{R}^2} \frac{u^2(y)}{|x-y|}\,dy\ . \tag{5}$$

Here B is a linear self-adjoint operator with domain D_B, semi-bounded from below, whose spectrum consists of a positive continuous spectrum and a negative point spectrum with eigenvalues

$$\lambda_n = -2/n^2 \qquad n = 1, \ldots\ . \tag{6}$$

The nonlinear operator $C(u)$ is homogeneous of degree three; that is, $C(\tau u) = \tau^3 C(u)$ for real τ. Further, its "linearization at the point u" is given by

$$dC_u h = h(x)\int_{\mathbb{R}^3} \frac{u^2(y)}{|x-y|}\,dy + 2u(x)\int_{\mathbb{R}^3} \frac{u(y)h(y)}{|x-y|}\,dy\ , \tag{7}$$

which, for $u \neq 0$, is a positive and symmetric operator when considered as defined on functions h in D_B. By a well known theorem [17,p.56] the operator $A(u)$ has a potential $\Phi_A(u)$ given by

$$\Phi_A(u) = \frac{1}{2}(Bu,u) + \int_0^1 (C(\tau u),u)\,d\tau = \frac{1}{2}(Bu,u) + \frac{1}{4}(C(u),u)\ . \tag{8}$$

The minimal critical energy E on the unit sphere is defined by

$$E = \min_{\substack{u \in D_B \\ \|u\|=1}} \Phi_A(u) \;/\; \tfrac{1}{2}\|u\| \quad , \tag{9}$$

or equivalently

$$E = \min_{\substack{u \in D_B \\ \|u\|=1}} \Phi(u) \quad , \tag{10}$$

where $\Phi = 2\Phi_A$ is given by

$$\Phi(u) = (Bu,u) + \tfrac{1}{2}(C(u),u) \quad . \tag{11}$$

It will be shown later that this minimum in fact exists. Obviously the class of minimizing vectors can be enlarged by introducing the operator $\sqrt{B+2}$. If $\tilde{u}$ is a normalized minimizing vector, then it must satisfy $A(\tilde{u}) = \tilde{\lambda}\tilde{u}$ by the principle of Euler-Lagrange. Thus the minimal critical energy is related to the eigenvalue by

$$E = \tilde{\lambda} - \tfrac{1}{2}\,(C(\tilde{u}),\tilde{u}) \quad . \tag{12}$$

Finally, we remark that the result of Reeken [10] uses a constructive argument to prove the existence of a positive eigenfuction $\bar{u}(x) > 0$ satisfying

$$A(\bar{u}) = \bar{\lambda}\bar{u} \quad . \tag{13}$$

We will later show that Reeken's solution in fact attains the minimal critical energy E. The example in [1] shows that this result is non-trivial.

UPPER BOUNDS

The problem of obtaining upper bounds to E leads to a generalization of the Rayleigh-Ritz method to the case of nonlinear operators. By (10) any normalized trial vector $\tilde{u}$ gives upper bounds according to

$$E \leq \Phi(\tilde{u}) \quad . \tag{14}$$

If $\tilde{u}$ is obtained by expansion in the first k elements of a complete orthonormal set $\{v_i\}_1^\infty$, then $\tilde{u} = \sum_{i=1}^{k} \alpha_i v_i$, where $1 = \sum_{i=1}^{k} \alpha_i^2$. This leads to the nonlinear algebraic problem of finding optimal $\alpha_i, \ldots, \alpha_k$. We will later use results of Weyer to show that the method of convex programming is applicable.

LOWER BOUNDS

We now turn to the arguments given in [13] and [2]. There the idea is to introduce a scalar product

$$[v,w] = \iint_{\mathbb{R}^3} \frac{v(x)w(y)}{|x-y|}\, dxdy \tag{15}$$

and remark that in terms of this scalar product the quadratic term $\frac{1}{2}(C(u),u)$ in Φ is simply

$$\frac{1}{2}(C(u),u) = \frac{1}{2} \iint_{\mathbb{R}^3} \frac{u^2(x)u^2(y)}{|x-y|}\, dxdy = \frac{1}{2}[u^2,u^2] \quad . \tag{16}$$

Following [14] and letting $z(x)$ be a fixed vector, we rewrite the inequality $\frac{1}{2}[u^2-z^2,u^2-z^2] \geq 0$ as

$$\frac{1}{2}[u^2,u^2] \geq [u^2,z^2] - \frac{1}{2}[z^2,z^2] \; . \tag{17}$$

For sufficiently regular z we can write (17) in the form

$$\frac{1}{2}(C(u),u) \geq (L_z u,u) - \frac{1}{2}(C(z),z) \quad , \tag{18}$$

where L_z is a linear operator for each fixed function $z(x)$. It has the form $L_z u = q_z(x)u$, where $q_z(x)$ is the operator of multiplication by the function

$$q_z(x) = \int_{\mathbb{R}^3} \frac{z^2(y)}{|x-y|}\, dy \quad ; \tag{19}$$

further, L_z has the property that for the vector $u = z$,

$$L_z z = C(z) \tag{20}$$

and (18) becomes an identity. Another possible expression for L_z is given in [2].

To obtain our lower bounds we introduce the functional

$$\psi(u,z) = (Bu,u) + (L_z u,u) - \frac{1}{2}(C(z),z) , \tag{21}$$

which satisfies

$$\Phi(u) \geq \psi(u,z) . \tag{22}$$

Thus

$$\begin{aligned} E = \min_{\substack{u\varepsilon D_B \\ \|u\|=1}} \Phi(u) &\geq \min_{\substack{u\varepsilon D_B \\ \|u\|=1}} \psi(u,z) \\ &= \min_{\substack{u\varepsilon D_B \\ \|u\|=1}} \{(Bu,u) + (L_z u,u)\} - \frac{1}{2}(C(z),z)\} . \end{aligned} \tag{23}$$

If we denote the first eigenvalue of the linear operator $B + L_z$ by $\lambda_1(z)$ then (23) gives the lower bound

$$E \geq \lambda_1(z) - \frac{1}{2}(C(z),z) . \tag{24}$$

We can replace $\lambda_1(z)$ by any lower bound and refer the reader to [18] for methods of calculating lower bounds to eigenvalues. Reasonable candidates for z might be the Rayleigh-Ritz trial vector $\tilde{u}$ or the nth Hartree iteration vector u_n obtained from the iteration scheme $Bu_n + u_n q_{u_{n-1}} = \lambda u_n$.

EXISTENCE OF E

Until now we have not proven the existence of E . However, a simple proof follows from the special choice $z(x) = \overline{u}(x)$, where $\overline{u}(x)$ is the positive solution of Reeken satisfying (13). In fact, by the minimal characterization of the first eigenvalue of the operator $B + q_{\overline{u}}$, we have

$$\overline{\lambda} = \min_{\substack{u\varepsilon D_B \\ \|u\|=1}} \{(Bu,u) + (q_{\overline{u}}u,u)\} = (B\overline{u},\overline{u}) + (C(\overline{u}),\overline{u}) . \tag{25}$$

Thus (23) can be written as

$$\min_{\substack{u\epsilon D_B\\ ||u||=1}} \Phi(u) = E \geq \min_{\substack{u\epsilon D_B\\ ||u||=1}} \{(Bu,u) + (L_{\overline{u}}u,u)\} - \frac{1}{2}(C(\overline{u}),\overline{u}) \tag{26}$$

$$= (B\overline{u},\overline{u}) + \frac{1}{2}(C(\overline{u}),\overline{u}) = \Phi(\overline{u}) .$$

Thus $E = \Phi(\overline{u})$ and the minimum in (10) is actually attained.

UPPER BOUNDS BY CONVEX PROGRAMMING AND SOME RESULTS OF J. WEYER

We now return to the problem mentioned earlier of obtaining upper bounds $\Phi(\widetilde{u}) \geq E$, $||\widetilde{u}|| = 1$, by expanding $\widetilde{u}$ as $\widetilde{u} = \sum_{i=1}^{k} \alpha_i v_i$, where $\{v_i\}_1^k$ are the first k elements of an orthonormal set of functions. Our problem is to find a systematic method for determining the optimal constants $\{\alpha_i\}_1^k$. We show that they can be found by the method of convex programming. This does not seem to be obvious from inspection of the functional Φ. In fact our analysis depends on properties of the linearization of $A(u)$ and makes essential use of recent results of J. Weyer [19]. We first outline his theory and remark that the usefulness of monotone operators in the analysis of the Hartree operator was first pointed out by Gustafson and Sather [5].

As the work of Weyer is based on the theory of cyclically monotone operators, we first recall their definition. Let A be a single-valued nonlinear operator defined from a domain D_A into $\tilde{h}$ (this definition can be extended to multi-linear operators [11]).

Definition Suppose, for each integer $n > 0$ and for n arbitrary vectors $u_o, u_1, \ldots, u_{n-1}, u_n = u_o$ in D_A, that

$$\sum_{i=1}^{n} (A(u_i), u_i - u_{i-1}) \geq 0 , \tag{27}$$

$i=1,\ldots,n$. Then A is said to be cyclically monotone.

The cyclically monotone operators form an important subset of the familiar monotone operators $(n=2)$, defined by $(A(u_2) - A(u_1), u_2 - u_1) \geq 0$.

In [19] Weyer proves the following basic result.

Theorem: Let D_A be dense and linear in $\mathfrak{H}$ and $A : D_A \to \mathfrak{H}$. Further, suppose A possesses a hemi-continuous Gateaux derivative A'_u . Then A is cyclically monotone if and only if A'_u is positive and symmetric for all u in D_A .

Further he shows, for certain classes of operators, that if A'_u is positive and self-adjoint at one point $u = u_o$ in D_A then the corresponding operator A is maximal cyclically monotone.
The proof of Weyer's theorem is based on applications of known results in the theory of monotone and cyclically monotone operators. It is indirect in the sense that it passes through the potential corresponding to A . At one point Weyer makes use of a result which goes back to Minty [8]; namely, that A is monotone if and only if its Gateaux derivative A'_u is positive on D_A for all u in D_A . Thus the cyclically monotone restriction of monotone operators is in some sense equivalent to the symmetry of the derivatives.
The usefulness of the above theorem is readily seen when it is combined with the following theorem of Rockafellar [11,12].

Theorem: Let $A : D_A \to \mathfrak{H}$ with D_A linear and dense. Then

a) There exists a convex functional ψ with $A \subset \partial\psi$, the subgradient of ψ , if and only if A is cyclically monotone.

b) There exists a proper lower semi-continuous convex functional ψ with $\partial\psi = A$ if and only if A is maximal cyclically monotone.

Physically interesting operators usually arise from potentials. The importance of the above theorems lies in the relationship between the positivity and symmetry (or self-adjointness) of the derivatives A'_u of a nonlinear operator $A(u)$ and the convexity (proper lower semi-continuity, uniqueness, and convexity) of potentials ψ . Thus, given a nonlinear differential expression for an operator of mathematical physics, it is natural to ask the question "what linear boundary conditions should one prescribe for the nonlinear differential expression in order that it be the unique gradient of a convex functional?" By the theorem of Rockafellar it is clear that we want boundary conditions so that $A(u)$ is a maximal cyclically monotone operator. But cyclical monotonicity is difficult to check directly. Fortunately the theorem of Weyer allows us to answer this question by investigating the positivity and self-adjointness of the derivative.

Weyer has applied his results to determine boundary conditions for the class of differential expressions given by

$$\bar{A}(u) = \sum_{i=1}^{k} (-1)^i \left(q_i(x)\Phi_i(u^{[i]})\right)^{[i]}, \quad k \geq 1 , \tag{28}$$

where q_i are positive functions and Φ_i are monotone functions. This class was suggested by some work of B. Werner and includes as a special case the equations for the vibrations of a nonlinear beam [6].

We remark that in the case of linear operators the maximal cyclically monotone operators are the self-adjoint operators [4]. A vast literature exists on the question of boundary conditions which assure the self-adjointness or essential self-adjointness of linear differential expressions, going back to original works of Friedrichs, Kato and Weyl.

In order to connect Weyer's theory with the minimization of a convex functional, we first introduce operators $\tilde{A}(u)$ defined by

$$\tilde{A}(u) = A(u) + 2Iu = Bu + 2u + C(u) , \tag{29}$$

where B and C are given in (4) and (5) respectively. Then we go back to equations (4) and (7) to write

$$\begin{aligned} \tilde{A}'_u h &= Bh + 2h + dC_u h \\ &= -\frac{1}{2}\nabla^2 h - \frac{2h(x)}{|x|} + 2h(x) + h(x)\int_{\mathbb{R}^3} \frac{u^2(y)}{|x-y|}\,dy \\ &\qquad + 2u(x)\int_{\mathbb{R}^3} \frac{u(y)h(y)}{|x-y|}\,dy , \end{aligned} \tag{30}$$

where $\tilde{A}'_u$ denotes the linearization of $\tilde{A}$ at the point u. Since B has the lower bound -2 by (6) and since dC_u is positive for each u, it follows at once that $\tilde{A}'_u$ is a positive operator. It is symmetric when defined on the domain D_B of $B+2I$. Furthermore, $\tilde{A}'_u$ is obviously self-adjoint at the point $u \equiv 0$, where it reduces to $B + 2I$.

Thus the results of Weyer are applicable and it follows that $\tilde{A}(u)$ is the unique subgradient of a convex functional $\frac{1}{2}\tilde{\Phi}(u)$. By the definition of $\tilde{A}$, $\tilde{\Phi}$ is given as

$$\tilde{\Phi}(u) = \Phi(u) + 2\|u\|^2 = (Bu,u) + 2(u,u) + \frac{1}{2}(C(u),u), \tag{31}$$

where $\Phi(u)$ is defined in (11).

We now investigate the problem

$$\tilde{E} = \min_{\substack{u\varepsilon D_B \\ \|u\|=1}} \tilde{\Phi}(u). \tag{32}$$

For any trial vector $\tilde{u} = \sum_{i=1}^{k} \alpha_i v_i$, where $(v_i, v_j) = \delta_{ij}$, we have $\tilde{E} \le \tilde{\Phi}(\sum_{i=1}^{k} \alpha_i v_i)$, or equivalently,

$$\tilde{E} \le \sum_{i,j=1}^{k} \alpha_i\alpha_j(\{B+2\}v_i,v_j) + \sum_{i,j,s,t=1}^{k} \alpha_i\alpha_j\alpha_s\alpha_t \iint_{\mathbb{R}^3} \frac{v_i(x)v_j(x)v_s(y)v_t(y)}{|x-y|}\,dxdy, \tag{33}$$

where

$$1 = \sum_{i=1}^{n} \alpha_i^2. \tag{34}$$

But $\tilde{\Phi}$ is a convex functional and the right-hand side of (33) must be a convex function of the n real variables $\alpha_1,\dots,\alpha_n$. Thus the well known methods of convex programming (see, for example, [7]) can be applied to minimize the right-hand side of (33) subject to the constraint (34). A numerical and theoretical investigation of this problem is presently being carried out by A. Bongers and T. Küpper of the University of Cologne.

It remains to be shown that the minimal solution in (32) is attained by the positive solution $\bar{u}(x)$ of Reeken. For this purpose we simply reapply the arguments of this article to the operator $\tilde{A}(u)$. Clearly $\bar{u}(x)$ satisfies $\tilde{A}(\bar{u}) = (\bar{\lambda}+2)\bar{u}$ and by replacing B by $B+2I$ in all the previous arguments the result follows immediately. Since $\tilde{E} = E+2$, we must subtract 2 from our upper bound for $\tilde{E}$ to obtain our upper bound for E.

REFERENCES

[1] Bazley,N., Reeken,M. and Zwahlen,B.: Math.Z. 123, 301 (1971).

[2] Bazley,N. and Seydel,R.: Chem.Phy.Letters 24, 128 (1974).

[3] Bazley,N. and Zwahlen,B.: Manuscripta Math. 2, 365 (1970).

[4] Brezis,H.: Operateurs maximaux monotones (North-Holland, Amsterdam, 1973).

[5] Gustafson,K. and Sather,D.: Rendiconti di Matematica 4, 723 (1971).

[6] Hu,K. and Kirmser,P.: J.Appl.Mech.,Trans.ASME, 461 (1971).

[7] Luenberger,D.: Introduction to linear and nonlinear programming, (Addison-Wesley, Reading,Mass., 1973).

[8] Minty,G.: Duke Math.J. 29, 341 (1962).

[9] Quirmbach,W., Diplomarbeit, Universität zu Köln (Wintersemester 1973/74).

[10] Reeken,M., J.Math.Phys. 11, 2505 (1970).

[11] Rockafellar,T.: Pacific J. Math. 17, 497 (1966).

[12] Rockafellar,T.: Pacific J. Math. 33, 209 (1970).

[13] Seydel,R.: Diplomarbeit, Universität zu Köln (Wintersemester 1973/74).

[14] Stakgold,I.: Boundary value problems of mathematical physics, Vol.II (MacMillan, New York, 1968).

[15] Stuart,C.: Arch.Rat.Mech.Anal. 51, 60 (1973).

[16] Stuart,C.: J. Math. Anal. and Appl. (to appear).

[17] Vainberg,M.: Variational methods in the study of nonlinear operators (Holden-Day, London, 1964).

[18] Weinstein,A. and Stenger,W.: Methods of intermediate problems for eigenvalues (Academic Press, New York, 1972).

[19] Weyer,J.: Diplomarbeit, Universität zu Köln (Wintersemester 1973/74).

[20] Wolkowisky,J.: Indiana Univ. Math. J. 22, 551 (1972).

Positive Invariance and a Wazewski Theorem

J. Bebernes

I. Classical Wazewski Theorem

Let $X \subset R^1 \times R^n$ be open, $V \subset X$ open, $f: X \to R^n$ continuous. Consider

$$x' = f(t,x) \tag{1}$$

Def. 1. A point $(t_0, y_0) \in \partial V \cap X$ is an egress point of V relative to (1) iff, for every solution $y(t)$ of (1) with $y(t_0) = y_0$, there exists $\varepsilon > 0$ such that $(t,y(t)) \in V$ on $t_0 - \varepsilon \le t < t_0$. An egress point is a strict egress point in case for every solution $y(t)$ of (1) with $y(t_0) = y_0$ there exists $\varepsilon > 0$ such that $(t,y(t)) \notin \bar{V}$ for $t_0 < t \le t_0 + \varepsilon$.

Def. 2. Let $A \subset B$ be subsets of a topological space T. A is a retract of B iff there exists a continuous map g on B into A such that $g|_A = I$.

Theorem 1. (Ważewski, 1947). Let S be the set of egress points of V and assume that all egress points are strict. Assume IVP's for (1) are unique. Let Z be a subset of $V \cup S$ such that

$$Z \cap S \text{ is a retract of } S, \text{ but } Z \cap S \text{ is not a retract of } Z$$

Then there exists $(t_0, y_0) \in Z \cap V$ such that the solution $y(t,t_0,y_0)$ of (1) exists on V on its right maximal interval of existence (RMIE).

II. Positive Invariance for Generalized Differential Equations

Let $F: X \to cc(R^n)$ where $cc(R^n)$ denotes collection of nonempty, compact, convex subsets of R^n be an uppersemicontinuous (USC) multifunction and consider

$$x' \in F(t,x) \quad \text{and} \tag{2}$$

$$Dx \subset F(t,x) \,. \tag{3}$$

Def. 3. A solution of the generalized differential equation (2) is a function $\varphi: I \to R^n$ which is absolutely continuous on each compact subinterval of I and

$$\varphi'(t) \in F(t,\varphi(t)) \quad \text{a.e. on } I \,.$$

<u>Def. 4</u>. For $E \subset X$ and $(t_0,x_0) = P_0$ an accumulation point of E, the <u>positive contingent</u> of E at P_0 is

$$D^+(E,P_0) = \left\{ y \in R^n : \exists \{(t_n,x_n)\} \subset E,\ (t_n,x_n) \to (t_0,x_0),\ t_n > t_0,\ \frac{x_n - x_0}{t_n - t_0} \to y \text{ as } n \uparrow \infty \right\}$$

If E is the graph of a function $\varphi: I \to R^n$, write $D^+\varphi(t_0)$.

<u>Def. 5</u>. A solution of the contingent differential equation (3) is a continuous function $\varphi: I \to R^n$ such that $\emptyset \neq D\varphi(t) \subset F(t,\varphi(t))$ for all $t \in I$.

<u>Lemma</u>. A function φ is a solution of (2) iff it is a solution of (3).

<u>Def. 6</u>. A set $E \subset X$ is <u>right admissible</u> with respect to (2) in case for each $(t_0,x_0) \in E$ there exists $\alpha(t_0,x_0) > 0$ such that (2) has a solution x on $[t_0, t_0+\alpha]$ with $x(t_0) = x_0$ and $(t,x(t)) \in E$ on $[t_0, t_0+\alpha]$.

<u>Def. 7</u>. A set $E \subset X$ is <u>positively weakly invariant</u> with respect to (2) in case for each $(t_0,x_0) \in E$ there is a solution $x(t)$ of (2) with $x(t_0) = x_0$, $(t,x(t)) \in E$ on RMIE.

<u>Lemma</u>. 1) $B \subset X$ PWI $\Rightarrow$ B is RA, but not conversely.

2) $B \subset X$ relatively closed is RA $\Leftrightarrow$ B is PWI.

<u>Question</u>. For $E \subset X$ relatively closed, $F: X \to cc(R^n)$ USC, what conditions on F and E imply that E is right admissible?

The following (sub) tangential conditions are necessary and sufficient:

(S_1) (Nagumo 1942) For every $(t_0,x_0) \in E$ and every $\varepsilon > 0$ there exists $(t_1,x_1) \in E$ such that $t_0 < t_1 < t_0 + \varepsilon$ and

$$d\left(\frac{x_1 - x_0}{t_1 - t_0}, F(t_0,x_0)\right) < \varepsilon .$$

Def. 8. A vector $v \in R^n$ is subtangential with respect to $S \subset R^1 \times R^n$ at $(t_0, x_0) \in S$ in case

$$\liminf_{h \to 0+} \frac{1}{h} d(x_0 + hv, S(t_0 + h)) = 0$$

where $S(\tau) \equiv \{y \in R^n : (\tau, y) \in S\}$

(S_2) (Yorke 1967) For every $(t_0, x_0) \in E$, there exists $v \in F(t_0, x_0)$ which is subtangential to E at (t_0, x_0).

(S_3) (Bebernes - Schuur 1970) For every $(t_0, x_0) \in E$,

$$D^+(E, t_0, x_0) \cap F(t_0, x_0) \neq \emptyset .$$

Theorem 2. $(S_1) \Leftrightarrow (S_2) \Leftrightarrow (S_3) \Leftrightarrow$ (RA) $\Leftrightarrow$ (WPI). (E relatively closed subset of X and F USC on X.)

Theorem 3. Assume E_1 and E_2 are nonempty relatively closed positively weakly invariant subsets of W with $E_1 \cup E_2 = W$. Then $E_1 \cap E_2$ is PWI.

III. A Ważewski Theorem for GDE. Let $X \subset R^1 \times R^n$ be open.

Def. 9. For $P_0 = (t_0, x_0) \in V \subset X$, the (positive) zone of emission relative to V is $E_V(P_0) \equiv \{(\tau, y) \in V: y = \varphi(\tau), \varphi(t)$ is a solution of (2) with $\varphi(t_0) = x_0$ and $(t, \varphi(t)) \in V$ on $[t_0, \tau], \tau \geq t_0\}$. If $A \subset V$,

$$E_V(A) \equiv \cup\{E_V(P) : P \in A\}$$

Let W be a relatively closed subset of X. For $P_0 \in W$, the trace of emission relative to W is defined to be $T_W(P_0) = E_W(P_0) \cap \partial W$ and, for $A \subset W$, $T_W(A) = E_W(A) \cap \partial W$.

Def. 10. A point $Q \in \partial W \cap W$ is a strict egress point relative to (2) if, for every solution $\varphi(t, Q)$,

$$c_\varphi = \sup\{t: (s, \varphi(s, Q)) \in \partial W \cap W, \ t_Q \leq s \leq t\} < \infty \quad \text{and}$$

there exists a sequence $\{t_n\}$, $t_n \downarrow c_\varphi$, with $(t_n, \varphi(t_n, Q)) \in X - W$. Denote the set of strict egress points by S.

<u>Def. 11</u>. A solution $\varphi(t,P)$, $P \in W$, of (2) "leaves W" iff there exists $t_1 \in$ RMIE such that $(t_1, \varphi(t_1,P)) \in X - W$.

<u>Theorem 4</u>. Let $P_0 \in W$. If (1) $T_W(P_0) \subset S$ and (2) all solutions of (2) through P_0 leave W, then $T_W(P_0)$ is compact and connected.

<u>Theorem 5</u>. Let $A \subset W$, $T_W(A) \subset S$ and assume for each $P \in A$, all solutions through P leave W. Then $T_W: A \to ck(W \cap \partial W)$ (compact connected subsets of $W \cap \partial W$) is USC.

<u>Theorem 6</u>. Let $Z \subset \text{int } W \cup S$ be connected. If $T_W(Z) \subset S$ and $T(Z)$ is not connected, then there exists $P_0 \in Z$ and a solution $\varphi(t,P_0)$ of (2) such that $(t, \varphi(t,P_0)) \in W$ on RMIE.

<u>Def. 12</u>. Let $A \subset B$ be subsets of a metric space Y. A is a (set-valued) retract of B iff there exists a map $H: B \to ck(A)$ which is USC and $x \in H(x)$ for all $x \in A$.

<u>Theorem 7</u>. Let Z be a subset of $(\text{int } W) \cup S$ such that $T_W(Z) \subset S$. If $Z \cap T(Z)$ is a retract of $T(Z)$ but not of Z, then there exists $P_0 \in Z$ and a solution $\varphi(t,P_0)$ of (2) such that $(t, \varphi(t,P_0)) \in W$ on RMIE.

References

1. J. Bebernes and J. Schuur, The Ważewski topological method for contingent equations, Ann. Mat. Para Appl. 87(1970), 271-280.

2. J. Bebernes and W. Kelley, Some boundary value problems for generalized differential equations, SIAM J. Appl. Math., 25(1973), 16-23.

3. M. Nagumo, Über die Lage der Integralkurven gewöhnlicher Differentialgleichungen, Proc. Phys.-Math. Soc. Japan 24(1942), 551-559.

4. R. Redheffer and W. Walter, Flow invariant sets and differential inequalities in normed spaces, to appear.

5. T. Ważewski, Une méthode topologique de l'examen du phénomène asymptotique relativement aux équations differentielles ordinaires, Rend. Accad. Lincei (8) 3(1947), 210-215.

6. J. Yorke, Invariance for ordinary differential equations, Math. Systems Theory 1(1967), 353-372.

An Application of the Hausdorff-Young Inequality to Eigenfunction Expansions

by

Harold E. Benzinger

1. Introduction. Let $a(x)$ be in $L^{\infty}[0,1]$, and let $\{u_k(x)\}$, $k = 1,2,\cdots$, denote the sequence of eigenfunctions of the boundary value problem

(1.1) $$u^{(2)} + a(x)u = \lambda u \quad , \quad 0 \leq x \leq 1,$$

(1.2) $$u(0) = 0, \quad u(1) = 0 .$$

If $a(x) \equiv 0$, the eigenfunctions are

(1.3) $$\varphi_k(x) = \sqrt{2} \sin k\pi x \quad , \quad k = 1,2,\cdots.$$

Since the problem of expanding an arbitrary function in its Fourier series of φ_k's is well understood, it is natural to ask if there is a general principle which implies that the general case $\{u_k\}$ inherits the expansion properties of the special case $\{\varphi_k\}$.

The first effort in this direction was due to G. D. Birkhoff [3] in 1917, who proved that if $\{\varphi_k\}$ and $\{u_k\}$ are any two orthonormal systems in $C[0,1]$, and if only $f(x) \equiv 0$ is orthogonal to all φ_k's, then only $f(x) \equiv 0$ is orthogonal to all u_k's provided the series

$$H(x,y) = \sum_{k=1}^{\infty} [\varphi_k(x) - u_k(x)]\varphi_k(y) \quad , \quad 0 \leq x,y \leq 1 \quad ,$$

has the properties i) $|H(x,y)| < 1$, $0 \leq x,y \leq 1$, ii) for each f in $C[0,1]$,

$$\sum_{k=1}^{\infty} \int_0^1 f(x)[\varphi_k(x) - u_k(x)]\, dx\, \varphi_k(y)$$

converges uniformly. Birkhoff then used this result to show that if $a(x)$ in (1.1) is real and of class $C^1[0,1]$, then only $f(x) \equiv 0$ is orthogonal to all eigenfunctions u_k.

In 1921, J. L. Walsh [5] proved that if $\{\varphi_k(x)\}$, $\{u_k(x)\}$ are two orthogonal sequences in $C[0,1]$, if $\{\varphi_k(x)\}$ is uniformly bounded, and if

$$u_n(x) - \varphi_n(x) = \sum_{k=1}^{\infty} c_{nk}\varphi_k(x) \quad , \qquad n = 1,2,\cdots,$$

$$\sum_{n=1}^{\infty} \Big(\sum_{k=1}^{\infty} |c_{nk}| \Big)^2 < \infty \quad ,$$

then for any f in $L^2(0,1)$, the Fourier series of f with respect to $\{u_k\}$ and with respect to $\{\varphi_k\}$ are uniformly and absolutely equiconvergent. Walsh then used this result to show that if $a(x)$ in (1.1) is real and continuous, then the above hypotheses are satisfied by the eigenfunctions of (1.1),(1.2) and the eigenfunctions (1.3).

We present here a simple abstract theorem on perturbation of bases in Banach spaces which allows us to conclude that since $\{\varphi_k\}$ is a basis for $L^p(0,1), 1 < p < \infty$, then so is $\{u_k\}$. Since $a(x)$ in $L^\infty[0,1]$ is possibly complex valued, our result holds for biorthogonal systems as well as orthonormal ones.

2. Perturbation of Bases. Let B denote a complex Banach space with dual space B^*. Let $\{\varphi_k\}$ denote a (Schauder) basis for B, with dual basis $\{\psi_k\}$ in B^*. A sequence $\{u_k\}$ in B is <u>strongly linearly independent</u> if the equation

$$\sum_{k=1}^{\infty} a_k u_k = 0$$

implies that $a_k = 0$ for all k.

Theorem. Let $\{u_k\}$ be a strongly linearly independent sequence in B, and let $\{\varphi_k\}$, $\{\psi_k\}$ be as above. Then $\{u_k\}$ is a basis for B if

i) there exists $r, 1 < r < \infty$, such that if f is in B, then $\{(f,\psi_k)\}$ is in ℓ^r,

ii) $\{\|u_k - \varphi_k\|\}$ is in ℓ^s, where $rs = r + s$.

Proof. Note that if the map $B \longrightarrow \ell^r$ defined by $f \longrightarrow \{(f,\psi_k)\}$ is defined everywhere, then it is closed. Thus from i), there exists $m > 0$ such that for all f in B,

$$[\sum_{k=1}^{\infty} |(f,\psi_k)|^r]^{1/r} \leq m\|f\| . \tag{2.1}$$

Consider the operator $K : B \longrightarrow B$ defined by

$$Kf = \sum_1^{\infty}(f,\psi_k)(\varphi_k - u_k). \tag{2.2}$$

It is clear that K is defined for all f, since, using Hölders's inequality,

$$\sum_1^{\infty}\| (f,\psi_k)(\varphi_k - u_k)\| \leq [\sum_1^{\infty}|(f,\psi_k)|^r]^{1/r} [\sum_1^{\infty}\| \varphi_k - u_k\|^s]^{1/s} < \infty .$$

Also, K is compact, since if

$$K_N f = \sum_1^{N}(f,\psi_k)(\varphi_k - u_k) ,$$

then K_N is of finite rank, and

$$\|Kf - K_N f\| \leq \sum_{N+1}^{\infty} |(f,\psi_k)| \, \|\varphi_k - u_k\| \leq m[\sum_{N+1}^{\infty} \|\varphi_k - u_k\|^s]^{\frac{1}{s}} \|f\| ,$$

which converges to zero as N gets large. Let $A = I - K$. Clearly A is everywhere defined and

$$(2.3) \qquad Af = \sum_{1}^{\infty}(f,\psi_k)u_k \quad .$$

In particular, $A\varphi_k = u_k$, so if A is one-to-one and onto, then $\{u_k\}$ is a basis with dual basis $v_k = A^{*-1}\psi_k$. Since K is compact, A is onto if it is one-to-one. Since $\{u_k\}$ is strongly linearly independent, we see that $Af = 0$ implies $(f,\psi_k) = 0$ for all k, so $f = 0$.

The above theorem, in case that B is a Hilbert space and $\{\varphi_k\}$ is a complete orthonormal system, is due to N. Bari [1]. We note that a sequence $\{u_k\}$ is strongly linearly independent if there exists a sequence $\{v_k\}$ in B^* such that $(u_k,v_j) = \delta_{kj}$.

3. Eigenfunction Expansions. Using standard arguments, we see that the eigenfunctions of (1.1), (1.2) are given by

$$(3.1) \qquad u_k(x) = \varphi_k(x) + O(\tfrac{1}{k}) \quad , \quad k = 1,2,\cdots,$$

uniformly in x, $0 \leq x \leq 1$, where φ_k is given by (1.3).

From the theory of Fourier series, we know that the orthonormal system $\{\varphi_k\}$ is a basis for $L^p(0,1)$, $1 < p < \infty$, and that the Hausdorff-Young inequality is valid: for each f in $L^p(0,1)$, $1 \leq p \leq 2$, $\{(f,\varphi_k)\}$ is in ℓ^q, where $pq = p + q$.

Theorem. The sequence $\{u_k\}$ of eigenfunctions of (1.1),(1.2) is a basis for $L^p(0,1)$, $1 < p < \infty$.

Proof. Let $\{v_k\}$ denote the sequence of eigenfunctions for the adjoint problem:

(3.2) $$v^{(2)} + \overline{a(x)}v = \lambda v ,$$

(3.3) $$v(0) = 0, v(1) = 0 .$$

Since $(u_k, v_j) = \delta_{kj}$, we see that $\{u_k\}$ and $\{v_k\}$ are both strongly linearly independent. From the Hausdorff-Young inequality, we see that if f is in $L^p(0,1)$, $1 < p \leq 2$, then $\{(f,\psi_k)\}$ is in ℓ^r, where $r = q$. From (3.1), we see that $\{\| u_k - \varphi_k \|\}$ is in ℓ^s, where $s = p$. Thus $\{u_k\}$ is a basis for $L^p(0,1)$, $1 < p \leq 2$, and $\{v_k\}$ is a basis for $L^p(0,1)$, $2 < p \leq \infty$. Interchanging the roles of the problems (1.1),(1.2) and (3.2),(3.3), we extend these results to $1 < p < \infty$.

4. Remarks. In [2] we shall show that if $n \geq 1$ is a given integer, then the set of all n-th order Birkhoff regular boundary value problems can be decomposed into equivalence classes, each equivalence class containing a simple representative, and the theorem of §2 is applicable to the eigenfunction $\{u_k\}$ of any member of the equivalence class, and the eigenfunctions $\{\varphi_k\}$ of the simple representative.

The set of problems (1.1),(1.2) is such an equivalence class for $n = 2$. See [2] also for further references to the general theory of perturbation of bases.

In [4], the Bari theorem was applied to the problem (1.1),(1.2) in the case of real $a(x)$ for $p = 2$.

REFERENCES

1. N. Bari, Sur les systèmes complets de fonctions orthogonales, Mat. Sbornik 14(1944), 51-108.

2. H. E. Benzinger, Perturbation of bases, the Hausdorff-Young inequality, and eigenfunction expansions, in preparation.

3. G. D. Birkhoff, A theorem on series of orthogonal functions with an application to Sturm-Liouville series, Proc, Nat. Acad. of Sci. 3(1917), 656-659.

4. J. Kazdan, Perturbation of complete orthonormal sets and eigenfunction expansions, Proc. Amer. Math. Soc. 27(1971), 506-510. MR 42 # 6648.

5. J. L. Walsh, On the convergence of Sturm-Liouville series, Annals of Math., ser. 2, 24(1921), 109-120.

ON CENTRAL DISPERSIONS OF THE DIFFERENTIAL EQUATION $Y'' = q(t)Y$ WITH PERIODIC COEFFICIENTS

O. Borůvka

I. Introduction.

In the following lecture I shall deal with ordinary 2nd order linear differential equations of Jacobi's type

$$(q) \qquad y'' = q(t)y ,$$

mostly oscillatory.

Suppose that the coefficient q, the so-called carrier of the equation (q), is a continuous function in the interval $j = (-\infty, \infty)$. The equation (q) is called oscillatory if each of its integrals has an infinite number of zeros which accumulate towards both ends of j. The prototype of these oscillatory equations is the equation (-1): $y'' = -y$, $t \in j$, which we shall often deal with in what follows.

Let us consider, in particular, the equations (q) with periodic carriers q. The theory of these equations is governed by Floquet's theory. According to the latter, every equation (q) with a periodic carrier q is either disconjugate, i. e., without conjugate points, or oscillatory. To oscillatory equations (q), on the other hand, one may apply the theory of dispersions, based, as we know, on other principles than Floquet's theory. We may therefore expect that by combining the notions and the results of both theories we may arrive at a new approach to oscillatory equations (q) with periodic carriers, yielding new results. And this is exactly the leading idea of the following considerations.

II. Generalities.

Let me, first, introduce some basic notions and results from the theory of dispersions, needed in our study: 1. phases, 2. central dispersions, 3. inverse equations. ([1],[2])

All the equations (q) we shall deal with are oscillatory in the interval $j = (-\infty, \infty)$. A composite function, e. g., $a[b(t)]$ will often be written in the form $ab(t)$ or ab. Furthermore, we shall use the notation: $c_n(t) = t+n\pi$ $(n=0,\pm1,\ldots;\ t\in j)$ and instead of c_1 we shall generally only put c.

Let (q) stand for an arbitrary (oscillatory) equation.

1. Phases. Let us note, first, that by a basis of the equation (q) or of the carrier q we mean an ordered pair, u,v, of independent integrals of (q). As to the phases, more precisely: the first phases, we distinguish the phases of a given basis of (q) and the phases of the equation (q) or of the carrier q. By a phase of the basis u,v we mean every function in the interval j, $\alpha(t)$, which is continuous in j and, for $v(t) \neq 0$ satisfies the relation: $\tan\alpha(t) = u(t):v(t)$. By a phase of the equation (q) (of the carrier q) we mean any phase of some basis of (q). Every phase α of (q) has the following properties:

1. $\alpha(t) \in C_j^3$, 2. $\alpha'(t) \neq 0$ for $t \in j$, 3. $\lim\limits_{t\to\sigma.\infty}\alpha(t) = \sigma.\operatorname{sgn}\alpha'.\infty$ $(\sigma = \pm1)$

Every phase α of (q) uniquely determines its carrier in the sense of formula:

$$q(t) = -\{\tan\alpha, t\} \qquad (t\in j), \tag{1}$$

where the symbol { } denotes Schwartz's derivative of the function $\tan\alpha$ at the point t. The carrier q with the phase α is also denoted q_α.

α is called an upper or a lower phase if the property 2. is reinforced by the inequality $\alpha'(t) > 0$ and if, moreover, there holds $\alpha(t) > t$ or $\alpha(t) < t$ $(t \in j)$, respectively. Both the upper and the lower phases are called dispersion phases.

The phase α is called elementary if

$$\alpha(t+\pi) = \alpha(t) + \pi.\operatorname{sgn}\alpha' \qquad (t \in j),$$

i. e., $\alpha c = c_{\operatorname{sgn}\alpha'}\alpha$. Every elementary phase α is of the form

$$\alpha(t) = t.\operatorname{sgn}\alpha' + p(t) \qquad (t \in j)$$

where p is a periodic function with period π : $pc = p$.

The set formed of all the elementary phases, together with the operation given by composing the functions, forms a group called the group of elementary phases, $\mathcal{G}$.

An important part is played by phases of the equation (-1): $y'' = -y$. Each phase of the latter, $\varepsilon(t)$, may be expressed in the form

$$(2) \qquad \varepsilon(t) = {}_n\mathrm{Arc}\ \tan\left[C \cdot \frac{\sin(t+a)}{\sin(t+b)}\right],$$

where n is an integer and C;a,b constants such that $a \in (0,\pi]$, $b \in [0,\pi)$, $C\sin(b-a) \neq 0$. The expression on the right-hand side in (2) denotes the continuous function in the interval j uniquely determined by the conditions:

$$\varepsilon(-a-\overline{n-1}\,\pi)=0,\ \tan\varepsilon(t)=C.\sin(t+a):\sin(t+b) \text{ for } \sin(t+b)\neq 0.$$

ε is an upper phase if $C.\sin(b-a) > 0$ and, moreover, either $n = 1$ and between the constants C;a,b there are further relations, or $n \geqslant 2$; ε is a lower phase if, again, $C.\sin(b-a) > 0$ and, moreover, either $n = 0$ and between the constants C;a,b there are further relations, or $n \leqslant -1$.

The set of all phases of the equation (-1), together with the operation given by the composing of functions, forms again a group called the fundamental group $\mathfrak{E}$. The latter is a subgroup in $\mathfrak{L}$, $\mathfrak{E} \subset \mathfrak{L}$, hence all the phases of the equation (-1) are elementary.

An important result: Let α be an arbitrary phase of the equation (q). Then all the phases of the equation (q) are exactly the composite functions $\varepsilon\alpha(t)$, $\varepsilon \in \mathfrak{E}$.

2. Central dispersions. To every number n $(=0,\pm 1,\ldots)$ there corresponds the central dispersion (of the first kind) with the index n of the equation (q) or of the carrier q, denoted φ_n. It is a function in the interval j and its value $\varphi_n(t)$ $(t \in j)$ is determined as follows: If $n \neq 0$, then $\varphi_n(t)$ is the $|n|^{th}$ conjugate number (of the first kind) with the number t, greater or smaller than t according as $n > 0$ or $n < 0$. In case of $n = 0$, we put $\varphi_0(t) = t$. The dispersion φ_1, in particular, is called fundamental and is often denoted φ.

From the number of important properties of the dispersions φ_n I shall only introduce these:

Every dispersion φ_n has the above properties 1.-3. of phases while the property 2. is replaced by the inequality $\alpha'(t) > 0$; moreover, there holds $\varphi_{n-1}(t) < \varphi_n(t)$ $(t \in j)$. We see that φ_n is an upper or a lower phase of a convenient equation according as $n \geq 1$ or $n \leq -1$.

Every phase α of the equation (q) is connected with the dispersion φ_n by Abelian relation

$$\alpha\varphi_n(t) = \alpha(t) + n.\pi\,.\operatorname{sgn}\alpha' \qquad (t \in j), \tag{3}$$

more briefly: $\alpha\varphi_n = c_{n.\mathrm{sgn}\alpha'}\alpha$.

Hence follows the expression of the dispersion φ_n by α :

(4) $$\varphi_n(t) = \alpha^{-1}c_{n.\mathrm{sgn}\alpha'}\alpha(t) \qquad (t \in j)$$

Every integral y of (q) as well as its derivative y′ is, by the dispersion φ_n, transformed into itself in the sense of formula:

(5) $$\begin{aligned} y\varphi_n(t) &= (-1)^n\ [\varphi_n'(t)]^{\frac{1}{2}} \cdot y(t), \\ y'\varphi_n(t) &= (-1)^n\ [\varphi_n'(t)]^{-\frac{1}{2}} \cdot y'(t), \quad \text{if } y(t)=0. \end{aligned}$$

If $q(t) < 0$, $t \in j$, then there holds, for $n \geqslant 1$,

(6) $$\varphi_n'(t) = \frac{q(t_1)}{q(t_3)} \cdot \frac{q(t_5)}{q(t_7)} \cdot \ldots \cdot \frac{q(t_{4n-3})}{q(t_{4n-1})}$$

with convenient numbers $t_{2\nu-1}$ ($\nu=1,\ldots,2n$). The latter separate the zeros $(t=)a_0 < a_1 < \ldots < a_n (= \varphi_n(t))$ of every integral y of (q), which vanishes at the number t, and the zeros $b_1 < \ldots < b_n$ of its derivative y′, lying in the interval $(t, \varphi_n(t))$:

$$(t=)a_0 < t_1 < b_1 < t_3 < a_1 < t_5 < \ldots < t_{4n-3} < b_n < t_{4n-1} < a_n (= \varphi_n(t)).$$

By means of the central dispersions φ_n we define the distance functions of the equation (q) or of the carrier q by the relation

$$d_n(t) = \varphi_n(t) - t \qquad (n=0,\pm1,\ldots;\ t \in j)$$

with the evident meaning: $d_n(t)$ is the distance of the number $\varphi_n(t)$ from t. Clearly, $d_n(t) \gtreqless 0$ according as $n \gtreqless 0$. d_1 (>0) is the basic distance function; it is often denoted by d.

3. Inverse equations. Let me now introduce a new notion, namely that of the inverse equations with regard to (q).

A differential equation $(\bar{q})$ is called inverse of the equation (q) if it has a phase $\bar{\alpha}$ which is the inverse function of some phase α of (q): $\bar{\alpha} = \alpha^{-1}$.

From this definition there follows a number of properties of inverse equations; I shall introduce only those we shall need in what follows.

Symmetry: If $(\bar{q})$ is inverse of (q), then (q) is inverse of $(\bar{q})$.

The carrier of the equation inverse of (q), with the phase $\bar{\alpha}$, is given by the formula

$$q_{\bar{\alpha}}(t) = -1 - [1+q\bar{\alpha}(t)]\,(\bar{\alpha}(t))'^{2}$$

The set of all (oscillatory) equations (q) is decomposed into nonempty disjoint subsets, called blocks, so that the latter form a decomposition U of the set in question. To each block $u \in U$ there exists exactly one inverse block, $u^{-1} \in U$, with the characteristic property: Any two equations $(q) \in u$, $(\bar{q}) \in u^{-1}$ are inverse of each other.

The proof of this theorem lies deep in the algebraic theory of oscillatory equations (q) and cannot be introduced here for want of space.

III. Study of the problem.

Well, applying the above notions, I may now proceed to the main subject of my lecture dealing, according to the title, with the properties of the central dispersions of the oscillatory equations (q) with periodic carriers q: $qc = q$, $t \in j$. The set of these equations will be denoted A_p.

Let, first, $(q) \in A_p$. Then there holds: if y is an integral of the equation (q), then the function yc is an integral of (q) as well. Hence we easily deduce that to every phase α of (q) there exists a phase ε of the equation (-1), $\varepsilon \in \mathfrak{F}$, such that $\alpha c = \varepsilon\alpha$. Consequently α is a solution of the equation

$$(7) \qquad \alpha(t+\pi) = \varepsilon\alpha(t) \qquad (t \in j)$$

or

$$(\bar{\varphi}_{\operatorname{sgn}\alpha'} \equiv)\ \alpha c \alpha^{-1}(t) = \varepsilon(t).$$

The function $\bar{\varphi}_{\operatorname{sgn}\alpha'}$ is, by (4), the fundamental dispersion of $(q_{\alpha^{-1}})$ or the function inverse of this dispersion, according as $\alpha' > 0$ or $\alpha' < 0$.

Consequently:

$\varepsilon(t) \in \mathfrak{F}$ is a dispersion phase, upper or lower, according as $\alpha' > 0$ or $\alpha' < 0$.

The fundamental dispersion of the inverse equation $(q_{\alpha^{-1}})$ and, therefore, every central dispersion of the latter, lies in the fundamental group $\mathfrak{F}$.

These results suggest the question whether the above properties of phases and central dispersions of the equations of the class A_p or of the corresponding inverse equations are characteristic of the equations of A_p. The answer is affirmative in the sense of the following theorems:

The equation (q) belongs to the class A_p if and only if each of its phases α satisfies the equation (7) with a dispersion phase $\varepsilon(t) \in \mathfrak{F}$.

The equation (q) belongs to the class A_p if and only if all the central dispersions of each inverse equation of (q) lie in the fundamental group $\mathcal{G}$.

Furthermore, one may, in this connection, prove the theorems:

The fundamental dispersions of the equations inverse of the equations of the class A_p are exactly all the upper phases from the fundamental group $\mathcal{G}$.

All the equations (q) of the same block simultaneously either belong or do not belong to the class A_p.

Remember that the class A_p consists of oscillatory equations (q) with periodic carriers: $q(t+\pi) = q(t)$, $t \in (-\infty,\infty)$.

2. Into our considerations there have entered, as an important element, the central dispersions of the equations inverse of those from A_p.

In this connection there arises the question concerning the properties of the central dispersions of the equations of the class A_p themselves. The central dispersions of the equations $(q) \in A_p$ have, in fact, the following remarkable property:

All central dispersions of every equation $(q) \in A_p$ are elementary.

In other words:

All distance functions of every equation $(q) \in A_p$ are periodic.

Indeed, let $(q) \in A_p$ and φ be the fundamental dispersion of (q).

We easily ascertain that the proof need not be given but for the dispersion φ.

Well, let α (e. g. $\alpha' > 0$) be a phase of (q). From (7) and (3) there follows

$$\alpha c \varphi = \varepsilon \alpha \varphi = \varepsilon c \alpha = c \varepsilon \alpha = c \alpha c,$$

consequently, $c\varphi = (\alpha^{-1} c\, \alpha) c = \varphi c$ so that $\varphi c = c \varphi$ and the proof is accomplished.

3. In this place I have the opportunity to mention the class of the equations (q) characterized by the fact that their fundamental dispersions are elementary. Let the class of these equations be denoted by A, so that $(q) \in A \Longleftrightarrow \varphi c = c \varphi$. We have just seen that $A_p \subset A$.

The equations $(q) \in A$ may be characterized by the following geometric property: Let C be an arbitrary integral curve of the equation $(q) \in A$, whose parametric expression is given by some of the bases of $(q) \in A$. Let O stand for the origin of the coordinates and $\overrightarrow{OP(t)}$ for the radius vector from the point O to the point $P(t) \in C$ determined by the value of the parameter (time) t. Then the oriented areas traced out by the radius vector $\overrightarrow{OP(t)}$ and the opposite radius vector $\overrightarrow{OP(\varphi(t))}$ in the time from t to $t+\pi$ are the same.

The equations (q) with elementary fundamental dispersions occur even in other connections, e. g., if there is a question of determining pairs of equations (q) with interchangeable fundamental dispersions.

The theory of the equations $(q) \in A$ is extensive, so I cannot - for want of time - deal with it in detail but shall confine myself to a few remarks.

The theory of the equations $(q) \in A$ is fundamentally analogous to the theory of the equations $(q) \in A_p$, the part of the group $\mathfrak{G}$ in the latter being taken over by the group of the elementary phases, $\mathfrak{E}$.

In particular, there holds:

The equation (q) belongs to the class A if and only if each of its phases α satisfies the equation $\alpha(t+\pi) = h\alpha(t)$ $(t \in j)$ with a dispersion phase $h(t) \in \mathfrak{E}$.

The class A is closed with regard to the operation of forming inverse equations.

The equation (q) belongs to the class A if and only if the carriers $q(t)$, $q(t+\pi)$ have the same fundamental dispersion.

If $(q) \in A$, then there exist, in every interval $[t, \varphi(t))$, at least four mutually different numbers t_1, t_2, t_3, t_4 such that $q(t_i+\pi) = q(t_i)$ $(i = 1,2,3,4)$.

Note that, by the first theorem, the class A is wider than A_p: $A_p \subset A \neq A_p$.

4. Now we arrive at the last point of my lecture, where we return to the equations (q) with periodic carriers: $(q) \in A_p$. It will be a question of expressing real periodicity factors (characteristic roots) of the equations $(q) \in A_p$ by means of the values of the derivatives of central dispersions and, furthermore, of estimating the absolute values of the periodicity factors by means of the extreme values of the function $|q|$. ([3])

Well, let $(q) \in A_p$ be an arbitrary equation whose periodicity factors s_σ $(\sigma = \pm 1)$ are real. Denote

$$A = s_1 + s_{-1} ,$$

so that the characteristic equation corresponding to (q) is, by Floquet's theory, $s^2 - A.s + 1 = 0$ and, according to the supposition, we have $|A| \geq 2$.

Let s be one of the roots s_σ . Then there exists a nontrivial integral y_0 of the equation (q), with the property

(8) $$y_0 c = s.y_0 .$$

The equation (q) being oscillatory, the function y_0 has at least one zero x: $y_0(x) = 0$. From (8) then follows $y_0(x+\pi) = 0$. We see that the point $x+\pi$ is right conjugate with the point x. So we have, for a natural n:

(9) $$\varphi_n(x) = x + \pi .$$

Let u,v denote the basis of the equation (q), determined by the initial values:

(10) $$u(x)=1, \quad u'(x)=0; \qquad v(x)=0, \quad v'(x)=1.$$

Then $A = u(x+\pi)+v'(x+\pi)$ whence, by (9),(5),(10), there follows

$$A = (-1)^n \left([\varphi_n'(x)]^{\frac{1}{2}} + [\varphi_n'(x)]^{-\frac{1}{2}} \right).$$

We see that the periodicity factors of (q) are:

(11) $$s_\sigma = (-1)^n [\varphi_n'(x)]^{\frac{\sigma}{2}} \qquad (\sigma = \pm 1).$$

Now suppose that the carrier q of the equation (q) is always different from zero: $q(t) < 0$ $(t \in j)$.

Then, by (11) and (6), we have:

(12) $$s_\sigma = (-1)^n \left[\frac{q(x_1)}{q(x_3)} \cdot \frac{q(x_5)}{q(x_7)} \cdot \ldots \cdot \frac{q(x_{4n-3})}{q(x_{4n-1})} \right]^{\frac{\sigma}{2}} \quad (\sigma = \pm 1)$$

with convenient numbers $x_{2\nu-1}$ $(\nu=1,\ldots,2n)$ separating the zeros $(x=)a_0 < a_1 < \ldots < a_n(=x+\pi)$ of the integral y_0 and the zeros $b_1 < \ldots < b_n$ of its derivative y_0', lying in the interval $(x,x+\pi)$:

$$(x=)a_0 < x_1 < b_1 < x_3 < a_1 < x_5 < \ldots < x_{4n-3} < b_n < x_{4n-1} < a_n(=x+\pi)$$

Denote

$$m = \min_{t\in j} |q(t)|, \qquad M = \max_{t\in j} |q(t)|$$

so that

$$0 < m \leq M.$$

n stands for the number of zeros of the function y_0 in the interval $[x,x+\pi)$ and therefore, by a classical theorem, satisfies the inequalities

$$\sqrt{m} \leq n \leq \sqrt{M}. \tag{13}$$

From (6) we have

$$\left(\frac{m}{M}\right)^n \leq \varphi_n'(x) \leq \left(\frac{M}{m}\right)^n$$

which, together with (11) and (13), yields

$$\left(\frac{m}{M}\right)^{\frac{1}{2}\sqrt{M}} \leq |s_\sigma| \leq \left(\frac{M}{m}\right)^{\frac{1}{2}\sqrt{M}}. \tag{14}$$

Thus we have arrived at the following result:

If the periodicity factors s_σ of the equation $(q) \in A_p$ are real $(\sigma = \pm 1)$, then their values are given by the values of the function φ_n' in a number $x \in j$, in the sense of formula (11). If, furthermore, the carrier q is always different from zero, then the numbers s_σ may be expressed by values of the carrier q in the sense of formula (12). In that case there hold, for absolute values of the numbers s_σ, the inequalities (14).

Note that the inequalities (14) may be employed to obtain information as to the absolute values of the periodicity factors for Hill's equation in case of instability.

IV. Final remark.

Let me now finish my lecture with a short look at the algebraic theory of (oscillatory) equations (q) and stress the connection of the classes A_p and A with other elements of the theory in question.

The basic notion of the algebraic theory of the equation (q) is the group of phases.

The group of phases, $\mathfrak{G}$, is the set of all phases of all the equations (q) with the group-operation given by the composing of functions. The unit of $\mathfrak{G}$, $\underline{1}$, is the function t ($\in$ j): $\underline{1}$ = t.

The increasing phases form, in $\mathfrak{G}$, an invariant subgroup $\mathfrak{G}_0$, with the index 2; the decreasing phases form a coset of the factor group $\mathfrak{G}/\mathfrak{G}_0$, denoted G_1.

The upper phases α are characterized by the inequality $\alpha(t) > t$ ($\in$ j); they form a subset in $\mathfrak{G}_0$, called the upper complex, K_1. The lower phases α are characterized by the inequality $\alpha(t) < t$ ($\in$ j); they too form a subset in $\mathfrak{G}_0$, called the lower complex, K_{-1}. The complexes K_σ ($\sigma = \pm 1$) are disjoint and consist of functions mutually inverse. K_1 is composed of the fundamental dispersions of the equations (q); K_{-1} consists of functions inverse of these dispersions. For $\xi \in \mathfrak{G}$ there applies: $\xi^{-1} K_\sigma \xi = K_{\sigma . \operatorname{sgn} \xi'}$.

The set of all phases of the equation (-1) forms a subgroup in $\mathfrak{G}$, namely the above mentioned fundamental group $\mathfrak{E}$. The latter generates, on $\mathfrak{G}$, the right decomposition $\mathfrak{G}/_r\mathfrak{E}$ and the left decomposition $\mathfrak{G}/_\ell\mathfrak{E}$. Each element of the former has the form $\mathfrak{E}\alpha$ ($\alpha \in \mathfrak{G}$); it is the set of all phases of the equation (q) with the carrier given by the formula (1). Each element of the latter is of the form $\alpha\mathfrak{F}$ $(\alpha \in \mathfrak{G})$; it is the set of the functions inverse of the phases of $(q_{\alpha^{-1}})$. Every element $\bar{u} \in \bar{U}$ of the least common covering $\bar{U} = [\mathfrak{G}/_r\mathfrak{F}, \mathfrak{G}/_l\mathfrak{F}]$ is the union of the phases of some equations (q) [4]. The latter form the block corresponding to the element $\bar{u}$. To $\bar{u}$ there exists exactly one inverse element $\bar{u}^{-1} \in \bar{U}$ composed of the functions inverse of the phases lying in $\bar{u}$. Every equation (q) from the block corresponding to $\bar{u}$ is inverse of every equation from the block corresponding to $\bar{u}^{-1}$ and vice versa.

The center of the group $\mathfrak{G}_0 \cap \mathfrak{F}$ is the infinite cyclic group $\mathfrak{Z} = \{c_n(t)\}$ $(c_n(t) = t+n\pi\ ;\ n=0,\pm1,\ldots)$. The group $\alpha^{-1}\mathfrak{Z}\alpha$ $(\alpha \in \mathfrak{G})$ consists of the central dispersions of the equation (q_α). The normalizer of $\mathfrak{Z}$ in $\mathfrak{G}$ is the group of elementary phases, $\mathfrak{E}$. There holds: $\mathfrak{E} \supset \mathfrak{F}$.

The class A consists of all the equations (q) characterized by the fact that the inner automorphisms of the group $\mathfrak{G}$, formed by their phases α, the so-called phase-automorphisms of the equations (q), transform the center $\mathfrak{Z}$ into its normalizer: $\alpha^{-1}\mathfrak{Z}\alpha \subset \mathfrak{E}$. The same class A consists of the equations (q) inverse of the equations of class A.

The class A_p is a part of A: $A_p \subset A$. The equations of the class A_p are characterized by the fact that the phase-automorphisms of the inverse equations transform the center $\mathfrak{Z}$ into the fundamental group $\mathfrak{F}$.

BIBLIOGRAPHY

O. Borůvka:

[1] Linear differential transformations of the second order. The English Universities Press Ltd, London 1971, XVI+254 p.

[2] Sur la périodicité de la distance des zéros des intégrales de l'équation différentielle y" = q(t)y. Tensor,N.S. 26 (1972), 121-128.

[3] Sur quelques compléments à la théorie de Floquet pour les équations différentielles du deuxième ordre. Annali di matem. p. ed appl. (in print).

[4] Grundlagen der Gruppoid- und Gruppentheorie. VEB Deutscher Verlag der Wiss., Berlin 1960, XII+198 S.

Generalized translation operators associated with a singular differential operator

by

B.L.J.Braaksma and H.S.V.de Snoo

0. Introduction. Delsarte [7] considered the following Cauchy problem

$$(0.1)\qquad \frac{\partial^2 u}{\partial x^2} + \frac{2p+1}{x}\frac{\partial u}{\partial x} = \frac{\partial^2 u}{\partial y^2} + \frac{2p+1}{y}\frac{\partial u}{\partial y},\ u(x,0) = f(x),\ \frac{\partial u}{\partial y}(x,0) = 0,\ x > 0,$$

where $p \geq -\frac{1}{2}$ and f is a suitable function. The solution u(x,y) is called (generalized) translation of f and denoted by $u(x,y) = T^y f(x)$. It turns out that translation defined in this way shares several properties with ordinary translation $f(x) \to f(x+y)$ on the real line.

Levitan [14],[15], then considered the Cauchy problem

$$\frac{\partial^2 u}{\partial x^2} - q(x)u = \frac{\partial^2 u}{\partial y^2} - q(y)u,\ u(x,0) = f(x),\ \frac{\partial u}{\partial y}(x,0) = 0,\ x > 0,$$

where q is a continuous function that satisfies certain growth properties at ∞. For such translations there is a vast literature (see Levitan [15]). Recent work with milder conditions on q has been done by Leblanc [12], [13] and Hutson and Pym [10], [11].

Bochner [1] showed that the translations of Delsarte can be used for harmonic analysis with Hankel transforms. Similar work for Fourier-Jacobi transforms has been done by Flensted-Jensen and Koornwinder [8], [9]. The work on Hankel and Fourier-Jacobi transforms was generalized by Chébli [3], [4], [5] who considered the Cauchy problem

$$(0.2)\qquad \frac{\partial}{\partial x}\left(A(x)A(y)\frac{\partial u}{\partial x}\right) - \frac{\partial}{\partial y}\left(A(x)A(y)\frac{\partial u}{\partial y}\right) = 0,\ u(x,0) = f(x),\ \frac{\partial u}{\partial y}(x,0) = 0,\ x > 0.$$

Chébli also gives a concise introduction to the ideas of Delsarte.

In this paper we will define translations associated with the differential equation $(L_x - L_y)u = 0$, where

$$(0.3)\qquad L_x u = \frac{\partial^2 u}{\partial x^2} + \frac{2p+1}{x}\frac{\partial u}{\partial x} - q(x)u$$

and similarly $L_y u$, $p \geq -\frac{1}{2}$. We present properties of the translation kernel and we derive norm estimates for the translation. These properties can be used for the study of convolutions and harmonic analysis associated with this equation.

1. Another motivation for translations. We begin with a second manner to introduce our generalized translations. Let $\omega_\lambda(x)$ be the solution of

(1.1) $$L_x\omega_\lambda(x) = -\lambda^2\omega_\lambda(x),\ \omega_\lambda(0) = 1,\ \omega_\lambda'(0) = 0.$$

Then under suitable conditions we have the following correspondence, according to Weyl:

(1.2) $$\tilde{f}(\lambda) = \int_0^\infty f(x)\omega_\lambda(x)x^{2p+1}dx \leftrightarrow f(x) = \int_S \tilde{f}(\lambda)\omega_\lambda(x)d\rho(\lambda),$$

where ρ is the spectral function, related to our eigenvalue problem and S its spectrum.

We want to introduce a convolution which is mapped by the transformation $f \to \tilde{f}$ into a product. If $f*g$ denotes this convolution, then we should have

(1.3) $$\widetilde{f*g} = \tilde{f}.\tilde{g}.$$

So formally

$$\begin{aligned}(f*g)(x) &= \int_S \tilde{f}(\lambda)\tilde{g}(\lambda)\omega_\lambda(x)d\rho(\lambda)\\ &= \int_0^\infty g(y)\{\int_S \tilde{f}(\lambda)\omega_\lambda(x)\omega_\lambda(y)d\rho(\lambda)\}y^{2p+1}dy\end{aligned}$$

and hence

(1.4) $$(f*g)(x) = \int_0^\infty g(y)T^yf(x)y^{2p+1}dy,$$

where

(1.5) $$T^yf(x) = \int_S \tilde{f}(\lambda)\omega_\lambda(x)\omega_\lambda(y)d\rho(\lambda) = \int_0^\infty f(z)K(x,y,z)z^{2p+1}dz$$

with

(1.6) $$K(x,y,z) = \int_S \omega_\lambda(x)\omega_\lambda(y)\omega_\lambda(z)d\rho(\lambda).$$

Then in analogy to the usual definitions of convolution and translation we may consider T^yf as the generalized translation of f over y. Formally it follows from (1.5) that T^yf is a solution of the Cauchy problem

(1.7) $$(L_x-L_y)u = 0,\ u(x,0) = f(x),\ u_y(x,0) = 0,\ x > 0.$$

If $f \in C^2(0,\infty)$ and u is a solution of this Cauchy problem, then $v(x,y) = u(x,y) - f(x)$ satisfies

$$(L_x-L_y)v = -L_xf,\ v(x,0) = v_y(x,0) = 0,\ x > 0.$$

Let $A(\xi,\eta;x,y)$ be the Riemann function of L_x-L_y. In the sections 2 and 3 we show that $A(\xi,\eta;x,y)$ and $\frac{2p+1}{\eta}A(\xi,\eta;x,y) - \frac{\partial A}{\partial\eta}(\xi,\eta;x,y)$ are continuous for $(\xi,\eta) \in \Delta_{xy}$, where Δ_{xy} is the triangle with vertices $(x-y,0)$, $(x+y,0)$ and (x,y), $0 \leq y \leq x$. Moreover, $A(\xi,0;x,y) = 0$ if $p > -\frac{1}{2}$, $x-y \leq \xi \leq x+y$.

Now we use Riemann's method to solve a Cauchy problem on the line $y = \varepsilon$, $\varepsilon > 0$. Let $\varepsilon \downarrow 0$. Then we obtain

$$(1.8)\qquad v(x,y) = \tfrac{1}{2} \iint_{\Delta_{xy}} A(\xi,\eta;x,y)\, L_\xi f(\xi)\,d\xi d\eta.$$

Applying Green's theorem and the differential equation of which the Riemann function is a solution, we deduce

$$(1.9)\qquad v(x,y) = \lim_{\varepsilon\downarrow 0} \tfrac{1}{2} \int_{\partial\Delta_{xy,\varepsilon}} \{-f(\xi)\frac{\partial}{\partial\eta} A(\xi,\eta;x,y) + \frac{2p+1}{\eta} f(\xi)A(\xi,\eta;x,y)\}d\xi +$$
$$+ \{f'(\xi)A(\xi,\eta;x,y) - f(\xi)\frac{\partial}{\partial\xi} A(\xi,\eta;x,y) + \frac{2p+1}{\xi} f(\xi)A(\xi,\eta;x,y)\}d\eta,$$

where $\Delta_{xy,\varepsilon}$ is the triangle with vertices $(x-y+\varepsilon,\varepsilon)$, $(x+y-\varepsilon,\varepsilon)$ and (x,y). If $\xi \pm \eta = x \pm y$, then $d\xi = \mp d\eta$ and $\frac{\partial A}{\partial\eta}d\xi + \frac{\partial A}{\partial\xi}d\eta = \mp dA = \mp(p+\frac{1}{2})(\frac{1}{\xi} \mp \frac{1}{\eta})Ad\xi$.
Hence, if

$$(1.10)\qquad w(x,y,\xi) = \tfrac{1}{2} \lim_{\eta\downarrow 0} \{\frac{2p+1}{\eta} A(\xi,\eta;x,y) - \frac{\partial A}{\partial\eta}(\xi,\eta;x,y)\},$$

then

$$(1.11)\qquad v(x,y) = \int_{x-y}^{x+y} f(\xi)w(x,y,\xi)d\xi - f(x)$$

and

$$(1.12)\qquad u(x,y) = \int_{x-y}^{x+y} f(\xi)w(x,y,\xi)d\xi,$$

if $p > -\frac{1}{2}$. If $p = -\frac{1}{2}$, we have $A(\xi,0;x,y) = 1$ (cf. (2.1) and (2.9)) and so

$$u(x,y) = \tfrac{1}{2}\{f(x-y) + f(x+y)\} + \int_{x-y}^{x+y} f(\xi)w(x,y,\xi)d\xi.$$

If $p \neq 0$, then

$$w(x,y,\xi) = p\,\frac{\partial}{\partial\eta} A(\xi,0,x,y).$$

This suggests the following definition of the generalized translation $T^y f$ of a locally integrable function f on the positive x-axis if $p > -\frac{1}{2}$:

$$(1.13)\qquad T^y f(x) = \int_{x-y}^{x+y} f(\xi)w(x,y,\xi)d\xi, \text{ if } 0 \le y \le x$$

and

$$(1.14)\qquad T^y f(x) = T^x f(y), \text{ if } 0 \le x \le y.$$

If $p = -\frac{1}{2}$, then

$$(1.15)\qquad T^y f(x) = \tfrac{1}{2}\{f(|x-y|) + f(x+y)\} + \int_{|x-y|}^{x+y} f(\xi)w(x,y,\xi)d\xi.$$

Sufficient conditions on q such that this definition makes sense will be given in section 3, remark 1.

2. The Riemann function. Let L_{0x} denote L_x in the case $q \equiv 0$. The Riemann function $A_0(\xi,\eta;x,y)$ of $L_{0x} - L_{0y}$ may be deduced by a transformation from the Riemann function which Riemann calculated [17, p. 173] (cf. also [7 , p. 224])

$$A_0(\xi,\eta;x,y) = (\frac{\xi\eta}{xy})^{p+\frac{1}{2}}(1-z)^{-\frac{1}{2}-p}F(\tfrac{1}{2}+p,\tfrac{1}{2}+p;1;\frac{z}{z-1}),$$

where F denotes the hypergeometric function and

$$z = \{(x-y)^2 - (\xi-\eta)^2\}\{(x+y)^2 - (\xi+\eta)^2\}(16xy\xi\eta)^{-1}.$$

With $Z = (1-z)^{-1}$ we obtain

$$(2.1)\qquad A_0(\xi,\eta;x,y) = (\frac{\xi\eta}{xy})^{p+\frac{1}{2}}Z^{\frac{1}{2}+p}F(\tfrac{1}{2}+p,\tfrac{1}{2}+p;1;1-Z) =$$

$$= (\frac{\xi\eta}{xy})^{p+\frac{1}{2}}Z^{\frac{1}{2}-p}F(\tfrac{1}{2}-p,\tfrac{1}{2}-p;1;1-Z),$$

where

$$(2.2)\qquad Z = 16xy\xi\eta\{(x+y)^2 - (\xi-\eta)^2\}^{-1}\{(\xi+\eta)^2 - (x-y)^2\}^{-1}.$$

It is easily seen that

$$(2.3)\qquad \frac{\xi\eta}{xy} \leq Z \leq 1, \text{ if } 0 \leq y \leq x \text{ and } (\xi,\eta) \in \Delta_{xy}.$$

Let $Q(x,y)$ be a continuous function defined for $0 \leq y \leq x$, and let $A_Q(\xi,\eta;x,y)$ be the Riemann function of

$$(2.4)\qquad L_{0x} - L_{0y} - Q(x,y).$$

Then we have

$$(2.5)\qquad A_Q(\xi,\eta;x,y) = A_0(\xi,\eta;x,y) - \tfrac{1}{2}\iint_\Omega A_0(\xi,\eta;s,t)Q(s,t)A_Q(s,t;x,y)dsdt,$$

where $\Omega = \Omega(x,y,\xi,\eta)$ is the rectangle $\xi + \eta \leq s + t \leq x + y$, $x - y \leq s - t \leq \xi - \eta$ in the (s,t)-plane. This may be shown as follows: We have

$$(L_{0s} - L_{0t})A_0(\xi,\eta;s,t) = 0,$$

$$(L^*_{0s} - L^*_{0t})A_Q(s,t;x,y) = Q(s,t)A_Q(s,t;x,y),$$

if L^*_0 denotes the adjoint of L_0. Hence, by Green's theorem

$$(2.6)\qquad \iint_\Omega = -\int_{\partial\Omega} (A_Q\frac{\partial A_0}{\partial t} - A_0\frac{\partial A_Q}{\partial t} + \frac{2p+1}{t}A_0A_Q)ds + (A_Q\frac{\partial A_0}{\partial s} - A_0\frac{\partial A_Q}{\partial s} + \frac{2p+1}{s}A_0A_Q)dt.$$

In the same way as (1.10) was deduced from (1.9) we may deduce (2.5) from (2.6).

From (2.5) we now derive the following estimate for A_Q.

<u>Theorem 1.</u> Suppose $|Q(x,y)| \leq Q(y)$ if $0 \leq a \leq y \leq x \leq b$, where $Q(y)$ is integrable over $[a,b]$, and $Q(x,y)$ is measurable. Let

$$(2.7)\qquad R(\xi,\eta;x,y) = \frac{\xi\eta}{xy} \text{ if } p \geq \tfrac{1}{2}, \; = (\frac{\xi\eta}{xy})^{p+\frac{1}{2}}Z^{\frac{1}{2}-|p|} \text{ if } 0 < |p| \leq \tfrac{1}{2},$$

$$= (\frac{\xi\eta}{xy})^{\frac{1}{2}}Z^{\frac{1}{2}}(1+\log Z^{-1}) \text{ if } p = 0,$$

where Z is defined by (2.2). Let

(2.8) $$\rho(\eta,y) = \int_\eta^y tQ(t)dt.$$

Then there exists a constant M such that

(2.9) $$|A_Q(\xi,\eta;x,y) - A_0(\xi,\eta;x,y)| \leq R(\xi,\eta;x,y)\{-1 + \exp M\rho(\eta,y)\},$$

if $(\xi,\eta) \in \Delta_{xy}$, $a \leq \eta \leq y \leq x \leq b$.

<u>Proof</u>. From (2.1) and the properties of the hypergeometric function we may deduce that there exists a constant M_0 such that

(2.10) $$|A_0(\xi,\eta;x,y)| \leq M_0 R(\xi,\eta;x,y).$$

Next we show that if $Q_1 \in L(a,b)$ and ρ_1 is defined as in (2.8), then there is a constant M_1 such that

(2.11) $$\iint_{\Omega(x,y,\xi,\eta)} R(\xi,\eta;s,t)R(s,t;x,y)Q_1(t)dsdt \leq M_1 R(\xi,\eta;x,y)\rho_1(\eta,y)$$

with (ξ,η) as in the theorem and $p \neq 0$.
Then we construct the solution of (2.5) by means of the method of successive approximations. Using (2.11) we easily obtain (2.9) if $p \neq 0$.

First we prove (2.11) in the case $p \geq \frac{1}{2}$. If $(s,t) \in \Omega(x,y,\xi,\eta)$, then

(2.12) $$\xi + \eta - t \leq s \leq \xi - \eta + t, \; \eta \leq t \leq y, \; x - y + t \leq s \leq x + y - t.$$

Hence the lefthand side of (2.11) is at most

$$\int_\eta^y dt \int_{\xi+\eta-t}^{\xi-\eta+t} ds \, \frac{\xi\eta}{xy} Q_1(t),$$

which may be written as the righthand side of (2.11).

The proof in the case $p = -\frac{1}{2}$ is nearly the same. In the case $|p| < \frac{1}{2}$ we first need some estimates on Z. Let Z_1 and Z_2 denote Z if (ξ,η) and (x,y) respectively is replaced by (s,t) (cf. (2.2)). Put

$$u = (x+y)^2, \; v = (x-y)^2, \; u_1 = (s+t)^2, \; v_1 = (s-t)^2, \; u_2 = (\xi+\eta)^2,$$
$$v_2 = (\xi-\eta)^2.$$

Then

$$Z_1 Z_2 = 2^8 xy\xi\eta s^2 t^2 \{(u-v_1)(u_1-v)(u_1-v_2)(u_2-v_1)\}^{-1}.$$

Let σ_j and τ_j denote u_j and v_j when ξ and t are interchanged. Then $(u_1-v_2)(u_2-v_1) = (\sigma_1-\tau_2)(\sigma_2-\tau_1)$. On $\Omega(x,y,\xi,\eta)$ we have $u-v_1 \geq u-v_2$ and (cf. (2.12))

$$(u_1-v)(\sigma_1-\tau_2) = (s+t+x-y)(s+\xi+t-\eta)(s+t-x+y)(s+\xi-t+\eta) \geq$$
$$\geq s^2(\xi+\eta-x+y)(x-y+\xi+\eta) = s^2(u_2-v).$$

Hence

$$(2.13) \qquad Z_1 Z_2 \leq \frac{2^8 xy\xi\eta t^2}{(u-v_2)(u_2-v)(\sigma_2-\tau_1)} = 16Zt^2 \frac{1}{\sigma_2-\tau_1}.$$

So if I denotes the lefthandside of (2.11), then

$$I \leq 2^{2-4|p|}R(\xi,\eta;x,y) \iint_\Omega \left(\frac{t^2}{\sigma_2-\tau_1}\right)^{\frac{1}{2}-|p|} Q_1(t)dsdt, \text{ if } 0 < |p| < \tfrac{1}{2}.$$

Hence by (2.12)

$$(2.14) \qquad I \leq 2^{2-4|p|}R(\xi,\eta;x,y) \int_\eta^y dtQ_1(t)t^{1-2|p|} \int_{\xi+\eta-t}^{\xi-\eta+t} ds\{(t+\eta)^2-(\xi-s)^2\}^{|p|-\frac{1}{2}}$$

In the last integral we subsitute $s = \xi + (t-\eta)u$ and it follows that it is at most $C(t-\eta)^{2|p|}B(|p|+\frac{1}{2},|p|+\frac{1}{2})$. Hence (2.11) is valid for $0 < |p| < \frac{1}{2}$.

For the proof of (2.9) in case $p = 0$ we remark that $(A_Q-A_0)/A_0$ is an analytic function of p in a neighbourhood of $p = 0$. It is easily verified that its absolute value on a circle $|p| = \varepsilon > 0$ is at most $-1 + \exp M\rho(\eta,y)$ (with M independent of p), since R/A_0 is bounded on this circle and the proof of (2.9) is also valid for $|p| = \varepsilon$ if we replace p by Re p in (2.7). Hence by Cauchy's theorem (2.9) is valid for $p = 0$.

<u>Corollary 1</u>. Suppose $Q(x,y)$ is a measurable function defined for $0 < y \leq x$ and $|Q(x,y)| \leq Q(y)$ if $0 < y \leq x$, where $xQ(x) \in L(0,x_0)$ for all $x_0 > 0$. Then the Riemann function A_Q of (2.4) satisfies (2.9) if $(\xi,\eta) \in \Delta_{xy}$, $0 \leq y \leq x$, $p \geq -\frac{1}{2}$.

3. <u>The translation kernel</u>. From now on we assume that Q satisfies the conditions of corollary 1. Let w_Q be defined by (1.10) with A replaced by A_Q. From (2.5) we now derive an integral representation for w_Q. Let

$$(3.1) \qquad w_0(x,y,\xi) = \frac{2^{1-2p}\Gamma(p+1)}{\sqrt{\pi}\ \Gamma(p+\frac{1}{2})} \xi(xy)^{-2p}\{(x+y)^2 - \xi^2\}^{p-\frac{1}{2}}\{\xi^2 - (x-y)^2\}^{p-\frac{1}{2}}.$$

Then we may deduce from (2.1)

$$(3.2) \qquad \frac{2p+1}{\eta} A_0(\xi,\eta;x,y) - \frac{\partial}{\partial\eta} A_0(\xi,\eta;x,y) = 2w_0(x,y,\xi)(1 + O(\eta))$$

as $\eta \to 0$ uniformly for $(x,y) \in \Omega(x_0,y_0,\xi,\eta)$, if $0 < y_0 \leq x_0$. Hence, by section 1 the kernel for the Bessel translation i.e. the translation in the case $Q = 0$ considered by Delsarte ((0.1) and [7]) is w_0.

Now we use (2.5). This formula may also be written as

$$A_Q - A_0 = -\tfrac{1}{4} \int_{\xi+\eta}^{x+y} d\sigma \int_{x-y}^{\xi-\eta} d\tau A_0\left(\xi,\eta;\frac{\sigma+\tau}{2},\frac{\sigma-\tau}{2}\right)Q\left(\frac{\sigma+\tau}{2},\frac{\sigma-\tau}{2}\right)A_Q\left(\frac{\sigma+\tau}{2},\frac{\sigma-\tau}{2};x,y\right).$$

Hence

$$(3.3)\qquad \frac{2p+1}{2\eta} A_Q(\xi,\eta;x,y) - \tfrac{1}{2}\frac{\partial}{\partial\eta} A_Q(\xi,\eta;x,y) = w_0(x,y,\xi)(1 + O(\eta)) -$$

$$- \tfrac{1}{2} \iint_{\Omega} w_0(s,t,\xi)(1 + O(\eta))Q(s,t)A_Q(s,t;x,y)dsdt -$$

$$- \frac{1}{8}\Big(\int_{x-y}^{\xi-\eta} + \int_{\xi+\eta}^{x+y} \Big) A_0(\xi,\eta;s,t)Q(s,t)A_Q(s,t;x,y)dz,$$

where $s = \frac{\xi+\eta+z}{2}$, $t = \frac{\xi+\eta-z}{2}$, if $x - y \leq z \leq \xi - \eta$ and $s = \frac{\xi-\eta+z}{2}$, $t = \frac{-\xi+\eta+z}{2}$, if $\xi + \eta \leq z \leq x + y$. If $p > -\frac{1}{2}$, then $A_0 = 0$ for these values of s and t. If $p = -\frac{1}{2}$, $A_0 \equiv 1$, $w_0 \equiv 0$. So the lefthand side of (3.3) is continuous for $(\xi,\eta) \in \Delta_{xy}$, $0 \leq y \leq x$ and

$$(3.4)\qquad w_Q(x,y,\xi) = w_0(x,y,\xi) - \tfrac{1}{2} \iint_{\Omega(x,y,\xi,0)} w_0(s,t,\xi)Q(s,t)A_Q(s,t;x,y)dsdt$$

if $p > -\frac{1}{2}$ and

$$(3.5)\qquad w_Q(x,y,\xi) = -\frac{1}{8}\int_{x-y}^{x+y} Q\Big(\frac{\xi+z}{2},\frac{\xi-z}{2}\Big)A_Q\Big(\frac{\xi+z}{2},\frac{\xi-z}{2};x,y\Big)dz$$

if $p = -\frac{1}{2}$. Here we assume $x-y \leq \xi \leq x+y$.

As in section 1 we may show that if $u(x,y)$ is a solution of the Cauchy problem

$$(3.6)\qquad (L_{ox}-L_{oy})u = Q(x,y)u,\ u(x,0) = f(x),\ u_y(x,0) = 0,\ x > 0$$

with $p > -\frac{1}{2}$, then

$$(3.7)\qquad u(x,y) = \int_{x-y}^{x+y} f(\xi)w_Q(x,y,\xi)d\xi,\ 0 < y < x.$$

If $p = -\frac{1}{2}$, then

$$(3.8)\qquad u(x,y) = \tfrac{1}{2}\{f(x-y) + f(x+y)\} + \int_{x-y}^{x+y} f(\xi)w_Q(x,y,\xi)d\xi,\ 0 < y < x.$$

Hence the solution of the Cauchy problem (3.6) is unique.

Remark 1. In case $Q(x,y) = q(x) - q(y)$ this last observation together with (1.13) implies the so-called product formula

$$\omega_\lambda(x)\omega_\lambda(y) = \int_{x-y}^{x+y} w(x,y,\xi)\omega_\lambda(\xi)d\xi,\ 0 \leq y \leq x,$$

where ω_λ is defined by (1.1). This formula reduces to an addition formula for Bessel functions due to Sonine, cf. [19, p. 367] by taking $q(x) = 0$. Moreover we see that the definition of T in (1.13)-(1.15) makes sense if f is locally integrable and $Q(x,y) = q(x)-q(y)$ satisfies the conditions of Cor. 1.

Theorem 2. Assume $t^{2p+1}Q(t) \in L(0,1)$ if $-\frac{1}{2} < p < 0$, $t(\log t)Q(t) \in L(0,1)$ if $p = 0$, $tQ(t) \in L(0,1)$ if $p > 0$. Let $\tilde{w} = w_Q - w_0$. Let $\{(x+y)^2 - \xi^2\}\{\xi^2 - (x-y)^2\}$ be denoted by V.

Then

$$(3.9)\qquad |\tilde{w}(x,y,\xi)| \leq M\xi(xy)^{-1}\{-1 + \exp M\rho(0,y)\} \text{ if } p \geq \tfrac{1}{2},$$

$$(3.10)\qquad |\tilde{w}(x,y,\xi)| \leq Mw_0(x,y,\xi)\{-1 + \exp M\rho(0,y)\} \text{ if } 0 < p < \tfrac{1}{2},$$

$$(3.11)\qquad |\tilde{w}(x,y,\xi)| \leq M\xi^{2p+1}V^{-p-\frac{1}{2}} \int_0^y t^{2p+1}Q(t)dt \exp M\rho(0,y), \text{ if } -\tfrac{1}{2} < p < 0,$$

$$(3.12)\qquad |\tilde{w}(x,y,\xi)| \leq M\xi V^{-\frac{1}{2}}\left[\int_0^y tQ(t)dt + \int_0^y tQ(t) \log \frac{xy}{\xi t}dt\right] \exp M\rho(0,y), \text{ if } p = 0,$$

where M is a positive constant.

<u>Proof</u>. If $p > -\frac{1}{2}$, then (3.4), (2.9) and (2.10) imply

$$(3.13)\qquad |\tilde{w}(x,y,\xi)| \leq M \iint_{\Omega(x,y,\xi,0)} w_0(s,t,\xi)Q(t)R(s,t;x,y) \exp M\rho(t,y)dsdt.$$

The numbers M, M_1, ... denote positive constants which may be different in different formulas.

Let V_0 be V with (x,y) replaced by (s,t). If $p \geq \frac{1}{2}$, then by (3.1), (2.7) and (2.12)

$$w_0(s,t,\xi)R(s,t;x,y) \leq M(xy)^{-1}\xi(st)^{1-2p}V_0^{p-\frac{1}{2}} \leq M_1(xy)^{-1}\xi,$$

if $(s,t) \in \Omega(x,y,\xi,0)$. Hence by (3.13) and (2.12)

$$|\tilde{w}(x,y,\xi)| \leq M\xi(xy)^{-1} \int_0^y dt\ Q(t) \exp M\rho(t,y) \int_{\xi-t}^{\xi+t} ds.$$

From this we deduce (3.9).

If $|p| < \frac{1}{2}$ we use the estimate

$$(3.14)\qquad Z_1 \leq 16xystV^{-1},\quad (s,t) \in \Omega(x,y,\xi,0).$$

(Z_1 has been defined by (2.2) with (ξ,η) replaced by (s,t)). According to (2.7) we now have if $0 < |p| < \frac{1}{2}$

$$(3.15)\qquad R(s,t;x,y) \leq M(st)^{1+p-|p|}(xy)^{-p-|p|}V^{|p|-\frac{1}{2}}.$$

Since $\xi - t \leq s \leq \xi + t$, if $(s,t) \in \Omega(x,y,\xi,0)$, we have

$$(3.16)\qquad V_0 \leq 16s^2t^2.$$

Combining this with (3.13), (3.15) and (3.1) in case $0 < p < \frac{1}{2}$ we find

$$|\tilde{w}(x,y,\xi)| \leq Mw_0(x,y,\xi) \int_0^y dt\ Q(t)\ t\{\exp M\rho(t,y)\} \int_{\xi-t}^{\xi+t} V_0^{-\frac{1}{2}}sds.$$

Now we use the symmetry of V_1 in s, ξ and t and substitute $s^2 = (\xi-t)^2 + 4\xi tu$. Then we obtain (3.10).

In case $-\frac{1}{2} < p < 0$ we have by (3.13), (3.15) and (3.1)

$$(3.17)\qquad |\tilde{w}(x,y,\xi)| \leq M\xi V^{-p-\frac{1}{2}} \int_0^y dtQ(t)t\{\exp M\rho(t,y)\} \int_{\xi-t}^{\xi+t} sV_0^{p-\frac{1}{2}}ds.$$

With the same substitution for s as before we obtain (3.11). In case $p = 0$, (2.3) and (3.16) imply

$$Z_1^{-1} \leq \frac{xy}{st} \leq 4xyV_0^{-\frac{1}{2}}.$$

Hence (3.15) holds if the righthand side is multiplied by

$$\{1 + \log (xyV_0^{-\frac{1}{2}})\}.$$

Furthermore

$$\int_{\xi-t}^{\xi+t} sV_0^{-\frac{1}{2}} \log V_0^{-\frac{1}{2}} ds = -\tfrac{1}{2} \int_0^1 \frac{1}{\sqrt{u(1-u)}} \log \{4\xi t\sqrt{u(1-u)}\}du.$$

With (3.13), (3.15) and (3.1) now (3.12) follows.

<u>Theorem 3</u>. Let f be continuous on $[0,x_0]$. Assume that the conditions for Q in theorem 2 are satisfied. Let ρ and u be defined by (2.8) and (3.7) (in particular u may be a solution of the Cauchy problem (3.6) with $p > -\frac{1}{2}$, $f \in C^2[0,x_0]$).

Then there exist positive constants M_1 and M_2 independent of x_0 such that

$$(3.18)\qquad |u(x,y)| \leq \sup |f(\xi)|\{M_1 + M_2 \exp M\rho(o,y)\},\ \text{if } p > 0,$$

$$(3.19)\qquad (1 + \log^+ x)^{-1}(1 + \log^+ y)^{-1}|u(x,y)| \leq \sup |f(\xi)|(1 + \log^+ \xi)^{-1}. \\ .\{M_1 + M_2 \int_0^y tQ(t)(1 + \log^+ t^{-1})dt \exp M\rho(0,y)\},\ \text{if } p = 0,$$

$$(3.20)\qquad (1 + x^{-2p})^{-1}(1 + y^{-2p})^{-1}|u(x,y)| \leq \sup (1 + \xi^{-2p})^{-1}|f(\xi)|\ . \\ .\{M_1 + M_2 \int_0^y t^{2p+1}Q(t)dt \exp M\rho(0,y)\},\ \text{if } -\tfrac{1}{2} < p < 0.$$

Here the sup is taken over $x-y \leq \xi \leq x+y$, where $0 \leq y \leq x$ and $\log^+ t = \max (0,\log t)$. These estimates are best possible in a certain sense.

<u>Proof</u>. If $p > 0$ the proof is clear from (3.9) and (3.10). If $-\frac{1}{2} < p < 0$, we use (3.11). Now

$$\int_{x-y}^{x+y} (1 + \xi^{-2p})\xi^{2p+1}V^{-p-\frac{1}{2}}d\xi \leq \int_{x-y}^{x+y} [\xi\{(x+y)^2 - \xi^2\}^{-p-\frac{1}{2}}\{\xi^2 - (x-y)^2\}^{-\frac{1}{2}} + \\ + \xi V^{-p-\frac{1}{2}}]d\xi,$$

and (3.20) follows.

According to (3.12) we have in case $p = 0$

$$\left|\int_{x-y}^{x+y} \tilde{w}(x,y,\xi)f(\xi)d\xi\right| \leq M \sup \{|f(\xi)|(1 + \log^+ \xi)^{-1}\} \exp M\rho(0,y) \\ \int_{x-y}^{x+y} d\xi(1+\log^+ \xi)\xi V^{-\frac{1}{2}}[\int_0^y tQ(t)(1 - \log t)dt + \log (xy\xi^{-1}) \int_0^y tQ(t)dt].$$

Let

$$I = \int_{x-y}^{x+y} (1 + \log^+ \xi)\xi V^{-\frac{1}{2}} \log \frac{xy}{\xi} d\xi.$$

If $y \geq \frac{1}{2}x$, we substitute $\xi^2 = (x-y)^2 + 4xyu$. Then $\xi^2 \geq 4xyu$ and so

$$I \leq \tfrac{1}{4} \int_0^1 (1 + \log^+ 2x)\log^+ \frac{xy}{4u} \frac{du}{\sqrt{u(1-u)}} \leq M_1(1 + \log^+ x)(1 + \log^+ y).$$

If $y < \frac{1}{2}x$, then $\xi > \frac{1}{2}x$ and $\frac{xy}{\xi} \leq 2y$. Consequently

$$I \leq \int_{x-y}^{x+y} (1 + \log^+ 2x)\xi V^{-\frac{1}{2}} \log^+ 2y \, d\xi \leq M_1(1 + \log^+ x)(1 + \log^+ y).$$

Now (3.19) follows.

Let u_1 and u_2 be C^2-functions such that $u_1(x) = 1$, $u_2(x) = x^{-2p}$, if $0 < x \leq 1$ and $u_1(x) = x^{-2p}$, $u_2(x) = 1$ if $x \geq 2$. Then u_1 and u_2 are linearly independent solutions of a differential equation (0.3) if $p \neq 0$. Now $u_1(y)u_2(x)$ satisfies a corresponding Cauchy problem (3.6) with $Q(x,y) = q(x) - q(y)$, $f(x) = u_2(x)$, hence (3.7). If in the L.H.S. and R.H.S. of an estimate like (3.20) should occur $f(x)f(y)$ and $f(\xi)$ then we see $f(x) = (1+x^{-2p})^{-1}$. If $p = 0$, replace x^{-2p} by $\log x$.

<u>Remark 2</u>. In the special case $Q(x,y) = q(x) - q(y)$, where $|q(x) - q(y)| \leq Q(y)$, if $0 < y \leq x$ with q measurable and Q locally integrable on the positive axis we obtain estimates for the translation kernel $w(x,y,\xi)$ of section 1 if $p > -\frac{1}{2}$ (cf. (1.10), (1.13)). Estimates for the case $p = -\frac{1}{2}$ have been considered extensively by Hutson and Pym [11]. Therefore we do not consider these here.

Let

$$\psi(x) = 1, \text{ if } p > 0, \ \psi(x) = 1 + \log^+ x, \text{ if } p = 0, \ \psi(x) = 1 + x^{-2p}, \text{ if } -\tfrac{1}{2} \leq p < 0.$$

As in [11] we denote by $\Psi_\infty(R_+)$ the space of measurable functions on R_+ (the halfline $x \geq 0$) for which

$$||[f]||_\infty = \operatorname{ess\,sup}_{x \in R_+} |f(x)/\psi(x)| < \infty.$$

With the usual conventions $\Psi_\infty(R_+)$ is a Banach space. We define $\Psi_\infty(R_+^2)$ in a similar way using $\psi \otimes \psi$. Then we have the following theorem, which for $p = -\frac{1}{2}$ has been proved by Hutson and Pym [11] and which for $p > -\frac{1}{2}$ follows with theorem 3.

<u>Theorem 4</u>. Let $Q(x,y)$ be measurable for $0 \leq y \leq x$, $|Q(x,y)| \leq Q(y)$, if $0 \leq y \leq x$, where $xQ(x) \in L(1,\infty)$ and $xQ(x) \in L(0,1)$ if $p > 0$, $xQ(x) \log x \in L(0,1)$, if $p = 0$, $x^{2p+1}Q(x) \in L(0,1)$ if $-\frac{1}{2} \leq p < 0$.

Then the mapping $f(x) \to u(x,y)$ defined by (3.7) and (3.8) is a bounded mapping from $\Psi_\infty(R_+)$ in $\Psi_\infty(R_+^2)$. In particular if q is measurable on R_+ and $|q(x) - q(y)| \leq Q(y)$, $0 \leq y \leq x$, then the translation defined by (1.13) - (1.15) is bounded from $\Psi_\infty(R_+)$ to $\Psi_\infty(R_+^2)$.

Hutson and Pym [11] have shown that if q is of bounded variation on R_+ then there exists a positive decreasing function $\tilde{q}$ on R_+ such that $|q(y) - q(x)| \leq \tilde{q}(y) - \tilde{q}(x)$ a.e for $0 \leq y \leq x$, and we may take $Q(x) = \tilde{q}(x)$.
Norm estimates for the translation operator in certain L^n spaces with suitable measures analogous to those obtained by Hutson and Pym in the case $p = -\frac{1}{2}$ may be derived for arbitrary $p > -\frac{1}{2}$. These results will be published elsewhere.

4. Symmetry of the translation kernel. From now on we assume $Q(x,y) = q(x) - q(y)$. With w defined by (1.10) we denote

$$(4.1)\qquad \begin{aligned} K(x,y,z) &= z^{-2p-1}w(x,y,z) \text{ if } 0 \leq x - y \leq z \leq x + y,\\ K(x,y,z) &= z^{-2p-1}w(y,x,z) \text{ if } 0 \leq y - x \leq z \leq x + y,\\ K(x,y,z) &= 0 \text{ otherwise.}\end{aligned}$$

The symmetry of K in its three variables which is suggested by (1.6) and (3.1) will now be proved if $q \in L(0,a)$ for any $a > 0$. The proof depends on the product formula in remark 1 and the following inversion formula for an integral transform associated with the differential operator L, defined by (0.3).

Theorem 5. Let $q \in L(0,\infty)$, $p \geq -\frac{1}{2}$. Let $\sigma_\lambda(x)$ be the solution of $Lu = -\lambda^2 u$ such that

$$(4.2)\qquad \sigma_\lambda(x) \sim (\lambda x)^{-p-\frac{1}{2}}e^{i\lambda x},\ \frac{d}{dx}\sigma_\lambda(x) \sim i\lambda(\lambda x)^{-p-\frac{1}{2}}e^{i\lambda x},$$

as $x \to \infty$, $\operatorname{Im}\lambda \geq 0$. Let $W(\lambda)$ be the Wronskian of $(\lambda x)^{p+\frac{1}{2}}\omega_\lambda(x)$ and $(\lambda x)^{p+\frac{1}{2}}\sigma_\lambda(x)$.

Suppose f is a function on R_+ which is of bounded variation in a neighbourhood of t_0 and such that

$$(4.3)\qquad t^{p+\frac{1}{2}}e^{ct}f(t) \in L_1(0,\infty),$$

where t_0 and c are real numbers, $t_0 > 0$, $c \geq 0$.

Then $W(\lambda) \sim \alpha\lambda$, as $\lambda \to \infty$ on $\operatorname{Im}\lambda \geq 0$ where α is some constant and

$$(4.4)\qquad \frac{i}{\pi}\lim_{\mu\to\infty}\int_{ic-\mu}^{ic+\mu} d\lambda\,\frac{\lambda^{2p+2}}{W(\lambda)}\,\sigma_\lambda(t_0)\int_0^\infty f(t)\omega_\lambda(t)t^{2p+1}dt = \tfrac{1}{2}\{f(t_0 - 0) + f(t_0 + 0)\},$$

if the path of integration in the λ-plane is lying above the zeros of $W(\lambda)$.

This theorem is an extension of theorem 1 of de Snoo [18]. The proof can be found in [2] where also the relation of (4.4) to (1.2) is given. We remark that (4.4) in the case $q = 0$ has been proved by Cherry [6].

Let x_0 and y_0 be fixed positive numbers, $x_0 \neq y_0$. From the definition of K, (1.10) and (2.5) it is clear that $K(x_0,y_0,z)$ only depends on the values of q on $[0,x_0 + y_0]$. Now replace q by $\tilde{q}$ such that $\tilde{q} = q$ on $[0,x_0 + y_0]$ and $\tilde{q} = 0$ on $[x_0 + y_0 + 1,\infty)$, $\tilde{q}$ is continuous on $[x_0 + y_0, x_0 + y_0 + 1]$. The corresponding functions defined previously with q will also be provided with ~. Then $\tilde{K}(x_0,y_0,z) = K(x_0,y_0,z)$, $\tilde{\omega}_\lambda(z) = \omega_\lambda(z)$ if $0 \leq z \leq x_0 + y_0$. Moreover there is a constant $\alpha_1 \neq 0$ such that $\tilde{\sigma}_\lambda(z) = \alpha_1(\lambda z)^{-p}H_p^{(1)}(\lambda z)$ if $z \geq x_0 + y_0 + 1$.

Let $\tilde{\tau}_\lambda(x)$ be the solution of $\tilde{L}u = -\lambda^2 u$ such that
$\tilde{\tau}_\lambda(x) \sim \lambda^p x^{-p} J_{-p}(\lambda x)$, $\tilde{\tau}_\lambda'(x) \sim -2p\lambda^p x^{-p-1} J_{-p}(\lambda x)$, $x \to 0$ if p is not an integer.
Then $\tilde{\tau}_\lambda(x)$ is an even entire function of λ and $W\{(\lambda x)^{p+\frac{1}{2}}\tilde{\omega}_\lambda(x),(\lambda x)^{p+\frac{1}{2}}\tilde{\tau}_\lambda(x)\} = k\lambda^{2p+1}$, k a constant $\neq 0$.

Now

(4.5) $$\tilde{\sigma}_\lambda(x) = A_1(\lambda)\tilde{\omega}_\lambda(x) + A_2(\lambda)\tilde{\tau}_\lambda(x).$$

By considering values of $x > x_0 + y_0 + 1$ we infer that $A_j(\lambda) = \lambda^{-2p}\phi_j(\lambda) + \psi_j(\lambda)$, where $\phi_j(\lambda)$ and $\psi_j(\lambda)$ are entire functions, $j = 1,2$. Hence

(4.6) $$\tilde{W}(\lambda) = kA_2(\lambda)\lambda^{2p+1}.$$

This implies that $\tilde{W}(\lambda)$ has only a finite number of zeros with $\text{Im}\,\lambda \geq 0$. Let the zeros with $\text{Im}\,\lambda > 0$ be $\beta_1,\ldots,\beta_n$ and those on the axis $\lambda \geq 0$ be $\gamma_1,\ldots,\gamma_m$.

We now combine the product formula of remark 1 with theorem 5 where L is replaced by $\tilde{L}$ etc. and $f(t) = K(x_0,y_0,t)$ (cf.(4.1)). If $p > \frac{1}{2}$, then it follows from (3.4) and (3.1) that f is of bounded variation. Hence

(4.7) $$\frac{i}{\pi}\lim_{\mu\to\infty}\int_{-\mu}^{\mu}\tilde{\sigma}_\lambda(z_0)\omega_\lambda(x_0)\omega_\lambda(y_0)\frac{\lambda^{2p+2}}{\tilde{W}(\lambda)}\,d\lambda = K(x_0,y_0,z_0),$$

if $|x_0 - y_0| < z_0 < x_0 + y_0$. From (4.5) and (4.6) we deduce that the integrand in (4.7) equals

(4.8) $$\frac{\lambda}{k}\{\frac{A_1(\lambda)}{A_2(\lambda)}\tilde{\omega}_\lambda(z_0) + \tilde{\tau}_\lambda(z_0)\}\omega_\lambda(x_0)\omega_\lambda(y_0).$$

Therefore

$$K(x_0,y_0,z_0) = \frac{i}{\pi k}\int_0^\infty \lambda\{\frac{A_1(\lambda)}{A_2(\lambda)} - \frac{A_1(-\lambda)}{A_2(-\lambda)}\}\omega_\lambda(x_0)\omega_\lambda(y_0)\omega_\lambda(z_0)d\lambda + S(x_0,y_0,z_0),$$ where

$S(x_0,y_0,z_0)$ is equal to twice the sum of the residues of (4.8) in $\beta_1,\ldots,\beta_n$ and the sum of the residues in $\gamma_1,\ldots,\gamma_m$. If some $\gamma = 0$ we replace the corresponding residue by the limit for $\varepsilon \to 0$ of the integral over the half circle $|\lambda| = \varepsilon$, $0 \leq \arg\lambda \leq \pi$. Hence $K(x_0,y_0,z_0)$ is symmetric in x_0, y_0 and z_0 if $p>\frac{1}{2}$, $p\neq 1,2,\ldots$ Since K depends analytically on p if $\text{Re}\,p > -\frac{1}{2}$ we have

Theorem 6. The translation kernel K is symmetric in its three variables if $q \in L(0,1)$.

We may remark that a shorter proof can be given in case q is real by applying the spectral theorem.

5. Positivity of the translation. We now consider cases where the kernel K is positive. The simplest case occurs when q in (0.3) is monotonically decreasing and positive. More generally if $Q(x,y) \leq 0$ if $0 \leq y \leq x$ then the mapping $f \to u$ given by (3.7) is positive. This immediately follows from (2.5), (3.1) and (3.4).

Another case has been considered by Chébli [4] using a method of Protter and Weinberger [16] for maximum principles. Here we prove in the same way a more general result.

Theorem 7. Let $p(x)$ be monotonically decreasing and non-negative if $x > 0$ and $p(x) - \frac{\alpha}{x} \in L(0,1)$ for some $\alpha \geq 0$. Suppose $Q(x,y)$ is integrable and non-positive if $0 \leq y \leq x$. Let u be a solution of the Cauchy problem.

$$(5.1) \qquad \frac{\partial^2 u}{\partial x^2} - \frac{\partial^2 u}{\partial y^2} + p(x)\frac{\partial u}{\partial x} - p(y)\frac{\partial u}{\partial y} - Q(x,y)u = 0, \; u(x,o) = f(x),$$

$$u_y(x,0) = 0, \; x > 0.$$

If f is positive then $u(x,y)$ is positive if $0 \leq y \leq x$.

Proof. Define

$$A(x) = x^\alpha \exp \int_0^x \{p(t) - \frac{\alpha}{t}\}dt.$$

Then $p(x) = A'(x)/A(x)$. Put $a(x,y) = A(x)A(y)$. Now (5.1) is equivalent to

$$(5.2) \qquad (au_x)_x - (au_y)_y = Qau, \; u(x,0) = f(x), \; u_y(x,0) = 0.$$

Let u be the solution and let $u_\varepsilon = u + \varepsilon \exp(ky)$, $\varepsilon > 0$. Then we may deduce as in section 1

$$2a(x,y)u_\varepsilon(x,y) =$$

$$\iint_{\Delta_{xy}} [-Q(\xi,\eta)a(\xi,\eta)u_\varepsilon(\xi,\eta) + \varepsilon a(\xi,\eta)e^{k\eta}\{k^2 + kp(\eta) + Q(\xi,\eta)\}]d\xi d\eta \; +$$

$$\int_{(x-y,0)}^{(x,y)} a(\xi,\eta)u_\varepsilon(\xi,\eta)\{p(\xi)+p(\eta)\}d\xi + \varepsilon k\, A(0) \int_{x-y}^{x+y} A(\xi)d\xi \; +$$

$$\int_{(x+y,0)}^{(x,y)} a(\xi,\eta)u_\varepsilon(\xi,\eta)\,\{p(\eta) - p(\xi)\}d\eta + (au_\varepsilon)(x-y,0) + (au_\varepsilon)(x+y,0).$$

Suppose $f \geq 0$ and consider u_ε on a fixed triangle Δ. Choose $k^2 =$ ess sup $\{1 - Q(x,y)\}$ on Δ, $k > 0$. Suppose u_ε possesses a zero on Δ. Let (x_0,y_0) be a zero with minimal ordinate. Then $y_0 > 0$ and $u_\varepsilon > 0$ on $\Delta_{x_0y_0} \setminus \{(x_0,y_0)\}$. Since $p(\eta) - p(\xi) \geq 0$ and $Q(\xi,\eta) \leq 0$ on Δ we conclude $u_\varepsilon(x_0,y_0) > 0$ and we have a contradiction. Hence $u_\varepsilon > 0$ on Δ for any $\varepsilon > 0$ and so $u(x,y) \geq 0$ on Δ.

Corollary 2. The translation defined by (1.13) and (1.14) is positive if

$$(5.3) \qquad q(x) = r(x) + \tfrac{1}{4}s^2(x) + \tfrac{1}{2}s'(x) + \frac{2p+1}{2x}\,s(x),$$

where $r(x)$ and $\frac{2p+1}{x} + s(x)$ are monotonically decreasing, $p \geq -\frac{1}{2}$, $\frac{2p+1}{x} + s(x) \geq 0$, $r(x) \geq 0$, s is continuously differentiable for $x > 0$ and $s \in L(0,1)$. In particular $K(x,y,z) \geq 0$.

This follows by choosing $Q(x,y) = r(x) - r(y)$, $p(x) = \frac{2p+1}{x} + s(x)$ and replacing u by u $\exp -\frac{1}{2}\int_0^x s(t)dt$ in theorem 7. The corresponding theorem of Chébli follows by taking $r = 0$.

Remark 3. Under the conditions of corollary 2 it follows from a theorem of Sturm that $\omega_\lambda(x)$ is positive if λ is imaginary.

Positivity of the translation implies boundedness of a modified translation on suitably chosen L^n-spaces.

Theorem 8. Suppose q is real, $q \in L(0,1)$, $p > -\frac{1}{2}$ and the translation is positive. Suppose Ω is a positive function such that

$$T^y\Omega(x) \le \Omega(x)\Omega(y).$$

(For example $\Omega(x) = \omega_\lambda(x)$, λ imaginary, in the case of corollary 2).
Define

$$K_1(x,y,z) = \frac{K(x,y,z)}{\Omega(x)\Omega(y)\Omega(z)},\ d\mu_1(z) = z^{2p+1}\Omega^2(z)dz,$$

$$T_1{}^y f(x) = \int_{|x-y|}^{x+y} K_1(x,y,z)f(z)d\mu_1(z).$$

Then $||T_1{}^y||_{n,\mu_1} \le 1$ if $1 \le n \le \infty$ where the $L^n(0,\infty;d\mu_1)$ norm is taken.

Proof. The kernel $K_1(x,y,z)$ is symmetric by theorem 6 and $0 \le \int_0^\infty K_1(x,y,z)d\mu_1(z) \le$ ≤ 1.

Hence if n and m are conjugate indices $n > 1$, and $f \in L^n(0,\infty;d\mu_1)$, then

$$\int_0^\infty |T_1{}^y f(x)|^n d\mu_1(x) \le \int_0^\infty\int_0^\infty K_1(x,y,z)|f(z)|^n d\mu_1(z)\{\int_0^\infty K_1(x,y,z)d\mu_1(z)\}^{\frac{n}{m}}d\mu_1(x)$$

$$\le \int_0^\infty |f(z)|^n \int_0^\infty K_1(x,y,z)d\mu_1(x)d\mu_1(z) \le ||f||^n_{n,\mu_1}.$$

If $n = 1$ or ∞ the proof is immediate. Combining remarks 1 and 3 we get the special case $\Omega(x) = \omega_\lambda(x)$, λ imaginary.

Example. Flensted-Jensen and Koornwinder [9] considered the differential equation

$$\Delta^{-1}\frac{d}{dt}(\Delta f) + \rho^2 f = -\lambda^2 f,\ \Delta(t) = 2^{2\rho}(\text{sh}t)^{2\alpha+1}(\text{ch}t)^{2\beta+1},\ \rho=\alpha+\beta+1,\ \alpha\ge\beta\ge-\tfrac{1}{2}.$$

Substituting $f = \Delta^{-\frac{1}{2}}t^{\alpha+\frac{1}{2}}u$ we obtain $Lu = -\lambda^2 u$, where $p = \alpha$ and q is given by (5.3) with $r(x) = -\rho^2$, $s(x) = (2\alpha+1)(\text{cth}\ x - x^{-1}) + (2\beta+1)\text{th}\ x$.
Then

$$\omega_\lambda(x) = (x^{-1}\text{sh}\ x)^{\alpha+\frac{1}{2}}(\text{ch}\ x)^{\beta+\frac{1}{2}}F(\tfrac{1}{2}(\rho+i\lambda),\tfrac{1}{2}(\rho-i\lambda);\alpha+1;-(\text{sh}\ x)^2).$$

The factor of F is $\omega_{i\rho}$. These functions with λ imaginary may be taken as Ω. In [9] the case $\Omega = \omega_{i\rho}$ has been considered and the kernel K_1 is explicitly determined as a hypergeometric function.

6. Concluding remarks. We may define convolutions associated with L by means of

$$(f*g)(y) = \int_0^\infty (T^y f)(x)g(x)x^{2p+1}dx.$$

Using norm estimates for the translation like those in theorem 8 or those mentioned at the end of section 3 we may obtain norm estimates for the convolutions. Further results from harmonic analysis may be generalized such as Paley Wiener theorems (cf. Flensted-Jensen [8], Chébli [5]). These will be published elsewhere.

References

[1] S.Bochner, Sturm-Liouville and heat equations whose eigenfunctions are ultraspherical polynomials or associated Bessel functions, Proc. Conf. Diff. Eq., Maryland (1956).

[2] B.L.J.Braaksma and H.S.V.de Snoo, Internal report, Dept. of Math., University of Groningen (1974).

[3] H.Chébli, Sur la positivité des opérateurs de "translation généralisée" associés à un opérateur de Sturm-Liouville sur]0,∞[, C.R. Acad. Sc. Paris, 275, 601-604 (1972).

[4] H.Chébli, Sur la positivité des opérateurs de "translation généralisée" associés à un opérateur de Sturm-Liouville sur]0,∞[, Séminaire de Théorie Spectrale, Strasbourg, 1972-73.

[5] H.Chébli, Sur un théorème de Paley-Wiener associé à la décomposition spectrale d'un opérateur de Sturm-Liouville sur]0,∞[, Report Institut de Recherche Math. Avancée Strasbourg (1974).

[6] T.M.Cherry, On expansions in eigenfunctions, particularly in Bessel functions, Proc. London Math. Soc. 51, 14-45 (1949-50).

[7] J.Delsarte, Sur une extension de la formule de Taylor, Journal de Math. Pur. Appl. 17, 213-231 (1936).

[8] M.Flensted-Jensen, Paley-Wiener type theorems for a differential operator connected with symmetric spaces, Arkiv för Matematik 10, 143-162 (1972).

[9] M.Flensted-Jensen and T.H.Koornwinder, The convolution structure for Jacobi function expansions, Arkiv för Matematik 11, 245-262 (1973).

[10] V.Hutson and J.S.Pym, Generalized translations associated with a differential operator, Proc. London Math. Soc. 24, 548-576 (1972).

[11] V.Hutson and J.S.Pym, Measure algebras associated with a second order differential operator, Journ. Functional An. 12, 68-96 (1973).

[12] N.Leblanc, Classification des algèbres de Banach associées aux opérateurs différentiels de Sturm-Liouville, Journ. Functional An. 2, 52-72 (1968).

[13] N.Leblanc, Algèbres de Banach associées à un opérateur différentiel de Sturm-Liouville, Springer lecture notes no. 336, 40-50 (1973).

[14] B.M.Levitan, The application of generalized displacement operators to linear differential equations of the second order, Translation no. 59, Amer. Math. Soc. (1950).

[15] B.M.Levitan, Generalized translation operators and some of their applications, Jerusalem, 1964.

[16] M.H.Protter and H.F.Weinberger, Maximum principles in differential equations, Englewood Cliffs, 1967.

[17] B.Riemann, Gesammelte Mathematische Werke, Dover publ., New York, 1953.

[18] H.S.V.de Snoo, Inversion theorems for some generalized Laplace transforms I, II, Proc. Kon. Nederl. Akad. Wet. , Ser. A, 73, 222 - 244 (1970).

[19] G.N.Watson, A treatise on the theory of Bessel functions, 2nd ed., Cambridge, 1966.

MULTI-PARAMETER PROBLEMS

Patrick J. Browne

In this paper we shall report on recent results and directions in multi-parameter spectral theory. Since Atkinson's survey of the field in 1968 (see [1]) several new and interesting results have been obtained and recently an abstract functional analytic approach to the problem has been studied (see [4]). Specifically we are interested in the $k \times k$ system of second order ordinary differential equations

$$\frac{d^2 y_r(x_r)}{dx_r^{\,2}} + q_r(x_r) y_r(x_r) + \sum_{s=1}^{k} \lambda_s a_{rs}(x_r) y_r(x_r) = 0, \qquad (1)$$

$r = 1,2,\ldots,k$, where $\lambda_1,\ldots,\lambda_k$ are complex parameters and the functions $a_{rs}(x_r)$, $q_r(x_r)$ are real valued and continuous.

We introduce a definiteness condition on the functions $a_{rs}(x_r)$; viz.

$$\det\left(a_{rs}(x_r)\right) > 0. \qquad (2)$$

The first case to be considered was the "finite interval problem". Let $a = (a_1,\ldots,a_k)$, $b = (b_1,\ldots,b_k) \in R^k$, $a_r \le x_r \le b_r$, $r = 1,2,\ldots,k$ and consider the multi-parameter problem consisting of the differential equations (1) subject to the definiteness condition (2) and boundary conditions

$$y_r(a_r)\cos\alpha_r + y_r'(a_r)\sin\alpha_r = 0, \quad 0 < \alpha_r < \pi, \qquad (3)$$

$$y_r(b_r)\cos\beta_r + y_r'(b_r)\sin\beta_r = 0, \quad 0 \le \beta_r < \pi, \quad r = 1,2,\ldots,k. \qquad (4)$$

An eigenvalue $\lambda = (\lambda_1,\dots,\lambda_k)$ and eigenfunction $\prod_{r=1}^{k} f_r(x_r)$ for this problem is a k-tuple of (necessarily real) numbers and functions $f_r(x_r)$, $r = 1,2,\dots,k$ satisfying the equations (1) and conditions (3),(4).

In this setting it has been proved that the eigenvalues, repeated according to multiplicity, have no finite point of accumulation and that there is a set of eigenfunctions forming an orthonormal basis in the space $L^2([a,b])$ with respect to the weight function $|A|(x) = \det\big(a_{rs}(x_r)\big)$. That is to say, if λ^m, $m = 1,2,\dots$ is an enumeration of the eigenvalues repeated according to multiplicity there is a corresponding set of eigenfunctions $\psi(x,\lambda^m)$ such that

$$\int_{[a,b]} |A|(x)\psi(x,\lambda^m)\overline{\psi(x,\lambda^n)}\,dx = \delta_{mn},$$

$$\int_{[a,b]} |A|(x)|f(x)|^2 dx = \sum_m \left| \int_{[a,b]} |A|(x)f(x)\psi(x,\lambda^m)\,dx \right|^2$$

for all $f \in L^2([a,b])$.

For details of these results the reader may consult Faierman [7], Browne [2] and Sleeman [9].

The singular case in which each variable x_r is allowed to range over a half line (a_r,∞) has been studied by the author [3]. In this case the Parseval equality involves a Stieltjes integral with respect to a positively monotonic function of k real variables thus generalizing the well-known result for the 1 parameter (i.e. Sturm-Liouville) problem.

The doubly singular case in which each variable x_r ranges over $(-\infty,\infty)$ has as yet to be resolved. The Parseval equality and generalized eigenvector expansion for the corresponding Sturm-Liouville problem involves a 2×2 matrix measure--see for example Coddington and Levinson [5, pp. 246-252]. The following reformulation of results from the finite interval problem leads to the conjecture that the Parseval equality for the doubly singular case involves a matrix measure defined on R^k and of size $2^k \times 2^k$.

Let $a_r < 0 < b_r$, $r = 1,2,\ldots,k$ and let $\theta_r(x_r,\lambda)$, $\varphi_r(x_r,\lambda)$ be solutions of the system (1) satisfying $\theta_r(0,\lambda) = \varphi_r'(0,\lambda) = 0$, $\theta_r'(0,\lambda) = \varphi_r(0,\lambda) = 1$, $r = 1,2,\ldots,k$ for all λ. These functions will be real valued for real k-tuples λ since the coefficient functions in the system (1) are real valued.

For $\sigma \subset \{1,2,\ldots,k\}$, put $\sigma' = \{1,2,\ldots,k\} - \sigma$ and define

$$h^{\sigma}(x,\lambda) = \prod_{r\in\sigma} \theta_r(x_r,\lambda) \prod_{s\in\sigma'} \varphi_s(x_s,\lambda).$$

Each $h^{\sigma}(x,\lambda)$ is real valued for real k-tuples λ and there are 2^k such functions h^{σ}.

For the multi-parameter problem (1) over $[a,b]$ subject to the conditions (2),(3),(4) let λ^m, $m = 1,2,\ldots$, be an enumeration of the eigenvalues and $\psi(x,\lambda^m) = \prod_{r=1}^{k} \psi_r(x_r,\lambda^m)$ the corresponding $|A|(x)$-orthonormal eigenfunctions. The functions $\psi_r(x_r,\lambda^m)$ will be real valued and further there are real constants A_r^m, B_r^m such that

$$\psi_r(x_r,\lambda^m) = A_r^m\theta_r(x_r,\lambda^m) + B_r^m\varphi_r(x_r,\lambda^m), \quad r = 1,2,\ldots,k.$$

Thus

$$\begin{aligned}\psi(x,\lambda^m) &= \prod_{r=1}^{k} [A_r^m\theta_r(x_r,\lambda^m) + B_r^m\varphi_r(x_r,\lambda^m)] \\ &= \sum_{\sigma} \gamma_m^{\sigma} h^{\sigma}(x,\lambda^m)\end{aligned}$$

where γ_m^{σ} are (real) constants determined from A_r^m, B_r^m by expanding the above product. Specifically

$$\gamma_m^{\sigma} = \prod_{r\in\sigma} A_r^m \prod_{s\in\sigma'} B_s^m .$$

Here σ ranges through the 2^k subsets of $\{1,2,\ldots,k\}$.

For an interval $J \subset R^k$ whose closure is $[c,d]$ we define $\Omega_{\sigma\tau}(J) = \sum_m \gamma_m^{\sigma} \gamma_m^{\tau}$, the summation extending over those indices m for which $\lambda^m \in [c,d)$. The number of terms in this summation is finite as the eigenvalues have no finite point of accumulation. $\Omega_{\sigma\tau}$ is an additive function of intervals. Following the discussion in [8, §45.5, p. 246], for each $t \in R^k$ we define $n(t)$ to be the number of indices i for which $t_i < 0$ and the interval $I(t)$ as

$$I(t) = \{(s_1,\ldots,s_k) \mid \inf(t_i,0) \le s_i \le \sup(t_i,0), \quad i = 1,\ldots,k\}.$$

Finally we define

$$\rho_{\sigma\tau}(t) = (-1)^{n(t)} \, \Omega[I(t)].$$

Using McShane's concept of a function of bounded variation [8, §46.2, p. 248] and the discussion in [8, §45.5, p. 246] we claim that $\rho_{\sigma\tau}(t)$ has the properties:

(i) if $J \subset R^k$ is an interval whose closure is $[c,d]$, $\rho_{\sigma\tau}(t)$ is of bounded variation over J, its total variation over J being given by $\sum_m |\gamma_m^{\sigma} \gamma_m^{\tau}|$ the summation extending over those indices m for which $\lambda^m \in [c,d)$,

(ii) $\rho_{\sigma\tau}(t)$ is left continuous,

(iii) $\Delta_{\rho_{\sigma\tau}}(J) = \Omega_{\sigma\tau}(J)$ for all intervals $J \subset R^k$,

(iv) if $|\rho_{\sigma\tau}|(t)$ denotes the function constructed in like manner from the total variation $|\Omega_{\sigma\tau}|$ of $\Omega_{\sigma\tau}$, then $\rho_{\sigma\tau}(t)$ and $|\rho_{\sigma\tau}|(t)$ are zero if any of the numbers $t_1,\ldots,t_k$ are zero.

With respect to such a function $\rho_{\sigma\tau}(t)$ we may form Stieltjes integrals of the form $\int_{R^k} f(t)\, d\rho_{\sigma\tau}(t)$. This integration is clearly equivalent to summation of the sequence $f(\lambda^m)$ with respect to the weight sequence $\gamma_m^{\sigma}\, \gamma_m^{\tau}$.

For an interval $J \subset R^k$, it is readily seen from the definition of $\Omega_{\sigma\tau}(J)$ that $\left(\Omega_{\sigma\tau}(J)\right)$ is an Hermitian matrix. Further if (ξ_σ) is a 2^k-tuple of complex numbers we have

$$\begin{aligned}\sum_{\sigma,\tau} \Omega_{\sigma\tau}(J)\, \xi_\sigma\, \bar{\xi}_\tau &= \sum_{\sigma,\tau} \sum_m \gamma_m^{\sigma}\, \gamma_m^{\tau}\, \xi_\sigma\, \bar{\xi}_\tau \\ &= \sum_m \left|\sum_\sigma \gamma_m^{\sigma}\, \xi_\sigma\right|^2 \\ &\geq 0,\end{aligned}$$

showing that $\left(\Omega_{\sigma\tau}(J)\right)$ is positive semi-definite. For $t \in R^k$, the Hermiticity of $\left(\rho_{\sigma\tau}(t)\right)$ follows from the Hermiticity of $\Omega_{\sigma\tau}\left(I(t)\right)$ and the definition of $\rho_{\sigma\tau}(t)$.

Each function $\rho_{\sigma\tau}(t)$ generates a regular (signed) measure on the bounded Borel subsets of R^k, $\left(\rho_{\sigma\sigma}(t)\right.$ will generate a positive measure$\left.\right)$. Further the $\rho_{\sigma\tau}$-measure of an interval $J = [\lambda,\mu)$ is $\Omega_{\sigma\tau}(J)$. Details of these results may be found in [8, §§47-56, pp. 251-311]. The upshot of these remarks is that if we denote the measure generated by $\rho_{\sigma\tau}(t)$ as $d\rho_{\sigma\tau}$, then $(d\rho_{\sigma\tau})$ is a $2^k \times 2^k$ positive matrix measure on the bounded Borel subsets of R^k. The theory of matrix measures on R^1 is given in [6, pp. 1337-1350] and is readily extended to cover the case of R^k.

As a particular consequence of this observation we claim that if $\left(F_\sigma(t)\right)$ is a 2^k-tuple of bounded Borel functions defined on a bounded interval $J \subset R^k$ then

$$\int_J \sum_{\sigma,\tau} F_\sigma(t)\, \overline{F_\tau(t)}\, d\rho_{\sigma\tau}(t) \geq 0.$$

Now the completeness result for eigenfunctions over the interval $[a,b]$ states that for $f \in L^2([a,b])$,

$$\int_{[a,b]} |A|(x)\ |f(x)|^2\ dx = \sum_m \Big| \int_{[a,b]} |A|(x)\ f(x)\ \psi(x,\lambda^m)\ dx\Big|^2$$

$$= \sum_m \Big| \int_{[a,b]} |A|(x)\ f(x)\ \sum \gamma_m^{\sigma}\ h^{\sigma}(x,\lambda^m)\ dx\Big|^2.$$

For such a function $f \in L^2([a,b])$ let us define

$$F_{\sigma}(t) = \int_{[a,b]} |A|(x)\ f(x)\ h^{\sigma}(x,t)\ dx \qquad (5)$$

for each subset $\sigma \subset \{1,2,\ldots,k\}$. Here t ranges through R^k. Then incorporating our spectral matrix $(\rho_{\sigma\tau})$ and the transform (5) we may write the above Parseval equality as

$$\int_{[a,b]} |A|(x)\ |f(x)|^2\ dx = \int_{R^k} \sum_{\sigma,\tau} F_{\sigma}(t)\ \overline{F_{\tau}(t)}\ d\rho_{\sigma\tau}^{\ \delta}(t) \qquad (6)$$

where we have rewritten the spectral matrix as $(\rho_{\sigma\tau}^{\ \delta})$ to display the fact that the analysis is relevant to the interval $\delta = [a,b] \subset R^k$.

This reformulation of the Parseval equality is the k-dimensional analogue of the work of Coddington and Levinson [8, p. 246] used as a starting point for their discussion of the spectral theory of the Sturm-Liouville problem with two singularities.

The basic problem now in the k-dimensional setting is to let the interval δ expand to R^k and to investigate the limiting nature of the spectral matrix $(\rho_{\sigma\tau}^{\ \delta})$ and the Parseval equality (6). It would be of interest to discover what conditions are needed (if any) to ensure the existence of a limiting spectral matrix together with a corresponding limiting Parseval equality. As we mentioned earlier the existence of such a limiting matrix would generalize the well-known theory of the 1 parameter (i.e. Sturm-Liouville) problem.

REFERENCES

[1] F.V. ATKINSON, Multiparameter Spectral Theory, *Bull. Amer. Math. Soc.*, 74 (1968), 1-27.

[2] P.J. BROWNE, A Multi-parameter Eigenvalue Problem, *J. Math. Anal. Appl.*, 38 (1972), 553-568.

[3] P.J. BROWNE, A Singular Multi-parameter Eigenvalue Problem in Second Order Ordinary Differential Equations, *J. Differential Equations*, 12 (1972), 81-94.

[4] P.J. BROWNE, Multi-parameter Spectral Theory, *Indiana Univ. J. Math.*, to appear.

[5] E.A. CODDINGTON and N. LEVINSON, *Theory of Ordinary Differential Equations*, McGraw-Hill, New York, 1955.

[6] N. DUNFORD and J.T. SCHWARZ, *Linear Operators Part II*, Interscience, New York, 1963.

[7] M. FAIERMAN, The Completeness and Expansion Theorems Associated with the Multi-parameter Eigenvalue Problem in Ordinary Differential Equations, *J. Differential Equations*, 5 (1969), 197-213.

[8] E. MCSHANE, *Integration*, Princeton University Press, Princeton, 1944.

[9] B.D. SLEEMAN, Completeness and Expansion Theorems for a Two-parameter Eigenvalue Problem in Ordinary Differential Equations using Variational Principles, *J. Lond. Math. Soc.* (2), 6 (1973), 705-712.

MONOTONICITY WITH DISCONTINUITIES IN PARTIAL DIFFERENTIAL EQUATIONS

L. Collatz

Professor Dr. Helmut Heinrich
dedicated to his seventieth birthday

Summary

In this survey monotonicity theorems for linear and nonlinear partial differential equations of second order are collected, in which discontinuities in the coefficients of the differential equation or in lower and upper bounds v,w for the solution u may occur. Discontinuities at certain interfaces are considered. If v,w are continuous, but their normal derivatives not, one has interface conditions for the normal derivatives. The form of these conditions can be found from a quite general geometrical principle. Results for special types of differential equations are cited. For monotone operators of contractive type one gets often error bounds, for monotone operators of non contractive type the iteration procedure still gives a numerical method; for demonstration a nonlinear boundary value problem with several solutions is considered.

1. Introduction and the linear space C_q^k -

In the following we ask for inclusion theorems for the solutions of partial differential equations of second order, and for upper and lower bounds, which hold pointwise in the considered domains. We suppose the existence of the desired solutions and use properties of monotonicity in the pointwise sense, not in the sense of Minty (with inner products).

For functions with continuous partial derivatives up to order two, pointwise monotonicity was proved for very general types of nonlinear differential equations in the elliptic case by Redheffer [61] and in the parabolic case by Westphal-Nagumo [49], Redheffer [62], Nickel [62] a.o.

In the hyperbolic case monotonicity is not valid in the same generality, compare for instance Walter [61], Gloistehn [61], Sather [67], Hofmann [74] a.o.

In this paper we deal with discontinuities.
Discontinuities in boundary value problems with elliptic or parabolic differential equations have been considered more frequently during the last years. There are several attempts to generalize the theorems of monotonicity to cases in which the coefficients in the differential equations or the considered approximate solutions or their derivatives are not continuous. One has tried to use Sobolev spaces and more general spaces or to prove the monotonicity directly with the aid of the distribution-theory or by elementary considerations. The latter have the advantage of easier applicability.

We will use the linear space $C_q^k(D)$ of functions, which are defined in a given open domain D of the n-dimensional point space R^n, which are continuous in D and have continuous partial derivatives up to order k (k included) and piecewise continuous partial derivatives up to order q (q included). This space $C_q^k(D)$ is not complete with respect to the usual inner products or norms, but it occurs frequently in the numerical analysis (spline functions, F.E.M. = Finite Elements Method, a.o.) and in the applications, if discontinuities occur (nuclear physics, shock waves a.o.).

2. Functions of C_2^o as upper and lower bounds

It is quite natural to use continuous functions with discontinuous first derivatives as bounds for a desired solution. Let w_j $(j=1,\dots,m) \in C^2(D)$ be upper bounds for a function $u(x)$ with $x = \{x_1,\dots,x_n\}$; Then obviously

$$(2.1) \qquad \hat{w}(x) = \min_j (w_j(x)) \geq u(x)$$

is an upper bound for $u(x)$ and $\hat{w}$ belongs to $C_2^0(D)$ (in not exceptional cases). Analogeously

$$(2.2) \qquad v_\ell(x) \leq u(x) \quad (\ell=1,\dots,q), \quad v_\ell \in C^2(D)$$

has the consequence

$$(2.3) \qquad \hat{v}(x) = \max_\ell (v_\ell(x)) \leq u(x)$$

and we have normally $\hat{v} \in C_2^0(D)$.

<u>Example I</u> (Linear case) Let us consider the linear problem for a function $u(x,y)$

$$(2.4) \qquad -\Delta u = -\frac{\partial^2 u}{\partial x^2} - \frac{\partial^2 u}{\partial y^2} = 0 \text{ in } B \text{ , fig.1}$$

$B = \{(x,y), \; 0 < \operatorname{Min}(x,y) < 1\}$

with the given boundary values :

$$(2.5) \qquad \begin{cases} u(x,y) = \operatorname{Min}(x,y) = y & \text{for } y=0 \text{ and } y=1 \\ \lim\limits_{y \text{ fixed}, x\to\infty} [u(x,y)-y] = 0 & \text{for } 0<y<1 \\ \lim\limits_{x \text{ fixed}, y\to\infty} [u(x,y)-x] = 0 & \text{for } 0<x<1 \end{cases}$$

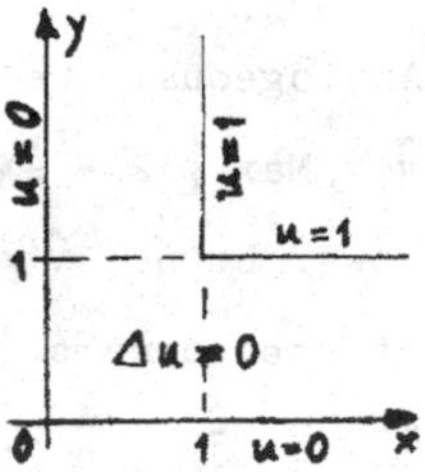

Fig.1

(ideal flow of a liquid around a corner)

Then the harmonic functions $w_1=x$, $w_2=y$, $w_3=xy$ are upper bounds for u and with (2.1) we have the new upper bound

$$\hat{w} = \mathrm{Min}(x,y,xy) \geq u(x) \ , \ \hat{w} \in C_2^o(B)$$

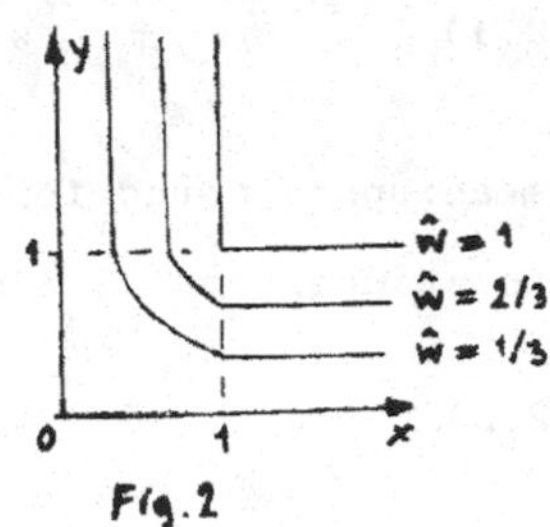

Fig.2

Fig.2 shows some lines $\hat{w}$=const. with discontinuities for the tangents at the interfaces $x=1$ and $y=1$.

Example II (Nonlinear case) Let be given the equation

$$(2.6) \qquad Tu = -\Delta u - 4 + u^2 = 0 \text{ in } B = \{(x,y), |x|<1, |y|<1\}$$

and the values at the boundary ∂B :

$$(2.7) \qquad u = 1 \text{ on } \partial B \ .$$

The functions $w_1=2$, $w_2= 1+\frac{3}{4}(2-\rho^2)$ with $\rho^2 = x^2+y^2$ satisfy $Tw_j \geq 0$ in B, $w_j \geq 1$ on ∂B $(j=1,2)$ and are therefore upper bounds for u; (2.1) gives then the upper bound $\hat{w} \geq u$. We have $\rho^2 = 2/3$ as interface; fig.3. w_1 is a better bound for $\rho^2 < 2/3$ and w_2 for $\rho^2 > 2/3$.

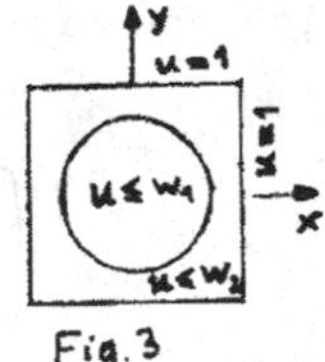

Fig.3

Analogeously we get
$\hat{v} = \mathrm{Max}\,\{\sqrt{2} - \frac{1}{2}\rho^2,\ 1 + (2\sqrt{3}-3)(1-x^2)(1-y^2)\} \in C_2^o$ as lower bound $\hat{v} \leq u$. The problem has singularities at the corners ((2.7) requires at the corners $\frac{\partial^2 u}{\partial x^2} = \frac{\partial^2 u}{\partial y^2} = 0$ in contradiction to (2.6)). Of course one could get better bounds by using for instance polynomials of higher degree.

3. The interface conditions for functions of the class C_1^o -

Now we ask for interface conditions, which one has to expect for upper bounds and analogeously for lower bounds. These conditions arise from a general geometrical principle and are independent of the special boundary value problems, which may be elliptic or not and linear or nonlinear, and of functions of one or more independent variables.

Let be $w_i(x), w_j(x) \in C_1^o(D)$ upper bounds for a function $u(x) \in C_1^o(D)$. Let us assume that there exist in D two subdomains D_i, D_j with boundaries $\partial B_i, \partial B_j$ and a smooth interface $J = \partial D_i \cap \partial D_j$ with the inner normals v_i, v_j; let be

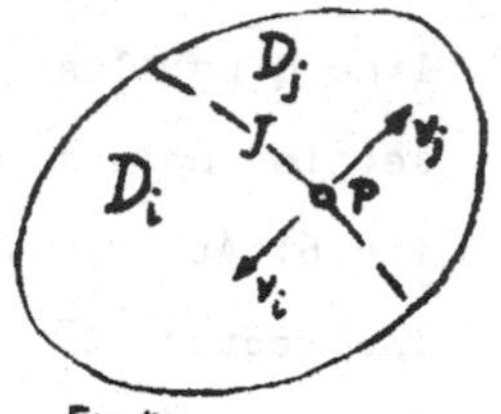

$$(3.1) \quad w_i < w_j \text{ in } D_i, \; w_i = w_j = \hat{w} \text{ on } J, \; w_i > w_j \text{ in } D_j, \quad w_i, w_j \in C^1(D_i \cup J \cup D_j).$$

fig.4,

Fig.4

Asking for local conditions at a point $P \in J$ we substitute in the neighbourhood of P the surfaces w_i=const., w_j= const. and J by (n-1)-dimensional planes (fig.5 for $R^n = R^3$).

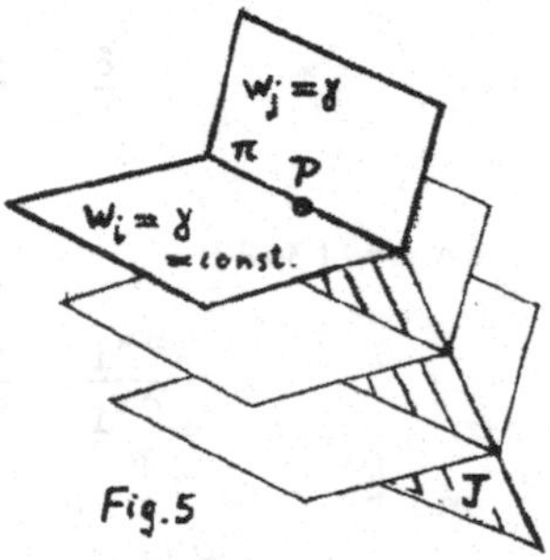

Fig.5

Let be $\hat{w}(P) = \gamma$. Then $w_i = w_j = \text{const.} = \gamma$ is a (n-2)-dimensional plane π containing the point P.

Projecting the R^n in the direction of this plan π, fig.6, we can identify π with the point P and the planes $w_\sigma = \gamma$, $(\sigma = i,j)$ are transformed into straight lines

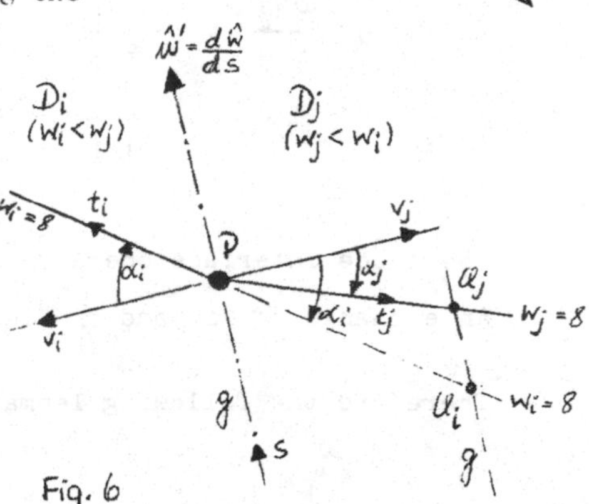

Fig. 6

with tangential vectors t_σ $(\sigma=i,j)$, fig.6. The plane J is carried over to a straight line g with the tangential vector $\hat{w}' = \frac{d\hat{w}}{ds}$; if we introduce the arc length s in the direction of increasing values of $\hat{w}$ on g. We introduce the angles α_σ between the inner normals v_σ and the tangential vectors t_σ $(\sigma=i,j)$ in clockwise direction, fig.6, where $|\alpha_\sigma| < \frac{\pi}{2}$; then we have $\alpha_j \leq \alpha_i$ with respect to $w_j < w_i$ in D_j , or

$$(3.2) \qquad \tan(\alpha_i) - \tan(\alpha_j) \geq 0 \quad .$$

One sees the fact $\alpha_j \leq \alpha_i$ by considering for instance a straight line q parallel to g in the domain D_j . We introduce the intersection points Q_σ of q with the lines w_σ = const. $(\sigma=i,j)$, fig.6. At Q_j we have $w_j = \hat{w} = \gamma$ and at Q_i we have $w_i = \gamma$, $w_j < \gamma$. The vector $\overrightarrow{Q_iQ_j}$ is directed to increasing w_j and opposite directed to increasing α_j . We have along the lines w_σ = const.

$$0 = \frac{\partial w_i}{\partial v_i} \cos \alpha_i + \hat{w}' \sin \alpha_i = \frac{\partial w_j}{\partial v_j} \cos \alpha_j - \hat{w}' \sin \alpha_j$$

or by (3.2)

$$-\frac{\partial w_i}{\partial v_i} - \frac{\partial w_j}{\partial v_j} = \hat{w}'(\tan \alpha_i - \tan \alpha_j) \geq 0$$

and by

$$\frac{\partial w_j}{\partial v_j} = -\frac{\partial w_j}{\partial v_i}$$

$$(3.3) \qquad S[w] = \frac{\partial w_j}{\partial v_i} - \frac{\partial w_i}{\partial v_i} \geq 0 ,$$

where the interface operator S is defined by the equation in (3.3); An exchange of i and j leaves S[w] invariant.

Therefore the following lemma holds:

<u>Lemma:</u> Assume a function $u \in C_q^o(D)$ with $q \geq 1$, where $D \subset R^n$ is an open domain, for instance a solution of a nonlinear boundary value problem, and two upper bounds w_i, w_j for u in D. Then $\hat{w} = \mathrm{Min}(w_i, w_j)$ is also an upper bound for u; if $w_i < w_j$ in D_i, $w_i > w_j$ in D_j, where D_i, D_j are open domains with a smooth interface $J = \partial D_i \cap \partial D_j$ and $D \supset (D_i \cup J \cup D_j)$. Then (3.3) holds at the interface J.

In the same way we get for lower bounds the interface conditions

$$\frac{\partial w_j}{\partial v_i} - \frac{\partial w_i}{\partial v_i} \leq 0 \tag{3.4}$$

We give two simple geometrical interpretations of the interface conditions:

Crossing the interface J, the graph of an upper, resp. lower bound as a function along inner normals has a vertex as shown in fig.7. In two dimensions the lines of constant value for upper (lower) bounds w have corners at the interface J directed to decreasing (increasing) w as shown in fig.8 and 9.

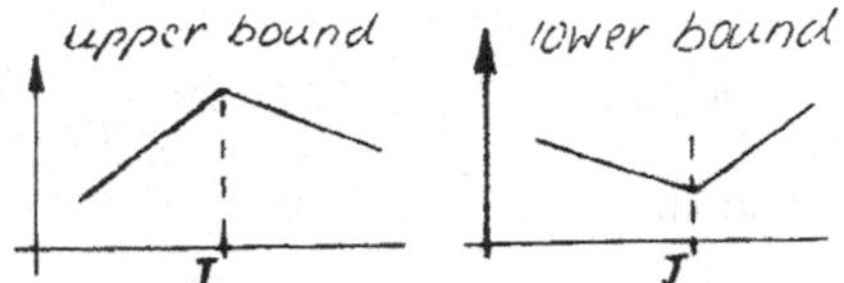

upper bound lower bound

J J

Fig. 7

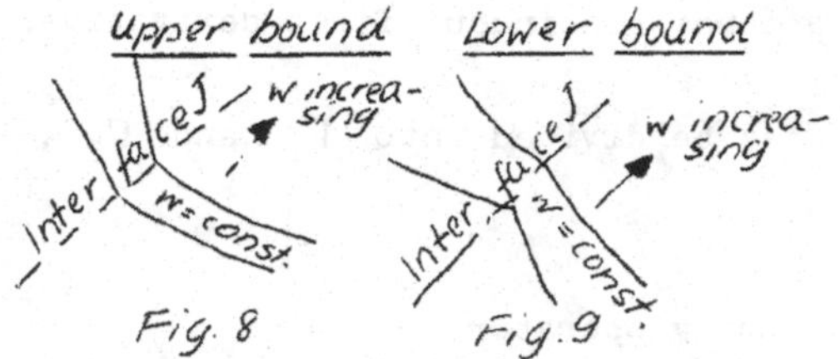

Fig. 8 Fig. 9

An example for this behaviour is given in fig. 2.

4. Mildly nonlinear elliptic boundary value problems

Let B be an open bounded domain $\subset R^n$ and B_i $(i=1,\dots,m)$ open subdomains of B with piecewise smooth boundaries ∂B_i, where

$$B_i \cap B_j = \phi \quad \text{for } i \neq j \quad \text{and}$$

$$\bar{B} = B \cup \partial B = \bigcup_i (B_i \cup \partial B_i) \tag{4.1}$$

The interfaces $J_{ij} = \partial B_i \cap \partial B_j$ (possibly empty) contain "regular points", in which the inner normals v_i and v_j exist, and those points which are not regular. Fig.10 shows some non regular points marked by small circles. Points at which interfaces meet ∂B, are non regular.

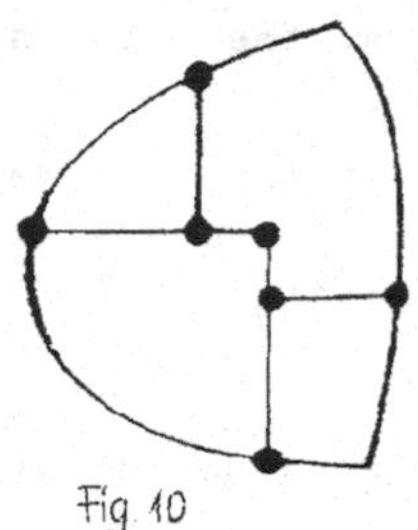
Fig 10

Let a function $u(x)$ satisfy the differential equation

$$Lu = 0 \quad \text{in } B_i \quad (i=1,..,m) \tag{4.2}$$

with

$$Lu = - \sum_{jk=1}^{n} a_{jk}(x) \frac{\partial^2 u}{\partial x_j \partial x_k} + \sum_{j=1}^{n} b_j(x) \frac{\partial u}{\partial x_j} + c(x,u(x)) \tag{4.3}$$

At the interfaces J_{ij} we suppose that u satisfies (4.1) or that u is continuous and has continuous first derivatives.

The boundary ∂B may be divided into Γ_1 and Γ_2, $\partial B = \Gamma_1 + \Gamma_2$, $\Gamma_1 \cap \Gamma_2 = \phi$.

We consider the boundary operator

$$Ru = \begin{cases} u(x) & \text{for } x \in \Gamma_1 \\ \alpha(x,u(x)) - \frac{\partial u}{\partial v} & \text{for } x \in \Gamma_2 \end{cases} \tag{4.4}$$

(v as inner normal)

We suppose a_{jk} and $b_j \in C(B)$, the matrix (a_{jk}) may be uniformly positive definite, $\frac{\partial c}{\partial u} \geq 0$ for $x \in \bar{B}$, $u \in \mathbb{R}$, $\frac{\partial \alpha}{\partial u} \geq 0$ for $x \in \Gamma_2$, $u \in \mathbb{R}$, $\frac{\partial \alpha}{\partial u} > 0$ in not regular points.

We consider the following linear subspace F of $C_2^0(\bar{B})$, which contains all functions $z \in C(\bar{B})$, for which the restrictions z_i from z to $\bar{B}_i$ satisfy

$$z_i \in C^2(\bar{B}_i \cap B) \cap C^1(\bar{B}_i - \Gamma_1) .$$

Then the general theorem of B. Werner [74] holds:

Theorem of Monotonicity (B. Werner): The vector (L,R,S) consisting of the nonlinear elliptic operator L in B_i , the boundary operator R on ∂B and the interface operator S in regular points of the interfaces J_{ij} , is of monotonic type, that means: if $(L,R,S)_{(v)} \geq (L,R,S)_{(w)}$ for functions v,w of the above defined subspace F , then

$$v \geq w \quad \text{in} \quad \bar{B} .$$

This theorem includes also possibilities of error estimations for many numerical procedures as iteration procedures, F.E.M. and others.

Example In this example occurs a discontinuity in the differential equation for the unknown function $u(x,y)$

$$-\Delta u = \begin{Bmatrix} 0 & \text{in} & B_1 \\ 1 & \text{in} & B_2 \end{Bmatrix}, \; u = 0 \text{ on } \partial B .$$

Here we have $0 < x < 1$ for B_1, $1 < x < 2$ for B_2, $0 < x < 2$ for B and $|y| < 1$ for all these domains, fig.11. We use finite elements with

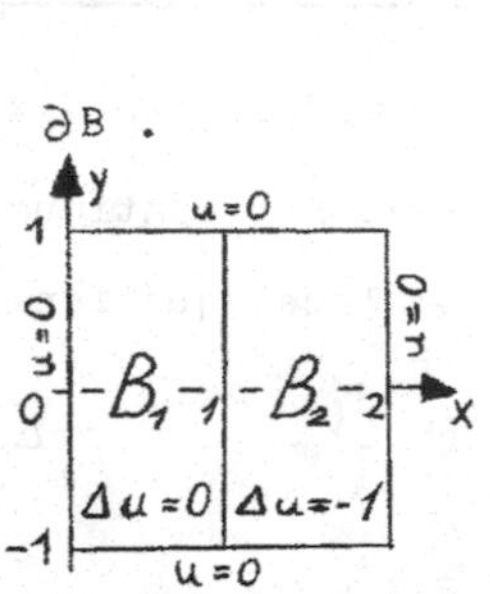

Fig.11

different analytic terms for the approximate solution $\hat{u}$ in the subdomains B_1, B_2 :

$$\hat{u} = \psi - \frac{1}{4}x + w \quad \text{with}$$

$$\psi = \begin{cases} 0 & \text{in } B_1 \\ \frac{1}{2}(x-1)^2 & \text{in } B_2 \end{cases}$$

$$w = \sum_{\nu=1}^{q} a_\nu \sin(\tfrac{\pi}{2}\nu x)\cosh(\tfrac{\pi}{2}\nu y)\ \frac{1}{\cosh\frac{\pi}{2}\nu}$$

We have the approximation problem $|u-\hat{u}| \leq \delta_q$ in B with δ_q = Min. The table gives some values of δ_q (I thank Mr. H. Günther und Mr. D. Schütt for the numerical calculation):

q	δ_q
1	0.065
2	0.0110
6	0.00140
9	0.00077•

For $q = 6$ are the values of the coefficients:

a_1=-0.25814, a_2=0.06450, a_3=-0.00940,

a_4=0.00002 , a_5=-0.00238, a_6=0.00320 .

5. Other monotonicity theorems for elliptic equations

W. Wetterling [68] has given lower and upper bounds, which are even not continuous. For instance for the Dirichlet-problem for the Laplace equation

$$(5.1) \qquad \Delta u(x) = 0 \ \text{in } B, \quad u = f(x) \ \text{on } \partial B$$

he introduces a modified difference operator for an approximate solution $\hat{u}(x)$

$$(5.2) \qquad D\,\hat{u}(x) = 20\,\hat{u}(x) - 4\sum_{j=1}^{4}\hat{u}(x+\xi_j) - \sum_{j=1}^{4}\hat{u}(x+\eta_j) \quad ,$$

where ξ_j are the vectors with components $(\pm h,0)$, $(0,\pm h)$ and η_j the vectors $(\pm h,\pm h)$ as in fig. 12 with h as meshsize; x means the vector x_1,x_2 in a x_1-x_2-plane. With certain additional assumptions has

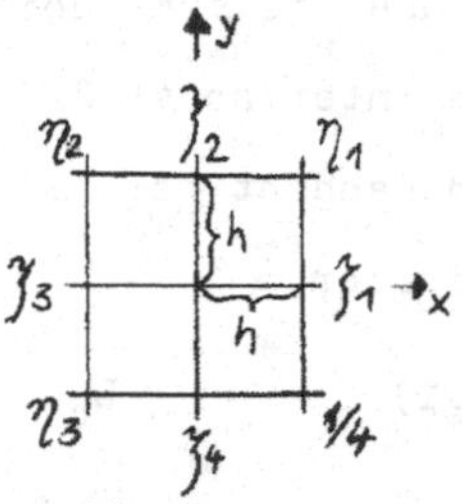

$$(5.3) \qquad D\,\hat{u} \leq 0 \text{ in } B\,, \quad \hat{u} \leq f \text{ on } \partial B$$

Fig. 12

the consequence $\hat{u} \leq u$ in B. He solves the interpolation problem, in order to find a polynomial $v(x)$ of degree 4 with $\Delta v = 0$ which has prescribed values at the eight points $x+\xi_j$, $x+\eta_j$. Then one calculates in the usual way the solution $\hat{u}$ of the finite difference method with the operator D and one can get exact (in general not continuous) bounds for u with the aid of the corresponding interpolating functions v for the inner grid points.

In another paper Wetterling [68a] gives generalizations for more general elliptic equations. He shows that one can get good numerical results even in more complicated problems with corners and he has treated many examples, for instance the problem, posed by Kantorowitsch-Krylov [61], fig. 13.

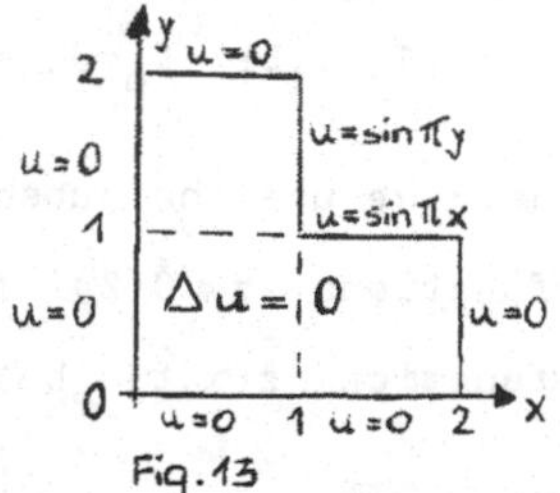

Fig. 13

F. Natterer and B. Werner [74] give a quite elementary proof for the monotonicity theorem of § 4 with the interface conditions (3.3) for the two-dimensional Dirichlet problem, which can be generalized to nonlinear equations. They use polar coordinates and therefore the proof is restricted to two dimensions.

6. Parabolic equations

We use the same notations as in § 4, but in (4.3) u, a_{jk}, b_j, c and the interfaces J_{ij} depend now on $x_1, \dots, x_n$ and on the independent variable t, usually the time, and we consider the operator

$$(6.1) \qquad L_t u = \frac{\partial u}{\partial t} + Lu \quad \text{in } Z$$

where

$$(6.2) \qquad Z = \{x, t,\ x \in B,\ 0 < t < T\} \text{ with a fixed } T > 0 .$$

Analogeously the boundary operator gets a slight modification

$$(6.3) \qquad R_t u = \begin{cases} u(x,0) & \text{for } x \in B \text{ and } t = 0 \\ Ru \quad (\text{as in } (4.4)) & \text{for } x \in \dot{B} \text{ and } 0 < t < T . \end{cases}$$

We suppose furthermore

$$(6.4) \qquad \frac{\partial c}{\partial u}(x,t,u) \geq 0 \text{ for } (x,t) \in Z,\ \frac{\partial \alpha}{\partial u}(x,t,u) > 0 \text{ for } x \in \Gamma_2,\ 0 < t < T,\ u \in \mathbb{R}$$

The cylinder Z may be divided into subcylinders by B dividing into B_i as in § 4 and the interval $[0,T]$ dividing into points t_k $(k=0,1,\dots,r;\ t_0=0,\ t_r=T)$ and open intervals (t_{k-1}, t_k) $(k=1,\dots,r)$.

Here we use the subspace F_t of $C_2^0(\bar{Z})$ which contains all functions $z \in C(\bar{Z})$ for which for all fixed $t=\hat{t}$ in $(0,T)$ the function $z(x,\hat{t})$ belongs to the subspace F of § 4 and for which for all fixed $x=\hat{x} \in B$ the function $v(\hat{x},t)$ belongs to $C^1(0,t_1]$, $C^1[t_{k-1},t_k]$ $(k=2,\dots,r-1)$, $C^1[t_{r-1},T)$. fig. 14

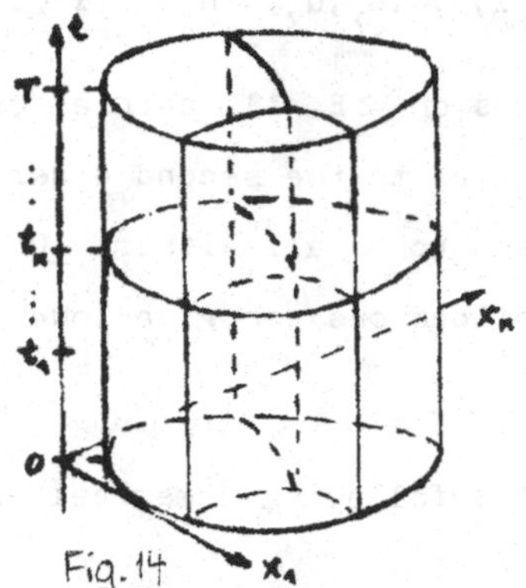

Fig. 14

The definition of regular points of the interfaces is not changed Then holds the general theorem

Theorem of Monotonicity (B. Werner [74]): The vector L_t, R_t, S consisting of the nonlinear parabolic operator L_t, the boundary operator R_t and the interface operator S in regular points (x,t) of the interfaces $x \in J_{ij}$, $0<t<T$, is of monotonic type, that means: if $(L_t,R_t,S)_{(v)} \geq (L_t,R_t,S)_{(w)}$ for functions v,w of the subspace F_t, then $v \geq w$ in $\bar{Z}$.

It is remarkable that we have on the hyperplanes $t = t_k$ $(k=1,\dots,r-1)$ no conditions about $\frac{\partial u}{\partial t}$, it is sufficient for u to be continuous with respect to t.

7. Nonlinear hyperbolic equations

Hofmann [74] proves that monotonicity holds in certain cases also for the derivatives of the unknown function u(x,y). He considers equations of the form

(7.1) $u_{xy} = f(x,y,u,u_x,u_y)$ in B $(x>0,y>0)$

with given boundary values on ∂B. All partial derivatives of f with respect to u,u_x,u_y up to the second order (included) are supposed to exist in B and to be all either ≥ 0 or to be all ≤ 0. With some further assumptions one can give lower and upper bounds for u,u_x,u_y.

I thank Dr.Hofmann for the following numerical example:

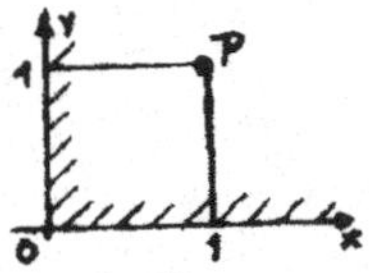

Let be given the equation

Fig. 15

(7.2) $u_{xy} - \frac{1}{4}uu_y - \frac{1}{2}x = 0$ in B $(0\leq x,y\leq 1)$, fig. 15.

with the initial values

(7.3) $u(x,0) = u(0,y) = 1$.

With simple calculations (without a computer) one gets the bounds

$$u_1(x,y) \leq u(x,y) \leq u_2(x,y)$$

$$\frac{\partial u_1}{\partial x} \leq \frac{\partial u}{\partial y} \leq \frac{\partial u_2}{\partial x} \qquad \text{in B}$$

$$\frac{\partial u_1}{\partial y} \leq \frac{\partial u}{\partial y} - \frac{\partial u_2}{\partial y}$$

with
$$u_1(x,y) = 1 + 8y(e^{\frac{1}{4}x} - (1 + \frac{1}{4}x))$$

$$u_2(x,y) = 1 + \frac{2y}{3}(44e^{\frac{1}{4}x} - x^2 - 11x - 44)$$

$$= u_1 + f(x,y)$$

Here $f(x,y) = \frac{y}{3}(64e^{\frac{1}{4}x} - (2x^2 + 16x + 64))$ is the difference between upper and lower bound, we have for instance

$$0 \le f(x,y) \le f(1,1) \le 0.056$$

$$0 \le \frac{\partial f}{\partial x}(x,y) \le \frac{\partial f}{\partial x}(1,1) \le 0.148$$

$$0 \le \frac{\partial f}{\partial y}(x,y) \le \frac{\partial f}{\partial y}(1,1) \le 0.056 .$$

For all details see the paper of Hofmann.

8. Iteration procedure for non contractive operators

The most theorems for convergence of a sequence v_n of elements with the iteration procedure

(8.1) $$v_{n+1} = Tv_n \quad (n=0,1,2,\ldots)$$

are given for operators T which are contractive or at least non expansive in a certain sense, compare for instance the theory of monotonically decomposible operators, J. Schröder [56], Collatz [66]. But for numerical calculation one can often use the procedure (8.1) also in case of non contracting operators, in cases, in which several solutions exist a.o., even when convergence-theorems are not known in the present stage.

Let us describe the method for a simple example.

Let be given the nonlinear boundary value problem for the unknown function $u(x,y)$ in the domain $B = \{(x,y), r^2=x^2+y^2 < 1\}$

(8.2) $$-\Delta u = x^2 + y^2 + u^3 \quad \text{in } B \quad u=1 \quad \text{on } \partial B .$$

The iteration (8.1) is here of the form

(8.3) $$-\Delta v_{n+1} = x^2 + y^2 + v_n^3 \quad \text{in } B \quad v_{n+1}=1 \quad \text{on } \partial B .$$

We look on a function v_1 of the type

$$v_1 = \sum_{\nu=1}^{q} a_\nu (1-r^{2\nu})$$

with $v_1 = v_0 = 1$ for $r=1$,

so that in (8.1) v_0 satisfies the boundary condition. Then we have the approximation problem:

(8.4) Approximate the function zero by $v_1 - v_0$.

We solve this problem approximately by looking on constants $a_\nu = a_{\nu(q)}$, so that $v_1 - v_0$ has an alternate of length (q+1). One starts for instance with $q=2$ (only two parameters a_ν), gets values, say $a_{\nu(2)}$, and takes the $a_{\nu(2)}$ as trial values for $q=3$, gets $a_{\nu(3)}$ and so on. I thank Mr. H. Günther and Mr. D. Schütt for calculation on a computer.

One observe that the problem (8.2) has 3 solutions for $q = 1,2,3$, the calculated values are given in the following table, in which the corresponding values of $\delta_{(q)} = \underset{x\in B}{\text{Max}} \mid v_1(x) - v_0(x) \mid$ are indicated; fig. 16 illustrates the approximate solutions one has got.

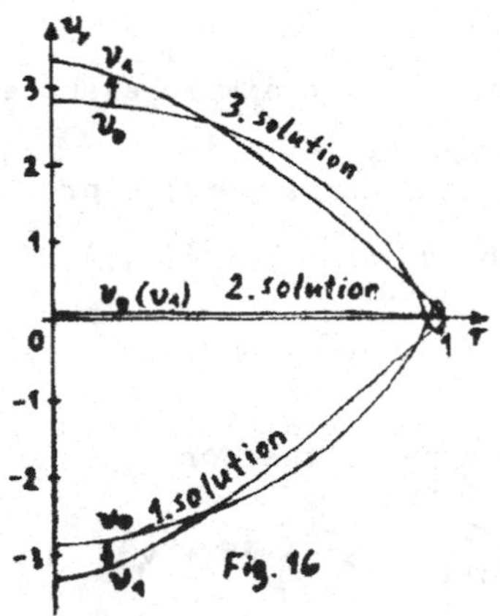

Fig. 16

calculation with a_1, a_2	a_1	a_2	a_3	$\delta_{(2)}$
1. solution	-5.1580	1.3520	-	1.063
2. solution	0.000033	0.062492	-	0.01274
3. solution	4.8658	-1.1540		1.022
calculation with a_1, a_2, a_3				$\delta_{(3)}$
1. solution	-5.8804	3.1919	-0.7375	0.5608
2. solution				
3. solution	5.4628	-2.8263	0.6769	0.5177

For $q=3$ the values of $\delta_{(3)}$ are approximately half the values of $\delta_{(2)}$

References

Agmon, S., L. Nierenberg and M.H. Protter [53]: A maximum principle for a class of hyperbolic equations and applications to equations of mixed elliptic-hyperbolic type. Comm. Pure Appl. Math. 6 (1953), 455-47o.

L. Collatz [52]: Aufgaben monotoner Art. Arch. Math. 3 (1952), 366-376.

L. Collatz [66]: Functional Analysis and Numerical Mathematics, Acad. Press 1966, 473 p.

L.. Collatz - W. Krabs [73]: Approximationstheorie, Tschebyscheffsche Approximation mit Anwendungen, Teubner, Stuttgart, 1973, 2o8 p.

H. H. Gloistehn [63]: Monotoniesätze und Fehlerabschätzungen für Anfangswertaufgaben mit hyperbolischer Differentialgleichung, Arch. Rat. Mech. Anal. 14 (1963), 384-4o4.

W. Hofmann [74]: Monotoniesätze für hyperbolische Anfangswertaufgaben und Einschließung von Lösungen, to appear.

I. W. Kantorowitsch - W. I. Krylow [61]: Näherungsmethoden der höheren Analysis, Berlin 1961, 611 p.

F. Natterer - B. Werner [74]: Eine Erweiterung des Maximumsprinzips für den Laplaceschen Operator, to appear.

K. Nickel [59]: Fehlerabschätzungen bei parabolischen Differentialgleichungen, Math. Z. 71 (1959), 268-282.

R. M. Redheffer [62]: Bemerkungen über Monotonie und Fehlerabschätzung bei nichtlinearen partiellen Differentialgleichungen, Arch. Rat. Mech. Anal. 1o (1962), 427-457.

R. M. Redheffer [62]: An Extension of Certain Maximum Principles, Mh. Math. Phys. 66 (1962), 32-42.

R. M. Redheffer [63]: Die Collatzsche Monotonie bei Anfangswertproblemen, Arch. Rat. Mech. Anal. 14 (1963), 196-212.

R. M. Redheffer [67]: Differentialungleichungen unter schwachen Voraussetzungen, Abhandl. Math. Sem. Univ. Hamburg, 31 (1967) 33-5o.

D. Sather [67]: Pointwise Bounds for the Solution of a Cauchy Problem, J. Diff. Equat. 3 (1967), 286-3o9.

J. Schröder [56]: Das Iterationsverfahren bei allgemeinerem Abstandsbegriff, Math. Z. 66 (1956), 111-116.

F. Stummel [69]: Rand- und Eigenwertaufgaben in Sobolevschen Räumen, Springer, Lecture Notes in Mathematics, Bd. 1o2, 1969, 386 S.

W. Walter [7o]: Differential and Integral Inequalities, Springer 197o, 352 p.

B. Werner [74]: Verallgemeinerte Monotonie bei schwach nichtlinearen elliptischen Differentialgleichungen, to appear.

H. Westphal [49]: Zur Abschätzung der Lösungen nichtlinearer parabolischer Differentialgleichungen, Math. Z. 51 (1949), 69o-695.

W. Wetterling [68]: Lösungsschranken beim Differenzenverfahren zur Potentialgleichung, Internat. Ser. Num. Math. 9 (1968), 2o9-222

W. Wetterling [68a]: Lösungsschranken bei elliptischen Differentialgleichungen, Internat. Ser. Num. Math. 9 (1968), 393-4o1.

ON A DEFICIENCY INDEX THEOREM OF W. N. EVERITT

A. Devinatz

1. We shall consider a fourth-order differential operator in the form

$$(p_0 y'')'' - (p_1 y')' + p_2 y \tag{1.1}$$

which is defined on the semi-axis $[0,\infty)$. W. N. Everitt in [3] proved the following:

If $p_0 = 1$, $p_1, p_2 \geq 0$, *and if there is a constant* K *so that* $p_1 \leq Kt^2(1 + p_2)^{½}$, *then the deficiency index of the minimal operator associated with* (1.1) *is* (2,2).

The object of this note is to generalize this theorem of Everitt. Toward this end we first introduce a function χ defined on $[a,\infty)$, $a \geq 0$, and having the following properties:

(a) $\chi \geq 0$, and is twice differentiable;

(b) $\int_a^t \chi \to \infty$ as $t \to \infty$;

(c) $\chi(t) + |\chi'(t)| + |\chi''(t)| \leq K \int_a^t \chi$, $K > 0$, $t \to \infty$.

In terms of this function χ we may now state the following result.

Theorem. *Suppose the coefficients of the differential (or more generally quasi-differential) operator* (1.1) *are non-negative, and* $1/p_0$, p_1 *and* p_2 *are locally integrable. If moreover there exist positive constants* K, m, *and* M, *and an* $\alpha < 1$ *so that*

(i) $\chi^2(t)p_1(t) \leq K(1 + p_2(t))^{\alpha}[\int_a^t \chi]^2$ *for all sufficiently large* t,

(ii) $0 < m \leq p_0(t) \leq M$, *wherever* $\chi(t) \neq 0$,

then the deficiency index of the minimal operator associated with (1.1) *is* (2,2).

We shall carry out the proof of this theorem in the next section. Before we do this we shall give some examples. If we take $p_0 = 1$ and $\chi = 1$, then we get Everitt's result for $\alpha = 1/2$. If we take $a = 2$, and $\chi(t) = 1/t$, then the condition (i) becomes $p_1(t) \leq K(t \log t)^2(1 + p_2(t))^{\alpha}$. If we take $\chi(t) = 1/(t \log t)$ then condition (i) becomes $p_1(t) \leq K(t \log t \log\log t)^2(1 + p_2(t))^{\alpha}$. Clearly we may carry out this procedure a finite number of times.

To get an example of a somewhat different nature we suppose that $\{I_m = (a_m,b_m)\}$

is a sequence of disjoint intervals with $a_m \to \infty$. We define

$$\chi_m(t) = \begin{cases} A_m(b_m - t)^3(t - a_m)^3, & t \in I_m \\ 0, & t \notin I_m. \end{cases}$$

We then take $\chi(t) = \sum_1^\infty \chi_m(t)$, where the constants A_m are to be chosen so that the conditions (b) and (c) on χ will be satisfied. If we suppose that for all sufficiently large m that $(b_m - a_m) \leq 1$, then an elementary calculation shows that in order for condition (c) to be satisfied it is sufficient that

$$A_m(b_m - a_m)^4 \leq K \sum_{k=1}^{m-1} A_k(b_k - a_k)^7$$

for all sufficiently large m. If we now, for example, take $A_k = 1/\{k(b_k - a_k)^7\}$, then condition (b) is clearly satisfied, and moreover we must have

$$(b_m - a_m)^3 \geq \frac{K}{m \log m},$$

where K is some generic constant. Thus if we take the length of I_m so that we have equality in the previous inequality, all of the conditions (a), (b) and (c) are satisfied for the function χ. Thus, in this situation we must have (ii) satisfied on the intervals I_m, and condition (i) will be satisfied if

$$p_1(t) \leq Km^{4/3}(\log m)^{4/3}(1 + p_2(t))^\alpha, \quad t \in I_m.$$

2. In this section we shall give a proof of our theorem. We prefer to work with quasi-differential operators instead of differential operators so as to avoid any differentiability assumptions on the coefficients. The quasi-derivatives of a function y with respect to the coefficients p_0, p_1, and p_2, provided they exist, are given by

$$y^{[1]} = y', \quad y^{[2]} = p_0 y'', \quad y^{[3]} = p_1 y' - (y^{[2]})', \quad y^{[4]} = p_2 y - (y^{[3]})'.$$

The quasi-differential operator corresponding to (1.1) is the quasi-derivative $y^{[4]}$. We refer the reader to [4] for a development of the theory of quasi-differential operators.

Let y be a real square integrable solution to the initial value problem

$$\begin{aligned} y^{[4]} &= 0, \\ y(a) &= y'(a) = 0. \end{aligned} \tag{2.1}$$

Our object is to show that $y = 0$. Toward this end we set

$$Q = p_0(y'')^2 + p_1(y')^2 + p_2y^2.$$

Using the conditions (2.1) and the definition of quasi-derivatives, an integration by parts yields

$$0 = \int_a^t yy^{[4]} = -y(t)y^{[3]}(t) - y'(t)y^{[2]}(t) + \int_a^t Q.$$

Thus we arrive at

$$Q = (yy^{[3]} + y'y^{[2]})'.$$

If we set $w(t,s) = \int_t^s \chi$, multiply the above equation through by $w(t,s)^\beta$, $\beta = 4/(1-\alpha)$, integrate with respect to t between a and s, and then integrate by parts using the initial conditions of (2.1), we get

$$\int_a^s Qw^\beta = \beta\int_a^s \{yy^{[3]} + y'y^{[2]}\}\chi w^{\beta-1}$$
$$= \beta\int_a^s \{yp_1y' - y(y^{[2]})' + y'y^{[2]}\}\chi w^{\beta-1}.$$

If we integrate the second term in the last integral by parts we get

$$\int_a^s Qw^\beta = \beta\int_a^s \{yp_1y' + 2y'y^{[2]}\}\chi w^{\beta-1}$$
$$+ \beta\int_a^s yy^{[2]}\chi'w^{\beta-1} - \beta(\beta-1)\int_a^s yy^{[2]}\chi^2w^{\beta-2}. \qquad (2.2)$$

Let us estimate the second term of the first integral on the right in (2.2). Using the definition of $y^{[2]}$ and Schwarz's inequality we get

$$\int_a^s y'y^{[2]}\chi w^{\beta-1} \leq \{\int_a^s p_0|y''|^2w^\beta\}^{1/2}\{\int_a^s p_0|y'|^2\chi^2w^{\beta-2}\}^{1/2}.$$

Using the assumption that $p_0(t) \leq M$ wherever $\chi(t) \neq 0$ we get

$$\int_a^s p_0|y'|^2\chi^2w^{\beta-2} \leq M\int_a^s |y'|^2\chi^2w^{\beta-2}$$
$$\leq -M[\int_a^s yy''\chi^2w^{\beta-2} + 2\int_a^s yy'\chi\chi'w^{\beta-2} - (\beta-2)\int_a^s yy'\chi^3w^{\beta-3}].$$

Using Schwarz's inequality for the first integral in the last line, using the assumption that $p_0(t) \geq m > 0$ wherever $\chi(t) \neq 0$, using integration by parts for the last two integrals, and using the inequality $(A+B)^{1/2} \leq A^{1/2} + B^{1/2}$ for non-negative A and B we get

$$\int_a^s y'y^{[2]}\chi w^{\beta-1} \leq K\{\int_a^s p_0|y''|^2w^\beta\}^{3/4}\{\int_a^s |y|^2\chi^4w^{\beta-4}\}^{1/4}$$
$$+ K\{\int_a^s p_0|y''|^2w^\beta\}^{1/2}\{\int_a^s |y|^2F\}^{1/2}, \qquad (2.3)$$

where

$$F = |(\chi^2)''|w^{\beta-2} + |\chi(\chi^2)'|w^{\beta-3} + |(\chi^3)'|w^{\beta-3} + \chi^4 w^{\beta-4}. \tag{2.4}$$

Since a bounded self-adjoint transformation may be added to a symmetric transformation without changing the deficiency index, we may, without loss of generality, suppose that $p_2 \geq 1$. We may also, without loss of generality, assume that $\alpha > 1/2$. Let us set $\gamma = (2\alpha - 1)/4$ and $q = 1/2(1 - \alpha)$. Then by Hölder's inequality we get

$$\int_a^s |y|^2\chi^4 w^{\beta-4} \leq \{\int_a^s p_2|y|^2 w^\beta\}^{\frac{1}{2}}\{\int_a^s |y|^2\chi^8 w^{\beta-8}\}^{\frac{1}{2}}$$

$$\leq \{\int_a^s p_2|y|^2 w^\beta\}^{\frac{1}{2}+2\gamma}\{\int_a^s |y|^2\chi^{8q}\, w^{\beta-8q}\}^{1/2q}.$$

Proceeding in the same way with the other terms in the sum (2.4) for F we find from (2.3) that

$$\int_a^s y'y^{[2]}\chi w^{\beta-1} \leq K\{\int_a^s Qw^\beta\}^{\frac{3}{4}+\gamma}\{\int_a^s |y|^2 G\}^{1/4q}, \tag{2.5}$$

where

$$G = \chi^{4q}w^{\beta-4q} + |(\chi^2)''|^{2q}w^{\beta-4q} + |\chi(\chi^2)'|^{2q}w^{\beta-6q} + \chi^{8q}w^{\beta-8q}.$$

Noting that $\beta = 8q$ it follows from the condition (c) that

$$G(t,s) \leq K\{w(a,t)^{4q}w(t,s)^{4q} + w(a,t)^{6q}w(t,s)^{2q} + w(a,t)^{8q}\}.$$

Since $|y|^2$ is integrable and $w(a,s) \to \infty$ as $s \to \infty$, it follows that

$$\int_a^s |y|^2 G = o(w(a,s)^\beta), \quad \text{as } s \to \infty.$$

Recalling that $\gamma = (2\alpha - 1)/4$, $\beta = 8q = 4/(1 - \alpha)$ we get from (2.5)

$$\int_a^s y'y^{[2]}\chi w^{\beta-1} = o\left(\{\int_a^s Qw^\beta\}^{(1+\alpha)/2}\, w(a,s)^2\right). \tag{2.6}$$

It follows by similar arguments that

$$\int_a^s yy^{[2]}\chi' w^{\beta-1} \quad \text{and} \quad \int_a^s yy^{[2]}\chi^2 w^{\beta-2}$$

are dominated by the right hand side of (2.6). Finally we have

$$\int_a^s yp_1y'\chi w^{\beta-1} \leq \{\int_a^s p_1|y'|^2 w^\beta\}^{\frac{1}{2}}\{\int_a^s p_1|y|^2\chi^2 w^{\beta-2}\}^{\frac{1}{2}}$$

$$\leq \{\int_a^s p_1|y'|^2 w^\beta\}^{\frac{1}{2}}\{\int_a^s p_2|y|^2 w^\beta\}^{\frac{1}{4}}\{\int_a^s |y|^2\frac{p_1^2}{p_2}\chi^4 w^{\beta-4}\}^{\frac{1}{4}}.$$

If we apply Hölder's inequality to the last integral with q and γ as before we get

$$\int_a^s |y|^2\frac{p_1^2}{p_2}\chi^4 w^{\beta-4} \leq \{\int_a^s p_2|y|^2 w^\beta\}^{4\gamma}\{\int_a^s |y|^2\frac{p_1^{2q}}{p_2^{2q-1}}\chi^{4q}\, w^{4q}\}^{1/q}.$$

Since $2q = 1/(1-\alpha)$, it follows from condition (i) that

$$\chi^{4q} \frac{p_1^{2q}}{p_2^{2q-1}} \leq Kw(a,t)^{4q}.$$

Thus we see that $\int_a^s yp_1y'\chi w^{\beta-1}$ is also dominated by the right hand side of (2.6).

If we use the preceding facts in (2.2) we get

$$\int_a^s Qw^\beta = o\left(\{\int_a^s Qw^\beta\}^{(1+\alpha)/2} w(a,s)^2\right), \quad s \to \infty.$$

If $Q \neq 0$, it follows from this that

$$\int_a^s Qw^\beta = o\left(w(a,s)^\beta\right), \quad s \to \infty. \tag{2.7}$$

Now $w(t,s)/w(a,s) = 1 - w(a,t)/w(a,s)$, so that if we divide both sides of (2.7) by $w(a,s)^\beta$ and allow $s \to \infty$ we see that since $Q \geq 0$,

$$\int_a^\infty Q = 0.$$

This contradicts our assumption that $Q \neq 0$. Hence we must have $y = 0$.

We have proved that any square integrable solution to the initial value problem (2.1) is identically zero. Thus, if y_1 and y_2 are solutions of (2.1) which satisfy the further initial conditions $y_1^{[2]}(a) = 0$, $y_1^{[3]}(a) = 1$, and $y_2^{[2]}(a) = 1$, $y_2^{[3]}(a) = 0$, then these solutions are not square summable. Further, since any linear combination of these solutions satisfies (2.1), no linear combination of these solutions is square integrable. Hence the deficiency number is at most 2. But since (1.1) is, by hypothesis, regular at 0, the deficiency number is at least 2. The proof is complete.

Remarks. Condition (ii) of the theorem was made for the purpose of simplicity of presentation. However, an examination of the proof shows that it can be relaxed. When the function χ vanishes off of a sequence of intervals, the result obtained is in the spirit of a recent result of W. Desmond Evans [2], although the results are not comparable. If the coefficient p_0 decreases rapidly on the disjoint intervals, then the lengths of the intervals can go to zero very fast. This is a phenomenon that also appeared in the work of Evans.

Using the same basic idea as used in this note, a number of other results can be obtained for fourth-order operators with positive coefficients. However, we shall present these on another occasion since lack of space prevents us from describing them here.

References

1. A. Devinatz, Positive definite fourth-order differential operators, J. London Math. Soc. (2), 6(1973), 412-416.

2. W. D. Evans, On non-integrable square solutions of a fourth-order differential equation and the limit-2 classification, To appear in J. London Math. Soc.

3. W. N. Everitt, Some positive definite differential operators, J. London Math. Soc. 43 (1968), 465-473.

4. M. A. Naimark, Linear Differential Operators, GITTL, Moscow 1954; English transl. Ungar, New York, part I, 1967; part II, 1968.

THE ESSENTIAL SELF-ADJOINTNESS OF SCHRÖDINGER OPERATORS

M. S. P. EASTHAM

1. Introduction

This lecture contains an account of work conducted jointly with W. D. Evans and J. B. McLeod concerning the essential self-adjointness of the minimal operator T_0 associated with the Schrödinger equation

$$\Delta \psi(x) + \{\lambda - q(x)\} \psi(x) = 0 \qquad (1.1)$$

in the Hilbert space $L^2(R^N)$ $(N \geqslant 1)$. Here, Δ denotes the Laplacian, $x = (x_1, \ldots, x_N)$ in cartesian coordinates, and $q(x)$ is real-valued. As a local condition on $q(x)$, it is assumed that $q(x)$ is continuous but it is expected that this can be relaxed to the requirement that $q(x)$ is only locally L^2 in R^N. Such a relaxation would be in the spirit of the recent papers by Simon [17] and Kato [11].

The domain $\mathcal{D}_0$ of T_0 consists of the functions f in $C_0^\infty(R^N)$. However, our method involves primarily, not T_0, but its extension T whose domain $\mathcal{D}$ consists of the functions f in $L^2(R^N)$ such that

(i) for $j = 1, \ldots, N$, $\partial f/\partial x_j$ is locally absolutely continuous as a function of x_j

(ii) $qf - \Delta f$ is in $L^2(R^N)$.

Then, for f in $\mathcal{D}$, $Tf = qf - \Delta f$. The operator T is the maximal operator associated with (1.1) if derivatives are understood in the classical, rather than the distributional, sense.

From a Green's Theorem, we have $T_0 \subset T \subset T_0^*$ and hence, taking closures, $\bar{T}_0 \subset \bar{T} \subset \bar{T}_0^*$. It follows that, if T_0 is essentially self-adjoint, then so is T. In the opposite direction, it is known that

(*) If T is symmetric, then T_0 is essentially self-adjoint.

In the proof of (*) in [8, p.189, Theorem 3], the continuity of $q(x)$ is assumed. In [10, p.299], it is stated that (*) holds if only $q(x)$ is locally L^2, but a published proof does not seem to be readily available.

One remark in the spirit of Titchmarsh's book [19] is appropriate. If it is assumed further that $q(x)$ has continuous first-order derivatives in R^N, then the general theory of [19, Chaps 11-12] applies and establishes the existence of a Green's function $G(x,\xi,\lambda)$ for (1.1). Then we have the result [13, Theorem 2] that

(**) $G(x,\xi,\lambda)$ is unique if and only if T is symmetric.

Additional related results in this spirit are in [6].

In one dimension, $N = 1$, the self-adjointness of T, rather than merely its symmetry, is linked to the limit-point, limit-circle classification of (1.1). In one dimension it is in fact customary to consider (1.1) on the half-line $[0,\infty)$ only, in which case it is necessary to add to the definition of the domain $\mathcal{D}$ the requirement that the functions f in $\mathcal{D}$ satisfy a homogeneous boundary condition

$$f(0)\cos\alpha + f'(0)\sin\alpha = 0.$$

Then we have the result, effectively proved in [8, §14.1], that

(ƒ) (1.1) is limit-point if and only if T is self-adjoint.

I give here, for reference below, the following limit-point result which is a particular case of the theorem in [3].

Theorem A Let there be a sequence of non-overlapping intervals (a_m, b_m) and a sequence of real numbers v_m such that

(i) $(b_m - a_m)^2 v_m \geqslant K$, where K ($>0$) is a constant,

(ii) $\sum v_m^{-1}$ is a divergent infinite series,

(iii) $q(x) \geqslant -kv_m^2(b_m - a_m)^2$ in (a_m, b_m), where k is a constant.

Then (1.1), with N = 1, is limit-point.

Theorem A itself was first proved by Ismagilov [9], who further considered the corresponding situation for differential operators of higher order. It was shown by Knowles [12] that the choice $v_m = (b_m - a_m)^{-2}$ covers the well-known criterion of Levinson [2, pp.229-30].

We point out that the condition (iii) on q(x) in Theorem A is imposed only in the intervals (a_m, b_m) and, outside these intervals, q(x) can be arbitrary. It was Hartman [7] who first drew attention to a condition of this type, and he dealt with the case $b_m - a_m$ = constant, $v_m = 1$.

In more than one dimension, we state first the result of Sears [16] (see also [19, §§22.15-16]), who expressed himself in terms of the uniqueness of the Green's function - see (**) above.

Theorem B Suppose that $q(x) \geqslant -Q(r)$, where $r = |x|$ and Q(r) is a real-valued, strictly positive function on $[0,\infty)$ and either

(i) Q'(r) exists and is continuous in $[0,\infty)$ with

$$Q'(r) = O\{Q^{3/2}(r)\} \quad (r \to \infty) \tag{1.2}$$

or (ii) Q(r) is non-decreasing and continuous in $[0,\infty)$.

Then T is symmetric if
$$\int^{\infty} Q^{-\frac{1}{2}}(r)\, dr = \infty.$$

The earlier result of Titchmarsh [18] dealt with the case $Q(r) = ar^2 + b$ $(a > 0,\ b > 0)$. In more recent papers [17], [11], [5] the local conditions on q(x) have been reduced to a minimum. We state here the result of Kato [11].

Theorem C Let $q(x) = q_1(x) + q_2(x)$, where $q_1(x)$ is locally L^2 and

(i) $q_1(x) \geqslant -Q(r)$, where $Q(r)$ is non-decreasing and $Q(r) = o(r^2)$ as $r \to \infty$,

(ii) $q_2(x)$ is $L^p(R^N)$, where $p \geqslant 2$ $(N \leq 3)$, $p > 2$ $(N = 4)$, and $p \geqslant \frac{1}{2}N$ $(N \geqslant 5)$.

Then T_0 is essentially self-adjoint.

In §2 below, we obtain the extension of Theorem A to $N > 1$. As with Theorem A, we find that the symmetry of T is assured if conditions are imposed on $q(x)$ only in a suitable sequence of annular regions centred on the origin. A consequence is that the conditions on $q(x)$ in Theorem B and on $q_1(x)$ in Theorem C need only be imposed in such annuli. It is anticipated that our method will also cover a similar situation involving $q_2(x)$ in Theorem C. However, at the time when this typescript has to be prepared for inclusion in the Conference Proceedings (February 1974), these details concerning $q_2(x)$ are not fully worked out. It is hoped that the complete result will be presented at the conference itself.

The non-symmetry of T for $N > 1$ appears to have been investigated only by Martin [13],[14]. This aspect of Martin's work is confined to the case where $q(x)$ is, in effect, a function of $|x|$. In particular, he showed that the Green's function is not unique when $q(x) = -|x|^{\gamma} + O(1)$ as $|x| \to \infty$, where $\gamma > 2$. In §3 below, we give a general class of functions $q(x)$ for which T is not symmetric. An example for $N = 2$ is $q(x) = -|x|^{\gamma}$ $(\gamma > 2)$ only in a sector of the plane with $q(x)$ arbitrary outside the sector.

2. The symmetry of T

Our method for considering the symmetry of T is based on that of [1] and it also has points of contact with that of [9]. For f and g in the domain $\mathcal{D}$ of T, we have the Green's formula

$$\int_{r \leq r_1} (\bar{g}Tf - fT\bar{g})\, dx = \int_{|\xi|=1} \left[f \frac{\partial \bar{g}}{\partial r} - \bar{g} \frac{\partial f}{\partial r} \right]_{r=r_1} r_1^{N-1}\, d\omega(\xi), \qquad (2.1)$$

where ξ is the unit vector defined by $x = r\xi$ and ω is the measure on the unit sphere in R^N. We write

$$[f,g](r) = \int_{|\xi|=1} \left(f \frac{\partial \bar{g}}{\partial r} - \bar{g} \frac{\partial f}{\partial r} \right) r^{N-1}\, d\omega(\xi). \qquad (2.2)$$

It follows from (2.1) and the definition of $\mathcal{D}$ that

$$[f,g](r_1) \to L, \qquad (2.3)$$

a finite limit, as $r_1 \to \infty$. Further, since the left-hand side of (2.1) tends to $(Tf,g) - (f,Tg)$ as $r_1 \to \infty$, we have the result that

(**) T is symmetric if and only if $L = 0$ in (2.3) for all f, g in $\mathcal{D}$.

We now require two lemmas as follows.

Lemma 1 Let $\Phi(r)$ be a real-valued $C^1[0,\infty)$ function with compact support and define $\phi(x) = \Phi(|x|)$ for x in R^N. Then there are positive absolute constants K_1 and K_2 such that

$$\int_{R^N} \phi^2 |\nabla f|^2 dx \leq K_1 \int_{R^N} \phi^2 f(Tf - qf)\, dx + K_2 \int_{R^N} |\nabla \phi|^2 f^2 dx \qquad (2.4)$$

for all real-valued f in $\mathcal{D}$.

In this proof the integrations are over R^N and we use the fact that $\phi(x)$ has compact support. By a Green's formula, we have

$$\begin{aligned}\int \phi^2 f (Tf - qf)\, dx &= -\int \phi^2 f \Delta f \, dx \\ &= \int \nabla(\phi^2 f) . \nabla f \, dx = \int (\phi^2 |\nabla f|^2 + 2\phi f \nabla\phi . \nabla f)\, dx \\ &\geq \int (\phi^2 |\nabla f|^2 - 2|\phi||\nabla\phi||f||\nabla f|)\, dx .\end{aligned}$$

Hence, for any $\epsilon > 0$,

$$\int \phi^2 f (Tf - qf)\, dx \geq (1-\epsilon)\int \phi^2 |\nabla f|^2 dx - \epsilon^{-1}\int |\nabla\phi|^2 f^2 dx ,$$

and (2.4) follows on choosing $\epsilon < 1$.

Lemma 2 Let $\phi(x)$ be as in Lemma 1. Then, for f and g in $\mathcal{D}$ and an arbitrary positive constant δ,

$$\int_0^\infty \Phi(r) |[f, g](r)| \, dr \leq \tfrac{1}{2}\delta \int_{R^N} \phi^2 (|\nabla f|^2 + |\nabla g|^2)\, dx + \\ + \frac{1}{2\delta}\int_{R^N} (|f|^2 + |g|^2)\, dx. \qquad (2.5)$$

By (2.2), we have

$$\begin{aligned}\int_0^\infty \Phi(r) |[f,g](r)| \, dr &\leq \int_{R^N} \phi \left| f \frac{\partial \bar{g}}{\partial r} - \bar{g}\frac{\partial f}{\partial r}\right| dx \\ &\leq \tfrac{1}{2}\delta \int_{R^N} \phi^2 \left(\left|\frac{\partial f}{\partial r}\right|^2 + \left|\frac{\partial g}{\partial r}\right|^2\right) dx + \frac{1}{2\delta}\int_{R^N} (|f|^2 + |g|^2)\, dx \\ &\leq \tfrac{1}{2}\delta \int_{R^N} \phi^2 (|\nabla f|^2 + |\nabla g|^2)\, dx + \frac{1}{2\delta}\int_{R^N} (|f|^2 + |g|^2)\, dx ,\end{aligned}$$

as required.

Theorem 1 Let there be a sequence of non-overlapping annular regions $A_m = \{x \in R^N \mid a_m \leq |x| \leq b_m\}$ and a sequence of real numbers v_m such that

(i) $(b_m - a_m)^2 v_m \geq K$, where K (>0) is a constant,

(ii) $\sum v_m^{-1}$ is a divergent infinite seies,

(iii) $q(x) \geq -kv_m^2(b_m - a_m)^2$ in A_m, where k is a constant.

Then T is symmetric.

Let $\rho_m = \frac{1}{3}(b_m - a_m)$ and let $\theta_m(r)$ be a real-valued $C^1[0,\infty)$

function satisfying

(1) $\theta_m(r) = \begin{cases} \rho_m & \text{in } [a_m+\rho_m, b_m-\rho_m] \\ 0 & \text{outside } [a_m, b_m], \end{cases}$

(2) $0 \leq \theta_m(r) \leq \rho_m$,

(3) $\theta_m'(r)$ is bounded independently of m.

Then, in Lemmas 1 and 2, we take $\Phi(r) = (b_m - a_m)\theta_m(r)$. By Lemma 1 and (iii) of Theorem1, we therefore have, for real-valued f,

$$\begin{aligned}\int_{A_m} \phi^2 |\nabla f|^2 dx &\leq k K_1 v_m^2 (b_m - a_m)^2 \int_{A_m} \phi^2 f^2 dx \\ &\quad + \tfrac{1}{2} K_1 \int_{A_m} \phi^2 \{f^2 + (Tf)^2\} dx + K_2 \int_{A_m} |\nabla\phi|^2 f^2 dx \\ &\leq K_3 \{v_m^2 (b_m-a_m)^6 + (b_m-a_m)^4 + (b_m-a_m)^2\} \int_{A_m} \{f^2 + (Tf)^2\} dx \\ &\leq K_3 \{v_m^2 (b_m-a_m)^6 + v_m (b_m-a_m)^4 + (b_m-a_m)^2 \int_{A_m} \{f^2 + (Tf)^2\} dx\end{aligned}$$

since we can assume without loss of generality that $v_m \geqslant 1$. In the above, K_3 is a new constant arising from condition (3) on θ_m'. Hence, by (i) of Theorem 1, we obtain

$$\int_{A_m} \phi^2 |\nabla f|^2 dx \leq K_4 v_m^2 (b_m - a_m)^6 \int_{A_m} \{f^2 + (Tf)^2\} dx,$$

where K_4 is a further new constant. Then Lemma 2 gives

$$\begin{aligned}(b_m - a_m) &\int_{a_m}^{b_m} \theta_m(r) |[f,g](r)| dr \\ &\leq \left(\tfrac{1}{2}\delta K_4 v_m^2 (b_m-a_m)^6 + \frac{1}{2\delta}\right) \int_{A_m} \{f^2 + g^2 + (Tf)^2 + (Tg)^2\} dx. \qquad (2.6)\end{aligned}$$

Considering (ﾉﾉ) now, it is clearly sufficient to prove that L = 0 in (2.3) for all <u>real-valued</u> f and g in $\mathcal{D}$. Then, if o(1) refers to $m \to \infty$, (2.6) and condition (1) on θ_m give

$$\begin{aligned}|L| \{1 + o(1)\} &\tfrac{1}{q} (b_m - a_m)^3 \\ &\leq \left(\tfrac{1}{2}\delta K_4 v_m^2 (b_m - a_m)^6 + \frac{1}{2\delta}\right) \int_{A_m} \{f^2 + g^2 + (Tf)^2 + (Tg)^2\} dx.\end{aligned}$$

The choice $\delta = v_m^{-1}(b_m - a_m)^{-3}$ gives

$$\tfrac{1}{q} |L| \{1 + o(1)\} v_m^{-1} \leq \left(\tfrac{1}{2}K_4 + \tfrac{1}{2}\right) \int_{A_m} \{f^2 + g^2 + (Tf)^2 + (Tg)^2\} dx.$$

Summation with respect to m gives a finite right-hand side and so (ii) of Theorem 1 implies that L = 0, as required.

The following two corollaries provide the extension of Theorem B mentioned towards the end of §1.

<u>Corollary 1</u> Let Q(r) be a continuous, strictly positive, non-decreasing function in $[0,\infty)$. Let there be a sequence of non-overlapping annular regions $A_m = \{x \in R^N \mid a_m \leq |x| \leq b_m\}$ such that

(α) $\{b_m - a_m\}$ is a non-increasing sequence,

(β) lim inf $(b_m - a_m) > 0$ as $m \to \infty$,

(γ) $q(x) \geqslant -Q(r)$ in A_m,

(δ) $\sum_{m=1}^{\infty} \int_{a_m}^{b_m} Q^{-\frac{1}{2}}(r)\, dr = \infty$.

Then T is symmetric.

In A_m, we have $\quad q(x) \geqslant -Q(r) \geqslant -Q(b_m) \geqslant -Q(a_{m+1})$. $\quad$ (2.7)

We define

$$v_m = \left(\int_{a_{m+1}}^{b_{m+1}} Q^{-\frac{1}{2}}(r)\, dr \right)^{-1}. \tag{2.8}$$

Then

$$\begin{aligned} v_m^{-1} = \int_{a_{m+1}}^{b_{m+1}} Q^{-\frac{1}{2}}(r)\, dr &\leqslant Q^{-\frac{1}{2}}(a_{m+1})(b_{m+1} - a_{m+1}) \\ &\leqslant Q^{-\frac{1}{2}}(a_{m+1})(b_m - a_m), \end{aligned} \tag{2.9}$$

by (α). Hence, by (2.7), $q(x) \geqslant -v_m^2(b_m - a_m)^2$ in A_m. Thus (iii) of Theorem 1 holds. By (2.8) and (δ), (ii) of Theorem 1 holds. By (2.9), $(b_m - a_m)^2 v_m \geqslant (b_m - a_m) Q^{\frac{1}{2}}(a_{m+1})$, and (i) of Theorem 1 holds on account of (β). This proves the corollary.

Corollary 2 Let $Q(r)$ be a continuous, strictly positive function in $[0,\infty)$. Let there be a sequence of non-overlapping annular regions $C_m = \{x \in R^N \mid c_m \leqslant |x| \leqslant d_m\}$ such that

(α) $Q'(r)$ exists and is continuous in C_m with $|Q'(r)| \leqslant KQ^{3/2}(r)$ in C_m, where K is a constant,

(β) $\int_{c_m}^{d_m} Q^{\frac{1}{2}}(r)\, dr \geqslant K'$,

where K' (>0) is a constant,

(γ) $q(x) \geqslant -Q(r)$ in C_m,

(δ) $\sum_{m=1}^{\infty} \int_{c_m}^{d_m} Q^{-\frac{1}{2}}(r)\, dr = \infty$.

Then T is symmetric.

The proof is based on that of Knowles [12, Theorem 3]. We first divide $\bigcup_m [c_m, d_m]$ into non-overlapping intervals $[a_m, b_m]$ such that

$$\tfrac{1}{2}K' \leqslant \int_{a_m}^{b_m} Q^{\frac{1}{2}}(r)\, dr \leqslant K', \tag{2.10}$$

where K' is as in (β), and we denote by M_m the maximum value of $Q(r)$ in $[a_m, b_m]$. For any r and s in $[a_m, b_m]$, we have

$$\log \frac{Q(s)}{Q(r)} = \int_r^s \frac{Q'(u)}{Q(u)}\, du \leqslant KK',$$

by (α) and (2.10), giving $Q(s)/Q(r) \leqslant k$, where $k = \exp KK'$. Choosing s so that $Q(s) = M_m$, we obtain

$$M_m k^{-1} \leqslant Q(r) \leqslant M_m \tag{2.11}$$

in $[a_m, b_m]$. We now define

$$v_m = \left(\int_{a_m}^{b_m} Q^{-\frac{1}{2}}(r)\, dr\right)^{-1}.$$

Then, using (2.11),

$$v_m \geqslant (b_m - a_m)^{-1} M_m^{\frac{1}{2}} k^{-\frac{1}{2}}. \tag{2.12}$$

Hence, by (γ),

$$q(x) \geqslant -M_m \geqslant -kv_m^2 (b_m - a_m)^2$$

in A_m. Thus (iii) of Theorem 1 holds. Also,

$$\sum_m v_m^{-1} = \sum_m \int_{a_m}^{b_m} Q^{-\frac{1}{2}}(r)\, dr = \sum_m \int_{c_m}^{d_m} Q^{-\frac{1}{2}}(r)\, dr = \infty,$$

by (δ). Hence (ii) of Theorem 1 holds. Finally, by (2.12) and (2.10)

$$(b_m - a_m)^2 v_m \geqslant (b_m - a_m) M_m^{\frac{1}{2}} k^{-\frac{1}{2}} \geqslant \tfrac{1}{2} K' k^{-\frac{1}{2}}.$$

Hence (i) of Theorem 1 holds. This proves the corollary.

We note that, even in one dimension, the conditions in Corollary 2 are a relaxation of those in Theorem 5 of [1] as far as our equation (1.1) is concerned.

The final remark in this section concerns the annular regions A_m introduced in Theorem 1. Our method works just as well if the A_m are replaced by "bands" of more general shape encircling the origin. For example, we can replace A_m by the region B_m between two N-dimensional cubes: $B_m = \{x \in R^N \mid a_m \leq |x_j| \leq b_m \ (j = 1, \ldots, N)\}$. Some remarks on the physical significance of such regions are made in §4.

3. The non-symmetry of T

In this section we give our main result, Theorem 2 below, on the non-symmetry of T. The proof, which is analogous to that of the corresponding one-dimensional theorem in [4, §2], is based on the observation that, if the limit L in (2.3) is not zero for some f and g in $\mathcal{D}$, then by $(\not\alpha)$ T is not symmetric. For simplicity, we write out the proof in the two dimensional case, N = 2, and we denote polar coordinates in the plane by r, θ.

Theorem 2 Let N = 2. Let $\rho(r)$ be real-valued and defined in an interval $[r_0, \infty)$, where $r_0 > 0$, with the properties

(1) $\rho(r) > 0$ in $[r_0, \infty)$,

(2) $\rho''(r)$ exists and is continuous in $[r_0, \infty)$,

(3) $\rho^{-\frac{1}{4}}(r)$ and $\{\rho^{-\frac{1}{4}}(r)\}''$ are $L^2(r_0, \infty)$.

In a region $r \geqslant r_0$, $\alpha \leqslant \theta \leqslant \beta$, where $0 < \alpha < \beta < 2\pi$, let

$$q(x) = -\rho(r) \tag{3.1}$$

and let q(x) be arbitrary elsewhere. Then T is not symmetric.

Let $v(\theta)$ be real-valued in $[0,2\pi]$ with a continuous second derivative and such that $v(\theta) > 0$ in (α,β) and $v(\theta) = 0$ outside (α,β). Also, define

$$P(r) = r^{-\frac{1}{2}}\rho^{-\frac{1}{4}}(r). \tag{3.2}$$

In (2.3), we consider the functions f and g defined by

$$f(x) = g(x) = v(\theta)P(r)\exp\left(i\int_{r_0}^{r} \rho^{\frac{1}{2}}(t)\,dt\right)$$

for $|x| \geqslant r_0$, while $f(x)$ and $g(x)$ are arbitrary for $|x| < r_0$. Confining our attention to $|x| \geqslant r_0$, we have

$$\begin{aligned}[f,g](r_1) &= r_1\int_0^{2\pi}\left[f\frac{\partial\bar{f}}{\partial r} - \bar{f}\frac{\partial f}{\partial r}\right]_{r=r_1} d\theta \\ &= -2ir_1P^2(r_1)\rho^{\frac{1}{2}}(r_1)\int_0^{2\pi} v^2(\theta)\,d\theta \\ &= -2i\int_0^{2\pi} v^2(\theta)\,d\theta,\end{aligned}$$

by (3.2). Thus we have a non-zero limit L in (2.3). To complete the proof of the theorem, we must show that f is in $\mathcal{D}$, i.e., that f and Tf are in $L^2(R^2)$. Clearly f is in $L^2(R^2)$ by (3.2) and (3) of the theorem. Also, expressing Δ in polar coordinates, a calculation gives

$$\begin{aligned}|(Tf)(x)| &= \Big|\Big(v(\theta)[r^{-\frac{1}{2}}\{r^{\frac{1}{2}}P(r)\}'' + \tfrac{1}{4}r^{-2}P(r) + \\ &\qquad + ir^{-1}P^{-1}(r)\{rP^2(r)\rho^{\frac{1}{2}}(r)\}'] - r^{-2}v''(\theta)P(r)\Big)\Big| \\ &= \Big|\Big(v(\theta)[r^{-\frac{1}{2}}\{\rho^{-\frac{1}{4}}(r)\}'' + \tfrac{1}{4}r^{-5/2}\rho^{-\frac{1}{4}}(r)] - r^{-5/2}v''(\theta)\rho^{-\frac{1}{4}}(r)\Big)\Big|.\end{aligned}$$

It now follows from (3) of the theorem that Tf is in $L^2(R^2)$, and the proof is complete.

The conditions of the theorem are satisfied if $q(x) = -r^{\gamma}$, where $\gamma > 2$, and this is the example given by Martin [13, §4]. In fact, we only need $q(x) = -r^{\gamma}$ in a sector $\alpha \leq \theta \leq \beta$ of the plane.

By introducing a factor $1 + S(r)$ on the right-hand side of (3.2), we could prove a result corresponding to Theorem 2 of [4] in which (3.1) is replaced by $q(x) = -\rho(r) + S''(r)$. We omit the details.

A further comment on Theorem 2 is that, had we used cartesian coordinates x_1, x_2 rather than polar coordinates, we would have obtained the result that T is not symmetric if (3.1) is replaced by

$$q(x) = -\rho(x_1), \tag{3.3}$$

where $\rho(x_1)$ satisfies (1), (2), and (3) of Theorem 2, and (3.3) need only hold in a strip $x_1 \geqslant x_0$, $X \leqslant x_2 \leqslant Y$ with $q(x)$ arbitrary elsewhere. Again an example is $q(x) = -x_1^{\gamma}$ $(\gamma > 2)$ in the strip.

4. A physical interpretation

The conditions imposed in Theorems 1 and 2 can be given a physical interpretation along the lines of the remarks in [5] and [15, §1].

The Schrödinger equation (1.1) describes the motion of a particle moving under the influence of forces represented by q(x). For the purposes of these remarks, we take q(x) to represent a repulsive force from the origin. Then, in rough terms, the symmetry of T has the physical interpretation that the particle does not reach infinity in a finite time. That this interpretation is only rough and not wholly accurate was pointed out in [15]. Nevertheless, the presence of the annuli A_m for which (iii) of Theorem 1 holds has the rough meaning that the repulsive force is restricted in size in the A_m to such an extent that it is unable to repel the particle to infinity in finite time.

On the other hand, in Theorem 2, the presence of a sector in which (3.1) holds means that the particle is provided with a region in which the repulsive force is large and through which the particle can be repelled to infinity in a finite time. The same remarks apply to (3.3) in the semi-infinite strip.

References

1. F.V.Atkinson and W.D.Evans, Math. Z. 127 (1972) 323-32.
2. E.A.Coddington and N.Levinson, Theory of ordinary differential equations (McGraw-Hill, New York, 1955).
3. M.S.P.Eastham, Bull. London Math. Soc. 4 (1972) 340-4.
4. M.S.P.Eastham, Quart. J. Math. (Oxford) (2) 24 (1973) 257-63.
5. W.G.Faris and R.B.Lavine, Commun. Math. Phys. 35 (1974) 39-48.
6. D.P.Goodall and I.M.Michael, J. Lond. Math. Soc. 7 (1973) 265-71.
7. P.Hartman, American J. Math. 73 (1951) 635-45.
8. G.Hellwig, Differential operators of mathematical physics (Addison-Wesley, 1967).
9. R.S.Ismagilov, Soviet Math. 3 (1962) 279-83.
10. T.Kato, Perturbation theory for linear operators (Springer, 1966).
11. T.Kato, Israel J. Math. 13 (1972) 135-48.
12. I.Knowles, J. London Math. Soc. (to appear).
13. A.I.Martin, Quart. J. Math. (Oxford) (2) 5 (1954) 212-27.
14. A.I.Martin, ibid. 7 (1956) 280-6.
15. J.Rauch and M.Reed, Commun. Math. Phys. 29 (1973) 105-11.
16. D.B.Sears, Canadian J. Math. 2 (1950) 314-25.
17. B.Simon, Math. Ann. 201 (1973) 211-20.
18. E.C.Titchmarsh, Canadian J. Math. 1 (1949) 191-8.
19. E.C.Titchmarsh, Eigenfunction expansions, Part 2 (Oxford, 1958).

Nonlinear elliptic equations

D. E. Edmunds

1. Introduction

The Dirichlet problem for nonlinear elliptic equations seems to offer a perpetual challenge to the mathematician, for no sooner is the situation understood for given classes of equations and domains than new equations and different kinds of domains make it necessary to develop fresh techniques. The object of this paper is to provide some idea of the methods which may be used to handle the particular complexities which arise when we wish to study the Dirichlet problems in an unbounded domain for an elliptic equation with coefficients which may depend on the unknown function u in a violently nonlinear way, so that terms of the form exp u, for example, may be encountered.

To set the scene, let Ω be a domain in R^n and consider the equation of order $2k$:

$$\sum_{|\alpha|\leq k} (-1)^{|\alpha|} D^{\alpha} A_{\alpha}(x, \xi_k(u)(x)) = 0 ,$$

where $x = (x_i)$ is any point of Ω , $D^{\alpha} = \prod_{i=1}^{n} (\partial/\partial x_i)^{\alpha_i}$, $|\alpha| = \sum_{i=1}^{n} \alpha_i$, and $\xi_k(u)(x) = \{D^{\alpha}u(x) : |\alpha| \leq k\}$. Provided the coefficients A_{α} have no more than polynomial growth in u and its derivatives, and provided also that monotonicity and coercivity conditions involving the A_{α} are satisfied, the theory of monotone operators may be brought into play to establish the existence of a variational solution of the Dirichlet problem in Ω for this equation: this can be done whether Ω is bounded (see, for example, [3]) or unbounded ([1], [6], [8]), although in the latter case there are substantial extra complications brought about by the need to develop suitable embedding theorems for Sobolev spaces on unbounded domains. If the equation is strongly nonlinear in the sense that the A_{α} do not satisfy a polynomial growth condition there are, when Ω is bounded, two particular methods of approach available: 1) the theory of monotone operators acting on an Orlicz-Sobolev space (cf. [2], [5], [11], [15]); 2) the use of mappings of monotone type, involving operators which may not be everywhere defined on the Sobolev spaces which are involved (cf. [4], [12], [13]).

Here we consider the Dirichlet problem for strongly nonlinear equations in an unbounded domain, and provide a report on very recent work in this area ([7], [9]) together with some indication of fresh developments. It turns out that method 2) above can be adapted so that in suitable domains, such as infinite strips, equations of the form

$$(\mathcal{L}u)(x) \equiv \sum_{|\alpha|, |\beta| \leq 1} (-1)^{|\alpha|} D^{\alpha}(a_{\alpha\beta}(x) D^{\beta}u(x)) + q(x)\, p(u(x)) = f(x), \qquad (1)$$

for example, can be handled: here p is any continuous function such that $tp(t) \geq 0$ for all real t, and q is a non-negative function which is integrable over Ω. As for method 1), we prefer not to lie on the rather uncomfortable bed of Orlicz-Sobolev spaces, but choose rather to exploit the special position occupied by elliptic equations of order n acting in a domain in R^n: for these equations progress can be made by means of the possibility of embedding certain Sobolev spaces in Orlicz spaces, the latter spaces appearing merely as intermediaries. The embedding results which are obtained (cf. [7]) appear to have considerable intrinsic interest, quite apart from their usefulness in treating the Dirichlet problem.

2. Methods involving operators of monotone type

We adapt the procedures of Hess [12] so that equation (1) can be discussed. Let V and W be real, reflexive, separable Banach spaces such that $W \subset V$ and the natural injection i of W in V is continuous; denote the norms on V and W by $\|\cdot\|_V$ and $\|\cdot\|_W$; denote by V^* and W^* the topological duals of V and W, and let $i^* : V^* \longrightarrow W^*$ be the adjoint of i. In any Banach space we shall denote strong and weak convergence by "$\longrightarrow$" and "$\rightharpoonup$" respectively; the natural pairing between the space and its topological dual will be represented by $(\cdot,\cdot)$.

Let $D(T)$ be the domain of a map $T: D(T) \longrightarrow W^*$, and suppose that $W \subset D(T) \subset V$. We say that T is quasi-bounded if given any sequence (v_n) in W with $v_n \rightharpoonup v$ in V and $(Tv_n, v_n) \leq c\|v_n\|_V$ for all n, c being a positive constant, it follows that (Tv_n) is bounded in W^*. Moreover, T is said to be of type (M) with respect to V and W if:

(i) T is continuous from finite dimensional subspaces of W to the weak topology on W^*;

(ii) if $(v_n) \subset W$, $g \in V^*$, $v_n \rightharpoonup u$ in V, $Tv_n \rightharpoonup i^*g$ in W^*, and $\limsup_{n\to\infty} (Tv_n, v_n) \leq (g,u)$, then $u \in D(T)$ and $Tu = i^*g$.

These concepts are used in the following theorem, due to Hess [12]:

Theorem. Let T be quasi-bounded and of type (M) with respect to V and W, and suppose $(Tv,v)/\|v\|_V \longrightarrow \infty$ as $\|v\|_V \longrightarrow \infty$, $v \in W$. Then the range of T contains $i^*(V^*)$.

Some more notation will be needed. Let Ω be an unbounded domain in R^n, and let $\bar{\Omega}$ and $\partial\Omega$ be the closure and boundary respectively of Ω. Given any positive integer k and any real number p with $1 < p < \infty$, we shall denote by $H_0^{k,p}(\Omega)$ the completion of the space $C_0^\infty(\Omega)$ (of infinitely differentiable real-valued functions with compact support in Ω) with respect to the metric induced by the norm
$\|u\|_{k,p} = \sum_{i=0}^{k} \|D^i u\|_p$, where $\|D^i u\|_p = (\int_\Omega |D^i u(x)|^p dx)^{1/p}$ and $|D^i u(x)| = (\sum_{|\alpha|=i} |D^\alpha u(x)|^2)^{1/2}$.
Also, $H^{k,p}(\Omega)$ will stand for the completion with respect to the same metric of the set of all functions u in $C^k(\bar{\Omega})$ (the set of all k times continuously differentiable functions on $\bar{\Omega}$) with $\|u\|_{k,p} < \infty$. Furthermore, let $B(x,1)$ be the open ball in R^n with centre x and radius 1, and given any $\alpha > 0$ and any measurable function Q set

$$M_\alpha(|Q|) = \sup_{x\in\Omega} \int_{\Omega\cap B(x,1)} |Q(y)|w_\alpha(x-y)dy,$$

where

$$w_\alpha(x) = \begin{cases} |x|^{\alpha-n} & \text{if } \alpha < n, \\ 1 & \text{if } \alpha > n, \text{ or } \alpha = n \text{ and } |x| > 1, \\ 1 - \log|x| & \text{if } \alpha = n \text{ and } |x| \leq 1. \end{cases}$$

We can now set about the discussion of the Dirichlet problem for (1) in Ω. Let $V = H_0^{1,2}(\Omega)$ and $W = H_0^{1,s}(\Omega) \cap V$, where $s > n$: the space W is provided with the norm $\|\cdot\|_W = \max\{\|\cdot\|_{1,2}, \|\cdot\|_{1,s}\}$, when like V it becomes a reflexive, separable Banach space. Moreover, it can be shown [7] that $H_0^{1,s}(\Omega) \subset C(\Omega) \cap L^\infty(\Omega)$, and that there is a constant C such that for all u in $H_0^{1,s}(\Omega)$,

$$\|u\|_\infty \equiv \operatorname*{ess\,sup}_{x\in\Omega} |u(x)| \leq C\|u\|_{1,s}.$$

Let $p : R \longrightarrow R$ be continuous and such that $tp(t) \geq 0$ for all real t, and let q be a non-negative function in $L^1(\Omega)$. Set

$$V_1 = \{v \in V : \int_\Omega q(x)|p(v(x))|dx < \infty \text{ and } \int_\Omega q(x)\ v(x)\ p(v(x))dx < \infty\}.$$

Then given any u in V_1 and any w in W,

$$b(u,w) \equiv \int_\Omega q(x)\ p(u(x))\ w(x)dx$$

is well-defined, and $|b(u,w)| \leq \|qp(u)\|_{L^1} \|w\|_\infty$. Hence for each u in V_1 the map $w \longmapsto b(u,w)$ defines a continuous linear functional T_1u on W by the rule $(T_1u, w) = b(u,w)$ for all w in W. We thus have a map $T_1 : V_1 \longrightarrow W^*$, where $W \subset V_1 \subset V$.

Our object is now to prove that T_1 is quasi-bounded. The key step in the proof is the following lemma.

<u>Lemma</u>. Let (w_n) be a sequence in W such that $w_n \longrightarrow v$ in V and $\limsup_{n\to\infty}(T_1 w_n, w_n) < \infty$. Then $v \in V_1$ and $T_1 w_n \longrightarrow T_1 v$ in W^*.

The proof of this may be carried out in three stages : (i) observe that there is a subsequence (v_n) of (w_n) such that $v_n(x) \longrightarrow v(x)$ for almost all x in Ω ; (ii) use Fatou's lemma and the decomposition of Ω into sets where $|v(x)| > 1$ or ≤ 1 to show that $v \in V_1$, (iii) use Vitali's convergence theorem for L^1 convergence. The details may be found in [9] .

That T_1 is quasi-bounded now follows easily from the lemma. Next, let $T_0: V \longrightarrow V^*$ be bounded and linear, and set $T = i^* T_0 + T_1$, so that T is a mapping from V_1 to W^*. A simple argument shows that T is quasi-bounded and of type (M) with respect to V and W, so that by the Theorem, provided that $(Tw, w)/\|w\|_V \longrightarrow \infty$ as $\|w\|_V \longrightarrow \infty$, $w \in W$, it follows that given any f in V^*, there exists $u \in V_1$ such that $Tu = i^*f$.

Finally, we consider equation (1), where we suppose, in addition to the conditions already imposed upon p and q, that the $a_{\alpha\beta}$ are real, measurable, and such that:

(i) if $|\alpha| = |\beta| = 1$, $a_{\alpha\beta} \in L^\infty(\Omega)$;

(ii) if $|\alpha| = 1$ or $|\beta| = 1$, but $|\alpha| + |\beta| \leq 1$, $M_\mu(a^2_{\alpha\beta}) < \infty$ for some μ with $0 < \mu < 2$;

(iii) if $|\alpha| = |\beta| = 0$, $M_\mu(|a_{\alpha\beta}|) < \infty$ for some μ such that $0 < \mu < 2$.

Under these conditions it can be shown (cf. [6]) that there is a bounded linear map $T_0: V \longrightarrow V^*$ such that

$$a(u,v) \equiv \sum_{|\alpha|,|\beta|\leq 1} \int_\Omega a_{\alpha\beta}(x)\, D^\beta u(x)\, D^\alpha v(x)dx = (T_0 u, v)$$

for all u, v in V. Given any f in W^*, we shall say that u is a <u>variational solution</u> of the Dirichlet problem

$$\mathcal{L}u = f \text{ in } \Omega, \quad u = 0 \text{ on } \partial\Omega, \tag{2}$$

if $u \in V_1$ and $a(u,w) + b(u,w) = (f,w)$ for all w in W. In view of our observations above we have immediately the following result:

Theorem 1. Suppose that $\{a(w,w) + b(w,w)\} / \|w\|_V \longrightarrow \infty$ as $\|w\|_V \longrightarrow \infty$, $w \in W$. Then given any f in V^* there is a variational solution of (2).

We note that since $b(w,w) \geq 0$ for all $w \in W$, the condition mentioned in Theorem 1 will be automatically satisfied if there is a positive constant K such that for all w in W, $a(w,w) \geq K\|w\|_V^2$. This will be so if (i) there is a positive constant c such that for all x in Ω and all $\xi \in R^n$,

$$\sum_{|\alpha|=|\beta|=1} a_{\alpha\beta}(x)\, \xi^\alpha \xi^\beta \geq c|\xi|^2 ;$$

(ii) the Poincaré inequality holds in V: $\|v\|_{1,2} \leq c_1 \|Dv\|_2$ for all v in V, with c_1 independent of v;

(iii) the terms $a_{\alpha\beta}$ with $|\alpha| + |\beta| < 2$ are not too large, in a certain sense (see [6] for more details).

Condition (i) is a straightforward ellipticity assumption, while (ii) amounts to a restriction on the domain Ω: it is verified if Ω is an infinite cylinder, and even for more general unbounded domains (see [6]). For such domains Ω, Theorem 1 may be applied to establish the existence of a variational solution of the Dirichlet problem for

$$-\Delta u + que^u = f,$$

where q is any non-negative L^1 function.

To conclude this illustration of the use of the use of the theory of operators of monotone type we remark that by appropriately modified arguments ([9], [16]) is is possible to obtain comparable results for elliptic equations with quasilinear principal parts, and also to handle variational inequalities for strongly nonlinear operators. Moreover, some progress can be made with the corresponding problems posed in different spaces, with $V = H^{1,2}(\Omega)$ for example (see [10]). Higher order equations may also be treated.

3. Orlicz space methods.

As in §2 we let Ω stand for an unbounded domain in R^n. An Orlicz function is any map $\phi : R \longrightarrow R$ which is continuous, convex, even and such that

$$\lim_{t\to 0} \phi(t)/t = 0 , \quad \lim_{t\to\infty} \phi(t)/t = \infty .$$

The Orlicz class $L_\phi(\Omega)$ is the set of all functions (or more precisely, equivalence classes of functions, functions equal almost everywhere being identified) $u : \Omega \longrightarrow R$ such that

$$\int_\Omega \phi(u(x))dx < \infty .$$

The Orlicz space $L_{\phi*}(\Omega)$ is the linear hull of $L_\phi(\Omega)$ furnished with the norm

$$\|u\|_\phi = \inf \{k : \int_\Omega \phi(u(x)/k)dx \leqslant 1\} .$$

It can be shown that $L_{\phi*}(\Omega)$ is a Banach space which is, in general, neither reflexive nor separable. However, the closure $E_\phi(\Omega)$ of the bounded functions in $L_{\phi*}(\Omega)$ is both separable and contained in $L_\phi(\Omega)$, while $E_\phi(\Omega) = L_{\phi*}(\Omega)$ if and only if ϕ satisfies the so-called Δ_2-condition, that is, there exists k such that $\phi(2t) \leqslant k\phi(t)$ for all $t \geqslant 0$.

Let ϕ and ψ be any two Orlicz functions. If for all $\lambda > 0$, $\lim_{t\to\infty} \phi(\lambda t)/\psi(t) = \infty$, we shall write $\psi \prec \phi$ at ∞; if given any $\varepsilon > 0$ there is a number λ such that for all $t \geqslant 0$, $\psi(t) \leqslant \lambda\phi(\varepsilon t)$, we write $\psi \prec \phi$ at 0.

For most of these notions, and a detailed treatment of Orlicz spaces, we refer to [14].

An important contribution to the theory of Orlicz spaces was made some time ago by Trudinger [15]. He proved that if Ω is a bounded domain in R^n and k is a positive integer less than n, then $H_0^{k,n/k}(\Omega)$ may be continuously embedded in $L_{\phi*}(\Omega)$, where

$$\phi(t) = \exp(|t|^{n/(n-1)}) - 1 ;$$

moreover, for any Orlicz function ψ with $\psi \prec \phi$ at ∞, the embedding into $L_{\psi^*}(\Omega)$ is compact. These results were then used to establish the existence of solutions of eigenvalue problems of the form

$$Au + \lambda B(x,u) = 0, \ u \in H_0^{m,2}(\Omega),$$

where $n = 2m$, A is a linear, uniformly elliptic operator of order $2m$, $tB(x,t) > 0$ if $t \neq 0$, and

$$\sup_{x \in \Omega} B(x,t) \prec \exp(|t|^{n/(n-1)}) \text{ at } \infty .$$

What we shall do here is to indicate how this approach may be carried over to unbounded domains Ω in R^n: the full details may be seen in [7] . In place of the Orlicz function ϕ which turns up in Trudinger's work, it seems more appropriate for our work to deal with the function Φ given by

$$\Phi(t) = |t|^{n/k} \exp(|t|^{n/(n-k)}), \ t \in R ;$$

here k is an integer with $0 < k < n$. The embedding results which are most directly relevant to our needs are contained in the following theorem.

Theorem 2. (i) The space $H_0^{k,n/k}(\Omega)$ may be continuously embedded in $E_\Phi(\Omega)$; in fact, if Ψ is an Orlicz function satisfying the following condition :

$$\text{for some } \lambda > 0, \ \Psi(t) \leq \Phi(\lambda t) \text{ for all } t \geq 0, \tag{3}$$

then $H_0^{k,n/k}(\Omega)$ may be continuously embedded in $E_\Psi(\Omega)$.

(ii) Let Ψ be an Orlicz function satisfying (3) and such that $\Psi \prec \Phi$ at 0 and ∞. Then if

$$\limsup_{|x| \to \infty, x \in \Omega} \operatorname{meas}(\Omega \cap \{y \in R^n : |x-y| \leq 1\}) = 0,$$

the natural embedding of $H_0^{k,n/k}(\Omega)$ in $E_\Psi(\Omega)$ is compact.

(iii) Let Ψ be as in (ii), and let $Q : \Omega \longrightarrow R$ be such that, for some α, $0 < \alpha < n$,

$$\liminf_{R\to\infty} \sup_{q \geq n/k} M_\alpha(|Q_R|^q) = 0,$$

where $Q_R(x) = Q(x)$ if $x \in \Omega$, $|x| \geq R$, and $Q_R(x) = 0$ otherwise. Then multiplication by Q is a compact map of $H_0^{k,n/k}(\Omega)$ in $E_\Psi(\Omega)$.

The proof of this theorem is achieved by a fairly natural combination of the methods of [6] and [15].

It remains to show how Theorem 2 may be used to handle the Dirichlet problem for a strongly nonlinear elliptic equation in Ω, say of the form

$$\sum_{|\alpha|,|\beta|\leq m} (-1)^{|\alpha|} D^\alpha(a_{\alpha\beta}(x)D^\beta u) + b(x,u) = f(x), \; u \in H_0^{m,2}(\Omega), \tag{4}$$

where $n = 2m$. For this illustration it is sufficient to discuss the nonlinear term $b(x,u)$. We assume for simplicity that b is continuous on $\Omega \times R$, and that for all x in Ω and all $t \in R$, $|b(x,t)| \leq \Psi(t)$, where Ψ is an Orlicz function such that the hypotheses of Theorem 2(ii) are met, so that $H \equiv H_0^{m,2}(\Omega)$ is compactly embedded in $E_\Psi(\Omega)$. Note that this requires that Ω should be of a rather special form. Suppose also that Ψ satisfies the Δ_3 condition, that is, there is a number $\lambda > 1$ such that $t\Psi(t) \leq \Psi(\lambda t)$ for all $t \geq 0$.

Use of the Orlicz space version of Hölder's inequality now shows that for any u,v in H,

$$\left|\int_\Omega b(x,u(x))v(x)dx\right| \leq 2\|\Psi(u)\|_{\tilde{\Psi}} \|v\|_\Psi ,$$

where $\tilde{\Psi}$ is the Orlicz function complementary to Ψ.

Since Ψ satisfies the Δ_3 condition it can be proved that $\|\Psi(u)\|_{\tilde{\Psi}} < \infty$. It follows that given any u in H, the map $v \longmapsto \int_\Omega b(x,u(x))v(x)\,dx$ is a bounded linear functional on H, which means we may write

$$\int_\Omega b(x,u(x))\, v(x)\, dx = (Bu, v),$$

where $B : H \longrightarrow H^*$.

At this stage we make the additional assumption that $\tilde{\Psi}$ satisfies the Δ_2 conditions. It can then be shown that $u \longmapsto b(x,u)$ is a continuous map of E_Ψ into $L_{\tilde{\Psi}^*}$. This is all we need to show that B is continuous and compact. For let (u_i) be a bounded sequence in H. Then

$$\|Bu_i - Bu_j\|_{H^*} = \sup\{|(Bu_i - Bu_j, v)| : v \in H, \ \|v\|_H \leq 1\}$$
$$\leq \text{const.}\ \|b(x,u_i) - b(x,u_j)\|_{\tilde{\Psi}}.$$

Since H is compactly embedded in E_Ψ, there is a subsequence of (u_i), denoted again by (u_i) for simplicity, such that $u_i \longrightarrow u$ in E_Ψ. But $u \longmapsto b(x,u)$ is continuous from E_Ψ to $L_{\tilde{\Psi}^*}$, and it is therefore clear that (Bu_i) is a Cauchy sequence in H^*, so that the compactness of B follows. That B is continuous is very easy to see.

Our problem (4) may thus be reduced to one involving an abstract operator equation which, given uniform ellipticity of the linear terms, reasonable behaviour of the $a_{\alpha\beta}$ with $|\alpha| + |\beta| < 2m$, and the requirement that $tb(x,t) \geq 0$ for all x in Ω and all t in R, involves the sum of a monotone and a completely continuous map, the sum being coercive. Standard theory therefore assures us of the existence of a solution. Of course, this discussion presupposes that the conditions of Theorem 2(ii) are fulfilled, and this entails a very stringent condition on the domain Ω. However, by dint of an appeal to Theorem 2(iii) instead, a similar argument can be carried out for such domains as infinite strips, provided that b(x,u) can be written as $b_1(x)\,\Psi(u)$, where b_1 is of such a nature as to justify the use of Theorem 2(iii).

By related methods eigenvalue problems such as those discussed by Trudinger may be handled.

References

1. M. S. Berger and M. Schechter, Embedding theorems and quasilinear boundary value problems for unbounded domains, Trans. Amer. Math. Soc. 172 (1973), 261-274.

2. F. E. Browder, Nonlinear elliptic functional equations in nonreflexive Banach spaces, Bull. Amer. Math. Soc. 72 (1966), 89-95.

3. F. E. Browder, Existence theorems for nonlinear partial differential equations, Proc. Symp. Pure Math. 16, Amer. Math. Soc.: Providence, R.I. (1970).

4. F. E. Browder, Existence theory for boundary value problems for quasilinear elliptic systems with strongly nonlinear lower order terms, Proc. Amer. Math. Soc. Symp. Partial Differential Equations (1971), to appear.

5. T. Donaldson, Nonlinear elliptic boundary value problems in Orlicz-Sobolev spaces, J. Diff. Eqs. 10 (1971), 507-528.

6. D. E. Edmunds and W. D. Evans, Elliptic and degenerate-elliptic operators in unbounded domains, Ann. Scuola Norm. Sup. Pisa, to appear.

7. D. E. Edmunds and W. D. Evans, Orlicz spaces on unbounded domains, to appear.

8. D. E. Edmunds and J. R. L. Webb, Quasilinear elliptic problems in unbounded domains, Proc. Roy. Soc. Lond. A 334 (1973), 397-410.

9. D. E. Edmunds, V. B. Moscatelli and J. R. L. Webb, Strongly nonlinear elliptic operators in unbounded domains, Publ. Math. Bordeaux, to appear.

10. D. E. Edmunds and V. B. Moscatelli, Semi-coercive nonlinear problems, to appear.

11. J. P. Gossez, Nonlinear elliptic boundary value problems for equations with rapidly increasing coefficients, Trans. Amer. Math. Soc., to appear.

12. P. Hess, On nonlinear mappings of monotone type with respect to two Banach spaces, J. Math. pures et appl. 52 (1973), 13-26.

13. P. Hess, Variational inequalities for strongly nonlinear elliptic operators, J. Math. pures et appl. 52 (1973), 285-298.

14. M. A. Krasnosel'skii and Ya. B. Rutickii, Convex functions and Orlicz spaces, Noordhoff, Groningen, 1961.

15. N. S. Trudinger, On imbeddings into Orlicz spaces and some applications, J. Math. Mech. 17 (1967), 473-483.

16. J. R. L. Webb, On the Dirichlet problem for strongly nonlinear elliptic operators in unbounded domains, to appear.

A differential inequality with applications to subharmonic functions

Matts Essén

By a solution of a differential equation or a differential inequality, we mean a function such that h and h' are absolutely continuous, h'' is the a.e. existing derivative of h' and the equation or the inequality is satisfied a.e. The main result to be discussed here is

Theorem 1 : *Let* $p: (-\infty, 0] \to [0, \infty)$ *be a lower semi-continuous function which is locally bounded. Assume that there exists a solution of the inequality*

$$(1) \qquad h''(t) - p(t)^2 h(t) \geq 0, \qquad -\infty < t \leq 0,$$

such that $h(0) = 1$, $\lim_{t \to -\infty} h(t)$ *exists.*

Let t_o *be given,* $t_o < 0$. *If* $\inf_{t<0} p(t) > 0$, *there exists a nonnegative solution* z *of the boundary value problem*

$$(2) \qquad z''(t) - p^*(t)^2 z(t) = 0, \qquad z(0) = 1, \quad z(-\infty) = 0,$$

and it is true that

$$(3) \qquad h(t_o) \leq z(t_o).$$

Here p^* *is a measure-preserving, nondecreasing re-arrangement of* $p|_{[t_o,0]}$ *on* $[t_o,0]$ *and* $p^*(t) = \inf_{s<0} p(s)$, $t < t_o$.

If $\inf_{s<0} p(s) = 0$, *the statement above will still be true except that the conclusion* $z(-\infty) = 0$ *is replaced by* $z(t) = z(t_0)$, $t \leq t_0$.

Remark 1. It is fairly easy to obtain an estimate of $z(t_0)$. Inequality (3) gives us then an estimate of $h(t_0)$ which is useful in the applications.

Remark 2. When $\inf_{s<0} p(s) = 0$, the proof of the theorem known to me is complicated (cf. Essén [1]). This case is important in the applications. When $\inf_{s<0} p(s) > 0$, there is an elemantary proof of an estimate which can be used instead of (3).

As an example how these results can be applied, let us consider the following result of A.Beurling from 1933 which has been extended by J. Lewis [3]. If u is a subharmonic function in the unit disk, let

$$m(r) = \inf_{\varphi} u(re^{i\varphi}), \qquad M(r) = \sup_{\varphi} u(re^{i\varphi}).$$

Theorem A: *Let* λ *be given*, $0 < \lambda < 1$. *With u as above, assume that*

$$m(r) \leq \cos\pi\lambda \; M(r), \qquad r \in E,$$

where E is a union of intervals in $(0,1)$. *Then, if* $M(1) = \sup M(r)$, $0 < r < 1$,

$$M(r) \leq \text{Const.}\; M(1) \exp\left\{ -\lambda \int_{E\cap(r,1)} dt/t \right\} .$$

The bridge from Theorem 1 to Theorem A (and to other results on growth problems for subharmonic functions) is given by a result of F. Norstad [4] (also cf. Essén [1], sect. 5). The differential inequality studied in this note is classical: cf. e.g. Heins [2], p. 121 for a study of a method of Carleman using this concept. There are also well-known estimates of the growth of harmonic measures and new results on the growth of meromorphic functions which can be mentioned in this context.

References.

1. M. Essén, Lectures on the $\cos\pi\lambda$ -theorem. University of Kentucky 1973.

2. M. Heins, Selected topics in the classical theory of a complex variable. Holt, Rinehart and Winston, New York 1962.

3. J.L.Lewis, Some theorems on the $\cos\pi\lambda$ -inequality, Trans. of the AMS 167 (1972), 171-189.

4. F. Norstad, Convexity of the mean-value of certain subharmonic functions (in Swedish). Manuscript 1970.

INEQUALITIES ASSOCIATED WITH CERTAIN PARTIAL DIFFERENTIAL OPERATORS

W. N. Everitt

and

M. Giertz

Notation and definitions

Throughout this lecture we shall use the following notation and definitions:

G is an open set in real euclidean n-space R^n of vectors $x = (x_1, x_2, \ldots x_n)$, with $n \geq 2$;

L^2, L^2_{loc} and C^∞_o refer to the set G, that is, they denote spaces of functions which are, respectively, of integrable square on G, of integrable square on compact subsets of G and infinitely differentiable with compact support on G;

∂_k is used to denote partial differentiation with respect to x_k $(k = 1, 2, \ldots n)$ in the generalised sense, so that for each f in L^2, $\partial_k f$ is the distributional derivative of f with respect to x_k;

$\Delta = \sum_{k=1}^{n} \partial_k^2$ is the generalised Laplacian in n dimensions;

q is a real-valued function on G;

$Q(c) = Q(G, c)$ is the subset of L^2_{loc} defined, for each c in R^n, by $q \in Q(c)$ when

(i) $q \in L^2_{loc}$, (ii) $q(x) > 0$ $(x \in G)$,

(iii) for each $k = 1, 2, \ldots, n$ (1)

$$\int_G \{q \partial_k \varphi\} \leq c_k \int_G |q^{3/2} \varphi| \qquad (\varphi \in C^\infty_o);$$

$M[\cdot]$ is the differential expression defined by $M[f] = -\Delta f + qf$;

S is the minimal operator generated by $M[\cdot]$ in L^2, that is,

$$Sf = M[f] \qquad (f \in D(S) = C^\infty_o);$$

$$D_1(q) = \{f: f \in L^2 \text{ and } M[f] \in L^2\}. \tag{2}$$

Remarks on the definitions of $Q(c)$ and $D_1(q)$

The condition (iii) in the definition of $Q(c)$ in (1) takes the form $|\partial^k q| \leq \leq c_k q^{3/2}$ $(k = 1, 2, \ldots, n)$ when q is differentiable in the classical sense. When x tends to a finite boundary point b of G this condition forces $q(x)$ either to tend to a finite limit or to $+\infty$, in which case $q(x) \geq 4|c|^{-2} |x - b|^{-2}$. If $n = 1$

and G is an interval of the real line with a finite end point b, this latter result implies that $M[\cdot]$ is limit point at b as long as $c \leq 4/\sqrt{3}$, see [8] and also [5]. Thus when q is in $Q(4/\sqrt{3})$ it follows that S is essentially self-adjoint, that is $S^{*} = S^{**}$, if (and only if) q is unbounded at all finite boundary points. A similar result holds true in arbitrary dimensions: When $|c| < 2$ and q is in Q(c) and is unbounded near all finite boundary points, then S is essentially self-adjoint. This result is a consequence of a very general, and for that reason also very complicated, theorem by Jörgens [7], see also [9].

The effect of the condition (iii) in (1) is, see [1] and [6], that it restricts the oscillatory behaviour of q. It does not, however, restrict the rate of increase of q near the boundary of G. In fact, unless q is extremely oscillatory we have $q^{-3/2}\,\partial_k q(x) \to 0 \quad (x \to b)$ whenever $q(x) \to +\infty \quad (x \to b)$.

The set $D_1(q)$ defined by (2) is the domain of the adjoint S^{*} of the minimal operator. It is also the maximal domain of any operator generated in L^2 by $M[\cdot]$ and it contains the domains of all such operators.

Statement of the results

This lecture is concerned with inequalities of the form

$$a_0 \|qf\|^2 + \sum_{k=1}^{n} a_k \|q^{1/2}\,\partial_k f\|^2 + a_{n+1} \|\Delta f\|^2 \leq \|M[f]\|^2 \qquad (f \in D_1(q)), \quad (3)$$

where the norm is the quadratic norm in L^2 and $a_0, a_1, \ldots, a_{n+1}$ are non-negative real numbers, not all zero.

Inequalities of the type (3) have been studied in the one-dimensional case, see [2] and [4], but the methods previously used to establish them do not apply directly to the present situation. When $n = 1$ such inequalities do not exist if $M[\cdot]$ is regular or limit-circle at any boundary point of G, but they do exist in most cases when $M[\cdot]$ is limit-point at each point of the boundary of G, that is when S is essentially self-adjoint, [4] and [3]. When $n \geq 2$ the corresponding result is

Theorem 1

Let $G \subset R^n$ be open and $c \in R^n$ satisfy $|c| < 2$. Then there are statements of the form (3), with $a_0, a_1, \ldots, a_{n+1}$ nonnegative and not all zero, which hold true for any q in Q(G, c) which is unbounded near every finite boundary point of G.

One way to prove this theorem is to start by establishing the inequalities for all f in C_0^{∞}, the domain of the minimal operator S, and then extend their validity, by a closure argument, to all f in the domain of the closure

S^{**} of S. This gives the desired result since the conditions on q imply that S is essentially self-adjoint, so that $D(S^{**}) = D(S^{*}) = D_1(q)$.

When $|c| < 2$ we may find explicit values of the coefficients $a_o, a_1, \ldots, a_{n+1}$, in terms of c, as follows:
First choose a δ in R^n such that $\min [\,|\delta|^2, c \cdot \delta] \leq 1$ and $2\delta_k \geq c_k$ $(k = 1, 2, \ldots, n)$, then choose a real number $\beta \geq c \cdot \delta$ satisfying $\min [1, 1 + \gamma] \leq \beta \leq \max [1, 1 + \gamma]$, where $\gamma = \delta \cdot (c - \delta)$, define $\alpha = (1 - \gamma)(\beta - \gamma)^{-1}$ and put

$$\begin{cases} a_o = \alpha(\beta - c \cdot \delta) \\ a_k = \alpha(2 - c_k \delta_k^{-1}) & (k = 1, 2, \ldots, n) \\ a_{n+1} = \begin{cases} 1 - (1 - \alpha)^2 (1 - \alpha\beta)^{-1} & (\alpha\beta < 1) \\ 1 & \alpha\beta = 1 . \end{cases} \end{cases} \tag{4}$$

(Here $a_k = 2\alpha$ when $c_k = 0$, even if $\delta_k = 0$).

Theorem 2

When $|c| < 2$ the statement (3) with coefficients given by (4) is valid for each q in Q(c) which is unbounded near every finite boundary point. Different choices of β and δ give different statements (3), and no such statement contains any of the others.

Many of the inequalities defined by (3) and (4) are sharp on Q(c) in the sense that they become false for some q in Q(c) if we replace any one of the coefficients by a larger number while keeping the others fixed. We shall give a few examples of such sharp inequalities in the next section.

The differential expression $M[\cdot]$ is said to be separated in L^2 if the two terms Δf and qf are both in L^2 whenever this is true of f and $M[f]$, that is, if $qf \in L^2$ $(f \in D_1(q))$. Thus we have the following corollary of Theorem 1:

Theorem 3

Let G be any (finite or infinite) open set in R^n and assume that q is in Q(G, c) with $|c| < 2$. Then $M[\cdot]$ is separated in $L^2(G)$.

Examples

We conclude this lecture by listing a few special examples of inequalities corresponding to particular choices of the parameters β and δ. All these examples are sharp on Q(c).

1. When $|c| < 2$ the choice $\delta = \frac{1}{2} c$, $\beta = 1 + |c|^2/4$ gives $\alpha = 1 - |c|^2/4$ and

$$(1 - |c|^2/4)\, \|qf\| \leq \|M[f]\| \qquad (f \in D_1(q)).$$

2. When $|c| \leq \sqrt{2}$, $\delta = \frac{1}{2} c$ and $\beta = 1$ give $\alpha = 1$ and

$$(1 - |c|^2/2) \, ||qf||^2 + ||\Delta f||^2 \leq ||M[f]||^2 \qquad (f \in D_1(q)).$$

3. When $\sqrt{2} \leq |c| < 2$, $\delta = \frac{1}{2} c$ and $\beta = \frac{1}{2}|c|^2$ give $\alpha = 4/|c|^2 - 1$ and

$$(4/|c|^2 - 1) \, ||\Delta f|| \leq ||M[f]|| \qquad (f \in D_1(q)).$$

4. When $|c| \leq 1$ we may choose $\delta = c/|c|$, $\beta = 1$ to obtain $\alpha = 1$ and

$$(1 - |c|) \, ||qf||^2 + (2 - |c|) \sum_{k=1}^{n} ||q^{\frac{1}{2}} \partial_k f||^2 + ||\Delta f||^2 \leq ||M[f]||^2$$

$$(f \in D_1(q)).$$

References

1. Everitt, W.N., Giertz, M: Some properties of the domains of certain differential operators, Proc. London Math. Soc. (3) 23 (1971) 301 - 324.

2. ——— : Some inequalities associated with certain ordinary differential operators, Math. Z. 126 (1972) 308 - 326.

3. ——— : An example concerning the separation property for differential operators, Proc. Royal Soc. Edinburgh Sect. A, 71, 14 (1972/73) 159 - 165.

4. ——— : Inequalities and separation for certain ordinary differential operators, Proc. London Math. Soc.

5. ——— : A Dirichlet-type result for ordinary differential operators, Math. Ann. 203 (1973) 119 - 128.

6. Giertz, M: On the solutions in $L^2(-\infty, \infty)$ of $y'' + (\lambda - q(x))y = 0$ when q is rapidly increasing, Proc. London Math. Soc. (3) 14 (1964) 53 - 73.

7. Jörgens, K: Wesentliche Selbstadjungiertheit singulärer elliptischer Differentialoperatoren zweiter Ordnung in $C_0^\infty(G)$. Math. Scand. 15 (1964) 5 - 17.

8. Sears, D.B: On the solutions of a second order linear differential equation which are of integrable square, J. London Math. Soc. 24 (1949) 207 - 215.

9. Triebel, H: Erzeugung nuklearer lokalkonvexer Räume durch singuläre Differentialoperatoren zweiter Ordnung, Math. Ann. 174 (1967) 163 - 176.

The Expansion Theorem in Multi-Parameter Sturm-Liouville Theory

M. Faierman

1. Introduction. In the first decade of this century there appeared important papers by Hilbert [1,pp.262-267] and Dixon [2] concerning the expansion of an arbitrary function in a double series involving the eigenfunctions of a pair of simultaneous, two parameter Sturm-Liouville systems. In this paper we wish to describe the methods employed by these authors in dealing with this expansion, and in view of more recent developements [3], to compare the results obtained by utilizing these methods. We shall therefore be concerned here with the simultaneous two-parameter systems

(1a) $$y_1'' + (\lambda A_1(x_1) - \mu B_1(x_1) + q_1(x_1))y_1 = 0, \quad 0 \le x_1 \le 1, \quad ' = d/dx_1,$$

(1b) $$y_1(0) \cos \alpha_1 - y_1'(0) \sin \alpha_1 = 0, \quad 0 \le \alpha_1 < \pi,$$
$$y_1(1) \cos \beta_1 - y_1'(1) \sin \beta_1 = 0, \quad 0 < \beta_1 \le \pi,$$

and

(2a) $$y_2'' + (-\lambda A_2(x_2) + \mu B_2(x_2) + q_2(x_2))y_2 = 0, \quad 0 \le x_2 \le 1, \quad ' = d/dx_2,$$

(2b) $$y_2(0) \cos \alpha_2 - y_2'(0) \sin \alpha_2 = 0, \quad 0 \le \alpha_2 < \pi,$$
$$y_2(1) \cos \beta_2 - y_2'(1) \sin \beta_2 = 0, \quad 0 < \beta_2 \le \pi,$$

where, for simplicity, we shall assume that for $r = 1,2$, A_r, B_r, and q_r are real-valued analytic functions in $0 \le x_r \le 1$, with both A_r and B_r positive in this interval. Furthermore, we shall also assume that $\Delta = (A_1B_2 - A_2B_1) > 0$ in I^2, where I^2 is the Cartesian product of the intervals $0 \le x_1 \le 1$, $0 \le x_2 \le 1$.

By an eigenvalue of the system (1,2) we shall mean a pair of numbers, (λ^*, μ^*), such that for $\lambda = \lambda^*$ and $\mu = \mu^*$, (1a) (resp. (2a)) has a non-trivial solution satisfying (1b) (resp. (2b)). If $y_1(x_1, \lambda^*, \mu^*)$ (resp. $y_2(x_2, \lambda^*, \mu^*)$) denotes such a solution, then the product

$\prod_{r=1}^{2} y_r(x_r, \lambda^*, \mu^*)$ is called an eigenfunction of the system (1,2) corresponding to (λ^*, μ^*). Let L_Δ^2 denote the Hilbert space constructed from those functions which are absolutely square-integrable on I^2 and with inner product $(f,g) = \iint_{I^2} \Delta f\bar{g}\, dx_1 dx_2$. Then by the multiplicity of an eigenvalue of the system (1,2) we shall mean the dimension of that subspace of L^2_Δ generated by the corresponding eigenfunctions. Utilizing the results given in [4,p.551, problem 16 and pp. 160-168] and [5,pp.248-251], we may show that the eigenvalues of the system (1,2) form a countably infinite subset of E^2 (Euclidean 2-space), with each eigenvalue having multiplicity 1, and with eigenfunctions corresponding to distinct eigenvalues being orthogonal in L^2_Δ. Furthermore, if p_1 and p_2 are any pair of non-negative integers, then there is exactly one eigenvalue of the system, say (λ^*,μ^*), such that $\phi_r(x_r,\lambda^*,\mu^*)$ has precisely p_r zeros in $0<x_r<1$ for $r = 1,2$, where ϕ_r is the solution of (ra) satisfying $\phi_r(0,\lambda,\mu) = \sin\alpha_r$, $\phi_r'(0,\lambda,\mu) = \cos\alpha_r$. We note for later use that the set of eigenvalues of the system (1,2) may be denoted by $\{(\lambda_{j,k},\mu_{j,k})\}_{j,k=0}^{\infty}$, where $\phi_1(x_1,\lambda_{j,k},\mu_{j,k})$ has precisely j zeros in $0<x_1<1$ and $\phi_2(x_2,\lambda_{j,k},\mu_{j,k})$ has precisely k zeros in $0<x_2<1$.

2. <u>The Method of Hilbert</u>. In this section we shall assume that $\alpha_r = 0$, $\beta_r = \pi$ for $r = 1,2$. Then by multiplying (1a) by A_2y_2, (1b) by A_1y_1, and adding, we are led to consider the eigenvalue problem

(3a) $$A_2(x_2)z_{x_1x_1} + A_1(x_1)z_{x_2x_2} + q(x_1,x_2)z = -\mu\Delta(x_1,x_2)z \text{ for } (x_1,x_2)\in I^2,$$

(3b) $$z = 0 \text{ on the boundary of } I^2,$$

where $z_{x_j} = \partial z/\partial x_j$ and $q = A_2q_1+A_1q_2$. Observing that the partial differential operator on the right-hand side of (3a) is uniformly elliptic in I^2, we may construct the Green's function for the system (3), and hence we now have the theory for symmetric integral equations at our disposal. In particular we are assured that infinitely many eigenvalues $\mu_0,\mu_1,\ldots$, exist together with corresponding eigenfunctions $z_0,z_1,\ldots$.

For $n\geq 0$, consider the system (1) with μ fixed at μ_n. Denote the eigenvalues of this system by $\lambda_0,\lambda_1,\ldots$, and the corresponding eigenfunctions (normalized with respect to the weight A_1) by $\chi_0,\chi_1,\ldots$. Then we may expand z_n in terms of these eigenfunctions, that is,

(4) $$z_n(x_1,x_2) = \sum_{j=0}^{\infty}\eta_j(x_2)\chi_j(x_1),$$

where $\eta_j(x_2) = \int_0^1 A_1(x_1)z_n(x_1,x_2)\chi_j(x_1)dx_1$ (and note also that $\eta_j(0) =$

$\eta_j(1) = 0$). If we differentiate η_j twice with respect to x_2, make use of (3a), and integrate by parts, then it will be seen that $\eta_j(x_2)$ satisfies (2a) when $\lambda = \lambda_j$ and $\mu = \mu_n$. Hence it follows that for $j \geq 0$, $\chi_j \eta_j$ is an eigenfunction of the system (1,2) corresponding to (λ_j, μ_n). It is also clear that $\chi_j \eta_j$ is an eigenfunction of (3) corresponding to μ_n. Since μ_n has finite multiplicity, and since it is easily seen from the orthogonal properties of the eigenfunctions of the system (1,2) that the subset of L^2_Δ composed of those $\chi_j \eta_j$ for which η_j is not identically zero is a linearly independent set, it therefore follows that only a finite number of the η_j are not identically zero. Hence, in light of (4), it is clear that the subspace of L^2_Δ formed from the solutions of (3) corresponding to μ_n may be generated by a finite number of elements of the form $\chi_j \eta_j$. In this way we arrive at the following result.

THEOREM 1. Let $f \in C_2$ on I^2 and vanish on the boundary of this square. Then f may be developed as an infinite series involving the eigenfunctions of the system (1,2), with the series converging absolutely and uniformly to f on I^2.

Our main criticism with the method of Hilbert lies in its limitation. For this method does not appear to work for the general case where there are no restrictions on the α_r or β_r on account of the difficulty of establishing the existence of Green's function for the system (3) when (3b) is replaced by more general Sturm-Liouville type boundary conditions.

3. The Method of Dixon. For r = 1,2, let θ_r denote the solution of (ra) satisfying $\theta_r(1,\lambda,\mu) = \sin \beta_r$, $\theta_r'(1,\lambda,\mu) = \cos \beta_r$, and put $W_r = \phi_r\theta_r' - \phi_r'\theta_r$, where the ϕ_r are defined above. With (x_1,x_2) a point of I^2 and f a function integrable in this square, put

$$\Omega(\lambda,\mu) = \iint_{I^2} \Delta(t_1,t_2) f(t_1,t_2) \left(\prod_{r=1}^{2} \Phi_r(x_r,t_r,\lambda,\mu) \right) dt_1 dt_2,$$

where, not indicating λ and μ explicitly, $\Phi_r(x_r,t_r)$ equals $\phi_r(x_r)\theta_r(t_r)$ if $x_r \leq t_r$, and equals $\phi_r(t_r)\theta_r(x_r)$ otherwise. Then with each positive integer n sufficiently large, we wish to associate a certain point set lying in the space of the complex variables λ, μ, which we shall call the surface of integration and denote by S_n. Our aim is to choose S_n in such a way so that on one hand the integral $I(n) = (1/2\pi i)^2 \iint_{S_n} (\Omega/W_1W_2)d\lambda d\mu$ is a sum of a finite number of terms of a double series involving f and the eigenfunctions of the system (1,2) (which, hereinafter, will be referred to as the Sturm-Liouville series for f), and on the other hand to ensure that $I(n) \to f(x_1,x_2)$ as $n \to \infty$.

The choice of S_n is a matter of some difficulty. However, the general idea is that the surface will be composed of tuples (λ,μ) such that λ will travel about a circle, say K, of large radius and μ about a cycle which varies with λ. The circle and the cycles must be chosen so as to ensure that the W_j and P_j do not vanish on S_n, where $P_1 = \lambda A_1 - \mu B_1$, $P_2 = -\lambda A_2 + \mu B_2$. These requirements will allow us to define the real numbers λ_1, λ_2, where $\lambda_1 < 0 < \lambda_2$ and both these quantities are of the order of n^2; and K is determined by demanding that the segment $[\lambda_1, \lambda_2]$ be a diameter. Now let the values of μ for which $W_1 = 0$ when λ has any particular real value be $\mu_0(\lambda)$, $\mu_1(\lambda), \ldots$, taken in descending order, and let those for which $W_2 = 0$ be $\mu_0^*(\lambda)$, $\mu_1^*(\lambda), \ldots$, in ascending order(observe that when λ is held fixed at a particular value, $\mu_r(\lambda)$ (resp. $\mu_r^*(\lambda)$), $r \geq 0$, is the eigenvalue of the system (1) (resp. (2)) whose corresponding eigenfunction has precisely r zeros in (0,1)).It is not difficult to verify that $\mu_r(\lambda)$ (resp. $\mu_r^*(\lambda)$), $r \geq 0$, is analytic in $-\infty < \lambda < \infty$, and at each point of this interval its derivative is a mean among the values of A_1/B_1 (resp. A_2/B_2). Hence assuming that n is large enough so as to ensure that $\mu_0(\lambda_1) < \mu_0^*(\lambda_1)$, and denoting by $\Gamma(\lambda)$ the cycle in the μ-plane corresponding to $\lambda \in K$, we shall require that $\Gamma(\lambda_2)$ be a closed curve that is symmetric about the real axis, that contains the points $\mu_j(\lambda_2)$, $\mu_k^*(\lambda_2)$, $j = (N_1+1), \ldots, N_3$, $k = 0, \ldots, N_2$ (but no other zeros of the $W_r(\lambda_2,\mu)$) as well as the zeros of $P_2(t_2,\lambda_2,\mu)$, but excludes the zeros of $P_1(t_1,\lambda_2,\mu)$ (here the $N_j = N_j(n)$ are certain positive integers which increase and tend to ∞ with n). As λ moves along K in the positive sense from λ_2 to λ_1, $\Gamma(\lambda)$ will decompose into two disjoint, closed curves, which, when $\lambda = \lambda_1$, will be symmetric about the real axis, with one containing the points $\mu_j(\lambda_1)$, $j = (N_1+1), \ldots, N_3$, and the other the points $\mu_k^*(\lambda_1)$, $k = 0, \ldots, N_2$. Finally, when $\operatorname{Im} \lambda < 0$, $\Gamma(\lambda)$ is obtained by reflecting $\Gamma(\bar{\lambda})$ about the real axis.

After this brief description of the surface S_n, we shall now show that

$I(n)$ is the sum of a finite number of terms of the Sturm-Liouville series for f. To this end we shall make use of the fact that if $r \geq 0$ and $\lambda^* < \lambda^\#$ are any two real numbers, then $\mu_r(\lambda)$ (resp. $\mu_r^*(\lambda)$) can be defined as an analytic function in a region D of the λ-plane that contains the segment $[\lambda^*, \lambda^\#]$. We then have $W_1(\lambda, \mu_r(\lambda)) = 0$ (resp. $W_2(\lambda, \mu_r^*(\lambda)) = 0$) for $\lambda \in D$. Also, if we utilize the obvious fact that the eigenvalues of the system (1,2) and the zeros of the simultaneous equations $W_1 = 0$, $W_2 = 0$, are identical, denote by γ_r (resp. γ_r^*), $r \geq 0$, the curve in the (λ,μ)-plane determined by $\mu_r(\lambda)$ (resp. $\mu_r^*(\lambda)$), and let j,k be any two non-negative integers, then it is clear from the above results that γ_j intersects γ_k^* in precisely one point, namely at the eigenvalue of the system (1,2), $(\lambda_{j,k}, \mu_{j,k})$. Now proceeding on with the problem under discussion here, we shall deform S_n continuously into a certain suitable surface S_n^* in such a way as to avoid the zeros of the W_j; hence from [6] it will follow that $I(n) = I^*(n)$, where $I^*(n) = (1/2\pi i)^2 \iint_{S_n^*} (\Omega/W_1 W_2)\, d\lambda\, d\mu$. S_n^* will be composed of tuples (λ,μ), where λ will travel about an ellipse, say K^*, having $[\lambda_1, \lambda_2]$ as major axis and μ about a cycle, say $\Gamma^*(\lambda)$, varying with λ. The ellipse K^* will lie in the interior of a rectangle in which $\mu_j(\lambda)$, $\mu_k^*(\lambda)$ are analytic and $\partial W_1(\lambda, \mu_j(\lambda))/\partial\mu$, $\partial W_2(\lambda, \mu_k^*(\lambda))/\partial\mu$ do not vanish for $j = 0, \ldots, N_3$, $k = 0, \ldots, N_2$. For $\lambda \in K^*$, $\Gamma^*(\lambda)$ will contain the points $\mu_j(\lambda)$, $\mu_k^*(\lambda)$, $j = (N_1+1), \ldots, N_3$, $k = 0, \ldots, N_2$, but no other zeros of the $W_r(\lambda,\mu)$ (it is also not difficult to verify that there can be no coincidences amongs't these $\mu_j(\lambda)$ and $\mu_k^*(\lambda)$ when $\mathrm{Im}\,\lambda \neq 0$, and clearly this assertion remains valid when $\lambda = \lambda_1$). Hence by employing the above results and applying the Cauchy residue theorem twice in succession, we may show that

(5) $$I(n) = I^*(n) = s_n,$$

where

(6) $$s_n = s_n(x_1,x_2) = \sum_{j=0}^{N_1} \sum_{k=0}^{N_2} (f, \phi_{j,k}) \phi_{j,k}(x_1,x_2)/(\phi_{j,k}, \phi_{j,k}) ,$$

$(\ ,\ )$ denotes the inner product defined in the introduction, and $\phi_{j,k}(t_1,t_2) = \prod_{r=1}^{2} \phi_r(t_r, \lambda_{j,k}, \mu_{j,k})$. In a similar manner we may show that

(7) $$\Delta(x_1,x_2) \iint_{S_n} \left(\prod_{r=1}^{2} P_r(x_r, \lambda, \mu) \right)^{-1} d\lambda\, d\mu = (2\pi i)^2.$$

The double series from which the s_n are obtained by summation over rectangles (as exemplified by (6)) is precisely the Sturm-Liouville series for f about which we have previously spoken. Hence, in light of (5) and (6), the above assertion concerning I(n) now follows.

Finally, it remains only to investigate the convergency of I(n); and this may be achieved by utilizing certain approximate solutions of the differential equations (1a) and (2a). Indeed, if we make use of (5) and (7), then the following theorem may be established.

THEOREM 2. Let $f \in C_2$ on I^2. Let $f(0,t_2)$ (resp. $f(1,t_2)$) vanish in $0 \leq t_2 \leq 1$ if $\alpha_1 = 0$ (resp. $\beta_1 = \pi$). Let $f(t_1,0)$ (resp. $f(t_1,1)$) vanish in $0 \leq t_1 \leq 1$ if $\alpha_2 = 0$ (resp. $\beta_2 = \pi$). Then $s_n(x_1,x_2) \to f(x_1,x_2)$ as $n \to \infty$ for $(x_1,x_2) \in I^2$. Moreover, the convergence is uniform on any compact set contained in the interior of I^2.

To conclude, we remark that although theorem 2 does not assert that at the point (x_1,x_2) the Sturm-Liouville series for f converges to $f(x_1,x_2)$ (in the sense in which convergence is defined for the double Fourier series), the theorem nevertheless furnishes us with an eigenfunction expansion for f, and an expansion which is not limited to particular values of the α_j and the β_j.

REFERENCES

1. D. Hilbert, "Grundzüge Einer Allgemeinen Theorie der Linearen Integralgleichungen," Chelsea, New York, 1953.
2. A. C. Dixon, Harmonic expansions of functions of two variables, Proc. London Math. Soc. (2) 5 (1907), 411-478.
3. M. Faierman, An eigenfunction expansion for a two-parameter system of ordinary differential equations of the second order, (to appear).
4. F. V. Atkinson, "Discrete and Continuous Boundary Problems," Academic, New York, 1964.
5. E. L. Ince, "Ordinary Differential Equations," Dover, New York, 1956.
6. H. Poincaré, Sur les résidus des intégrales doubles, Acta Math. 9 (1887), 321-380.

On the differential equation $\sum_{r=0}^{n} c_r (pD_1 + qD_2)^r u = 0$

I. Fenyő

Let u be a distribution of $D'(R^2)$ which is a solution of

$$\sum_{r=0}^{n} c_r (pD_1 + qD_2)^r u = 0 \tag{1}$$

(D_1 and D_2 are the partial differential operators) and ψ an arbitrary Schwartz-test function of one variable. We denote by $\odot$ a mapping from $D'(R^2) \times D(R)$ into $D'(r)$ such that $(u \odot \psi, \varphi) = (u, \varphi \otimes \psi)$ (φ is an abritrary testfunction of $D(R)$). It is proved, that the distribution $u \odot \psi$ possess in every point x_0 a local value (in the sense of Lojasiewicz) and the value of $u \odot \psi$ in x_0 is the scalar product of a distribution $u(x_0) \in D'(R)$ with ψ (i.e. $u \odot \psi|_{x_0} = (u(x_0), \psi)$). Now the following problem can be posed. There are given n points of the real axis: $x_1, x_2, \ldots, x_n$ ($x_i \neq x_k$, $i \neq k$) and n distributions $f_1, \ldots, f_n$ (of $D'(R)$). We are looking for a distribution $u \in D'(R^2)$ which is a solution of (1) and for which $u(x_i) = f_i$ ($i = 1, 2, \ldots, n$) holds. It is proved that this problem has always a unique solution.

Reference

1. Fenyő, I. : On a partial differential equation in two variables, Demonstration Math. (Poland) Vol 1973.

FUNDAMENTAL SOLUTIONS FOR DEGENERATE PARABOLIC EQUATIONS

Avner Friedman

Section 1. Definitions

Let

$$Lu \equiv \frac{1}{2} \sum_{i,j=1}^{n} a_{ij}(x) \frac{\partial^2 u}{\partial x_i \partial x_j} + \sum_{i=1}^{n} b_i(x) \frac{\partial u}{\partial x_i}$$

be an elliptic operator in R^n, and consider the Cauchy problem

$$(1.1)\qquad \begin{aligned} Lu - \frac{\partial u}{\partial t} &= 0 \qquad (x \in R^n,\ t > 0), \\ u(x,0) &= f(x) \qquad (x \in R^n). \end{aligned}$$

Definition 1. Let $K(x,t,y)$ be a function such that, for any continuous function f with compact support, the function

$$(1.2)\qquad u(x,t) = \int_{R^n} K(x,t,y) f(y) dy$$

is a solution of (1.1). Then we call K a fundamental solution.

Let us assume:

(A) $(a_{ij}(x))$ is uniformly positive definite in R^n, and a_{ij}, b_i are bounded and uniformly Lipschitz continuous.

Then, for any continuous f with compact support, there exists a unique bounded solution of the Cauchy problem (1.1). The solution can be represented in the form (1.2), and the fundamental solution K can be constructed by the method of parametrix (see Friedman [2]). K is non-negative, and

$$|D_x^\alpha K(x,t,\xi)| \leq \frac{C}{t^{(n+|\alpha|)/2}} \exp[-\frac{c}{t}|x-y|^2] \qquad (|\alpha| \leq 2)$$

where C, c are positive constants.

When $(a_{ij}(x))$ is not uniformly positive definite, the construction of K breaks down. There is however a class of equations, with $(a_{ij}(x))$ degenerating throughout all of R^n, for which a fundamental solution still exists. These are the ultra-parabolic equations

$$\sum_{i,j=1}^{n} a_{ij}(x,y) \frac{\partial^2 u}{\partial x_i \partial x_j} + \sum_{i=1}^{n} b_i(x,y) \frac{\partial u}{\partial x_i} + \sum_{i=1}^{m} x_i \frac{\partial u}{\partial y_i} - \frac{\partial u}{\partial t} \qquad (m \leq n).$$

One can construct a fundamental solution by a variant of the parametrix method (see [4],[5],[6]).

We shall now present a probabilistic definition of a fundamental solution. Assume that there exists an $n \times n$ matrix $\sigma = (\sigma_{ij})$ such that $a_{ij} = \sum \sigma_{ik}\sigma_{jk}$, and consider the system of n stochastic differential equations

(1.3) $$d\xi(t) = b(\xi(t))dt + \sigma(\xi(t))dw(t),$$

where $b = (b_1,\ldots,b_n)$ and $w(t) = (w_1(t),\ldots,w_n(t))$ is n-dimensional Brownian motion. If $b(x)$, $\sigma(x)$ are $O(|x|)$ and Lipschitz continuous, then (see [3]) this system defines a Markov process with transition probability function $q(t,x,A)$. When also (A) holds, then

(1.4) $$q(t,x,A) = \int_A K(x,t,y)dy,$$

i.e., the fundamental solution constructed by the parametrix method is the density function of q.

We shall now assume:

(B) $a_{ij} \in C^3(R^n)$; $b_i \in C^2(R^n)$; $a_{ij}(x) = O(|x|^2)$ and $b_i(x) = O(|x|)$.

(C) $(a_{ij}(x))$ is positive definite if $x \in R^n \setminus S$, where S is a compact set.

Then one can always construct σ as above (see [1]). Consequently, one can write down (1.3), and the transition probability function q is then well defined.

<u>Definition 2</u>. Let (B), (C) hold. If there exists a density function $K(x,t,y)$ of the transition probability function $q(t,x,A)$ (see (1.4)) then we call K the <u>fundamental solution</u> of the parabolic operator $L - \partial/\partial t$.

This definition is consistent with definition 1 when (A) holds. We shall henceforth adopt this definition.

<u>Section 2</u>. <u>Existence</u>

Assume that S is the boundary ∂G of a bounded domain G with C^3 boundary. Denote by $\nu = (\nu_1,\ldots,\nu_n)$ the outward normal. We shall assume:

$$\sum_{i,j=1}^{n} a_{ij}\nu_i\nu_j = 0 \qquad \text{along } \partial G, \tag{2.1}$$

$$\sum_{i=1}^{n} \Big(b_i - \frac{1}{2}\sum_{j=1}^{n} \frac{\partial a_{ij}}{\partial x_j}\Big)\nu_i > 0 \quad \text{along } \partial G. \tag{2.2}$$

Condition (2.1) means probabilistically that the "normal diffusion" of (1.3) vanishes along ∂G. Condition (2.2) means that the "normal drift" at ∂G points into $R^n \setminus \bar{G}$. Thus ∂G is an "obstacle" to the diffusion from outside G, but not from inside G. We call ∂G a *strictly one-sided obstacle*.

Theorem 2.1. Let (B), (C) and (2.1),(2.2) hold. Then there exists a fundamental solution K, and

$$K(x,t,y) = 0 \quad \text{if } x \in R^n \setminus \bar{G},\ y \in G,$$
$$> 0 \quad \text{otherwise.}$$

Denote by $R(y)$ the distance from y ($\in R^n \setminus G$) to ∂G.

Theorem 2.2. Under the assumptions of Theorem 2.1, for any compact set $E \subset R^n \setminus \bar{G}$, $\delta > 0$, $T > 0$,

$$K(x,t,y) \le C \exp[-\frac{c}{t} \log R(y)] \tag{2.3}$$

if $x \in E$, $R(y) < \delta$, $0 < t < T$, *where* C, c *are positive constants depending on* E,δ,T. *If also*

$$\sum a_{ij} \frac{\partial R}{\partial x_i}\frac{\partial R}{\partial x_j} \ge \alpha R^2 \qquad (\alpha > 0)$$

in a neighborhood of ∂G, *then*

$$K(x,t,y) \ge \bar{C} \exp[-\frac{\bar{c}}{t} \log R(y)] \tag{2.4}$$

when $x \in E$, $R(y) < \delta$, $0 < t < T$, *provided* δ *is sufficiently small;* $\bar{C}$ *and* $\bar{c}$ *are positive constants.*

Section 3. Outline of proofs

We give a very brief outline of the proofs. More details will be given elsewhere. Consider the parabolic operator

$$Lu + \varepsilon\Delta u - \frac{\partial u}{\partial t} \qquad (\varepsilon > 0). \tag{3.1}$$

Since it is non-degenerate, one can construct a Green's function $G_{\varepsilon,m}(x,t,y)$ in a cylinder with base $|x| < m$. One can estimate $G_{\varepsilon,m}(x,t,y)$ and its derivatives when either x or y is away from S, uniformly with respect to both m and ε. Hence, by taking a subsequence, we obtain a limit function $K(x,t,y)$.

Using the probabilistic connection, one can prove that $K(x,t,y)$ is independent of the particular subsequence. Further, (1.4) holds whenever $x \notin \partial G$.

In order to prove (1.4) for $x \in \partial G$, the following result is established:

$$P_x\{\xi(t) \in R^n \setminus \bar{G} \text{ for all } t > 0\} = 1 \quad \text{if } x \in \partial G.$$

To prove Theorem 2.2, one gives an alternate construction of $K(x,t,y)$ (for x, y in $R^n \setminus \bar{G}$) as limit of Green's functions $\Gamma_{m,\varepsilon}(x,t,y)$ of (3.1) in cylinders $\{x; |x| < m, R(x) > 1/m\}$. The functions $\Gamma_{m,\varepsilon}$ can be estimated by comparison functions.

Section 4. Generalized fundamental solutions

When (2.2) does not hold, a fundamental solution does not exist in general. Consider, for instance,

$$u_t = x^2 u_{xx} + \lambda u_x .$$

If $\lambda \neq 0$ then a fundamental solution exists, by Theorem 2.1. If $\lambda = 0$ then a fundamental solution does not exist. However a *generalized* fundamental solution exists in the following sense:

If $x \neq 0$ then $K(x,t,y)$ is a function, whereas if $x = 0$ then $K(0,t,y)$ is a Dirac measure $\delta(y)$ concentrated at $y = 0$.

This result is a special case of a general theorem asserting the existence of a generalized fundamental solution K in case (2.2) is replaced by

$$\sum_{i=1}^{n} \left(b_i - \frac{1}{2} \sum_{j=1}^{n} \frac{\partial a_{ij}}{\partial x_j}\right) \nu_i = 0 \quad \text{along } \partial G.$$

References

1. M. I. Freidlin, On the factorization of nonnegative definite matrices, *Theory Probability Appl.*, 13 (1968), 354-356.

2. A. Friedman, *Partial Differential Equations of Parabolic Type*, Prentice-Hall, Englewood Cliffs, N.J., 1964.

3. I. I. Gikhman and A. V. Skorokhod, *Introduction to the Theory of Random*

Processes, W. B. Saunders, Philadelphia, 1969.

4. A. M. Ilin, On a class of ultraparabolic equations, Soviet Math., 5 (1964), 1673-1676.

5. I. M. Sonin, On a class of degenerate diffusion processes, Theory Probability Appl., 12 (1967), 490-496.

6. M. Weber, The fundamental solution of a degenerate partial differential equation, Trans. Amer. Math. Soc., 71 (1951), 24-34.

Singular Perturbations of a Vector Boundary Value Problem

P. Habets(+)

1. INTRODUCTION

Consider the nonlinear boundary value problem

$$\varepsilon\ddot{y} + f(t,y,\dot{y},\varepsilon) = 0$$

$$y(0)=\bar{\alpha}(\varepsilon),\ y(1)=\bar{\beta}(\varepsilon) \tag{1}$$

where $y \in \mathbb{R}^n$, $0\leq t\leq 1$, $0<\varepsilon\leq E_0$. It is natural to associate with this problem the reduced equation

$$f(t,y,\dot{y},0) = 0 \tag{2}$$

together with one of the boundary conditions, namely

$$y(1)=\bar{\beta}(0) \tag{3}$$

if $\frac{\delta f}{\delta \dot{x}}$ is positive. In this note we prove that if the terminal value problem (2)(3) has a solution $y=y_0(t)$, defined on $[0,1]$, then with appropriate assumptions, problem (1) has a solution $y=y_0(t,\varepsilon)$ satisfying

$$y=y_0(t) + O(\varepsilon + e^{-\bar{\mu}t/\varepsilon}),\ \dot{y}=\dot{y}_0(t) + O(\varepsilon+\varepsilon^{-1}e^{-\bar{\mu}t/\varepsilon})$$

for some $\bar{\mu} > 0$.

This problem was already investigated by A. Erdélyi [4] and more recently by K.W. Chang [2] by applying twice the Banach's fixed point theorem. In the present paper we produce a one-step proof which makes use of Schauder's theorem and weakens the continuity assumptions. A similar result is proved by the same method in [5]

(+) Chargé de recherches FNRS

2. THE MAIN RESULT

Let us suppose $f \in C^1$ and $y_0 \in C^2$. Then

$$x = y - y_0(t)$$

satisfies the problem

$$\varepsilon\ddot{x} + P(t,\varepsilon)\dot{x} + Q(t,\varepsilon)x = g(t,x,\dot{x},\varepsilon)$$

$$x(0) = \alpha(\varepsilon),\ x(1) = \beta(\varepsilon) \tag{4}$$

where $P(t,\varepsilon) = \dfrac{\delta f}{\delta \dot{x}}(t, y_0(t), \dot{y}_0(t), \varepsilon)$

$$Q(t,\varepsilon) = \frac{\delta f}{\delta x}(t, y_0(t), \dot{y}_0(t), \varepsilon)$$

$g(t,x,\dot{x},\varepsilon) = -f(t, y_0(t)+x, \dot{y}_0(t)+\dot{x}, \varepsilon) + P(t,\varepsilon)\dot{x} + Q(t,\varepsilon)x - \varepsilon\ddot{y}_0(t)$, $\alpha(\varepsilon) = \bar{\alpha}(\varepsilon) - y_0(0)$, $\beta(\varepsilon) = \bar{\beta}(\varepsilon) - \bar{\beta}(0)$. The following assumptions are essential

ASSUMPTIONS A_1: *The matrices* $P(t,\varepsilon) = P(t,0) + \mathcal{O}(\varepsilon)$ *and* $Q(t,\varepsilon)$ *are continuous with respect to* t, *bounded uniformly with* ε *and every eigenvalue of* $P(t,0)$ *has a real part* $\geq \mu > 0$ *for* $0 \leq t \leq 1$.

A_2 : *The vector function* g *is continuous with respect to* $(t,x,\dot{x})$

A_3 : $g(t,x,\dot{x},\varepsilon) = \mathcal{O}(\varepsilon + \|x\|^2 + \varepsilon^m \|x\| \|\dot{x}\| + \varepsilon^{1+m}\|\dot{x}\|^2)$

$0 \leq m \leq 1$, $\alpha = \mathcal{O}(1)$ *and* $\beta = \mathcal{O}(\varepsilon)$

A_4 : $\|g(t,x_1,y_1,\varepsilon) - g(t,x_2,y_2,\varepsilon)\| \leq K\rho(x_1,x_2,y_1,y_2)$ *where* $\rho(x_1,x_2,y_1,y_2)$ *is the larger of the two quantities*

$$\|x_1 - x_2\| \max(\|x_1\|, \|x_2\|, \varepsilon^m\|y_1\|, \varepsilon^m\|y_2\|)$$
$$\varepsilon^m\|y_1 - y_2\| \max(\|x_1\|, \|x_2\|, \varepsilon\|y_1\|, \varepsilon\|y_2\|)$$

Using these assumptions we can prove the following

THEOREM. *Let* A_1, A_2 *and* A_3 *hold. Then there exists* $\delta > 0$ *such that for* ε *small enough and* $\|\alpha(\varepsilon)\| \leq \delta$ *the problem* (4) *has at least one solution* x *satisfying*

$$x = O(\varepsilon + e^{-\mu t/8\varepsilon}), \quad \dot{x} = O(\varepsilon + \varepsilon^{-1} e^{-\mu t/8\varepsilon})$$

Further if $m \neq 0$, $\delta = \infty$.

If in addition A_4 *holds, this solution is locally unique.*

3. PROOF OF THE THEOREM.

Let us first introduce the following change of variables

$$\begin{bmatrix} u \\ v \end{bmatrix} = \begin{bmatrix} I-\varepsilon SR & \varepsilon S \\ -R & I \end{bmatrix} \begin{bmatrix} x \\ \dot{x} \end{bmatrix}, \quad \begin{bmatrix} x \\ \dot{x} \end{bmatrix} = \begin{bmatrix} I & -\varepsilon S \\ R & I-\varepsilon RS \end{bmatrix} \begin{bmatrix} u \\ v \end{bmatrix} \tag{5}$$

where the matrices $R = R(t,\varepsilon)$ and $S = S(t,\varepsilon)$ are solutions of the equations

$$\varepsilon\dot{R} = -P(t,\varepsilon)R - \varepsilon R^2 - Q(t,\varepsilon), \quad R(0,\varepsilon) = 0 \tag{6}$$

$$\varepsilon\dot{S} = S(P(t,\varepsilon) + \varepsilon R(t,\varepsilon)) + \varepsilon R(t,\varepsilon)S - I, \quad S(1,\varepsilon) = 0 \tag{7}$$

I is the identity and 0 the zero matrix. It has been proved in [2] (see also [1]) that for ε small enough the solutions $R(t,\varepsilon)$ and $S(t,\varepsilon)$ of (6) and (7) are defined for $t \in [0,1]$, bounded uniformly in ε and $S^{-1}(0,\varepsilon) \to P(0,0)$ as $\varepsilon \to 0$. Using transformation (5) we see that u,v satisfy the differential equations

$$\dot{u} = Ru + S\bar{g}(t,u,v,\varepsilon), \quad u(1) = \beta(\varepsilon),$$

$$\varepsilon\dot{v} = -(P+\varepsilon R)v + \bar{g}(t,u,v,\varepsilon), \quad \varepsilon v(0) = S^{-1}(0,\varepsilon)(u(0) - \alpha(\varepsilon)),$$

where $\bar{g}(t,u,v,\varepsilon) = g(t, u-\varepsilon Sv, Ru + v - \varepsilon RSv, \varepsilon)$. Equivalently u,v satisfy the integral equations

$$u(t) = T_a(u,v)(t) = U(t,1)\beta - \int_t^1 U(t,\tau)S(\tau)\bar{g}(\tau,u(\tau), v(\tau),\varepsilon)\, d\tau, \tag{8-a}$$

$$\varepsilon v(t) = \varepsilon\, T_b(u,v)(t) = V(t,0)S^{-1}(0)(T_a(u,v)(0) - \alpha(\varepsilon)) + \int_0^t V(t,\tau)\bar{g}(\tau, u(\tau), v(\tau), \varepsilon)d\tau, \tag{8-b}$$

where $U(t,\tau)$ and $V(t,\tau)$ are the principal matrix solutions of

$$\dot{u} = Ru \text{ and } \varepsilon\dot{v} = -(P + \varepsilon R)v$$

which are equal to I at $t=\tau$. Notice that by theorem 2[3] and for ε small enough,

$$\|V(t,\tau)\| \leq K \exp\left(-\frac{\mu}{8\varepsilon}(t-\tau)\right).$$

Equations (8) define the operator

$$T : C^\circ \times C^\circ \to C^\circ \times C^\circ : (u,v) \to (T_a(u,v),\ T_b(u,v))$$

where C° is the space of continuous function $x:[0,1] \to \mathbb{R}$ with norm $\|x\|_\infty = \sup_{0\leq t\leq 1} \|x(t)\|$. It follows from the uniform boundedness of $R(t,\varepsilon)$ and $S(t,\varepsilon)$ that the problem (4) has a solution x satisfying

$$x = O(\varepsilon + e^{-\mu t/8\varepsilon}),\quad \dot{x} = O(\varepsilon + \varepsilon^{-1} e^{-\mu t/8\varepsilon})$$

if and only if, for ε small enough, T has a fixed point in some set

$$B = \{(u,v) : u \in C^\circ,\ v \in C^\circ,\ \|u\| \leq k_1\varepsilon + k_2 e^{-\mu t/8\varepsilon},\ \|v\| \leq k_1\varepsilon + k_2\varepsilon^{-1} e^{-\mu t/8\varepsilon}\}$$

LEMMA 1. *Let* A_1, A_2 *and* A_3 *hold. Then there exist* $\delta>0$, $k_1>0$ *and* $k_2>0$ *such that for* ε *small enough and* $|\alpha(\varepsilon)|<\delta$

$$T : B \to B$$

Further if $m \neq 0$, *for any* α, *there exist* $k_1>0$ *and* $k_2>0$ *such that for* ε *small enough*

$$T : B \to B$$

Proof. Let K be a generic constant independant of k_1, k_2, u, v and ε. If $(u,v) \in B$ then

$$\|\bar{g}\| \leq K(\varepsilon + \|u\|^2 + \varepsilon^m \|u\|\,\|v\| + \varepsilon^{m+1}\|v\|^2)$$
$$\leq K((1+k_1^2\varepsilon)\varepsilon + \varepsilon^m k_1 k_2 e^{-\mu t/8\varepsilon} + \varepsilon^m k_2^2 \frac{e^{-\mu t/4\varepsilon}}{\varepsilon})$$

and

$$\|T_a(u,v)(t)\| \leq K\|\beta\| + K\int_t^1 \|\bar{g}\|\,d\tau$$
$$\leq K(1+k_1^2\varepsilon)\varepsilon + K_1\varepsilon^m(k_1k_2\varepsilon + k_2^2)\, e^{-\mu t/8\varepsilon}$$
$$\leq k_1\varepsilon + 2K_1\varepsilon^m k_2^2 e^{-\mu t/8\varepsilon} \leq k_1\varepsilon + k_2 e^{-\mu t/8\varepsilon}$$

for k_1 large enough, ε small enough and if $m=0$ $k_2<1/2K_1$. Similarly

$$\|T_b(u,v)(t)\| \leq K(\|T_a(u,v)(0)\| + \|\alpha\|)\, \varepsilon^{-1} e^{-\mu t/8\varepsilon} + K\int_0^t \|\bar{g}\|\, \varepsilon^{-1} e^{-\mu(t-s)/8\varepsilon}\, ds$$

$$\leqslant K(1+k_1^2\varepsilon)\varepsilon + K(k_1\varepsilon+\varepsilon^m k_2^2+\|\alpha\|)\ \varepsilon^{-1}\ e^{-\mu t/8\varepsilon} + K\ \varepsilon^m k_1 k_2 t\ \varepsilon^{-1} e^{-\mu t/8\varepsilon}$$

and since $t\varepsilon^{-1}e^{-\mu t/8\varepsilon} \leqslant \varepsilon^{-\frac{1}{2}}\ e^{-\mu t/8\varepsilon} + K\varepsilon^2$

$$\|T_b(u,v)(t)\| \leqslant K(1+k_1^2\varepsilon+\varepsilon^{m+1}k_1k_2)\varepsilon + K_2(k_1\varepsilon+\varepsilon^{m+\frac{1}{2}}\ k_1k_2\ +\varepsilon^m k_2^2\ +\|\alpha\|)\varepsilon^{-1}e^{-\mu t/8\varepsilon}$$
$$\leqslant k_1\varepsilon\ +\ k_2\ \varepsilon^{-1}e^{-\mu t/8\varepsilon}$$

for k_1 large enough, ε small enough and, if $m \neq 0$, k_2 large enough. If $m=0$, one must choose $k_2<1/2\ K_2$ and $\|\alpha\| < k_2/2K_2$ Q.E.D.

From lemma 1, T maps the closed bounded convex set B into itself and from the continuity assumptions A_1 and A_2 it is easy to verify that T is completely continuous. Hence Schauder's theorem [6] applies and the existence part of the theorem follows.

Introducing the norm

$$\|(u,v)\|_B=\max\left(\left\|\frac{u(t)}{\varepsilon+\exp(-\mu t/8\varepsilon)}\right\|_\infty,\left\|\frac{v(t)}{\varepsilon+\varepsilon^{-1}\exp(-\mu t/8\varepsilon)}\right\|_\infty\right)$$

we can prove in much the same way.

LEMMA 2. *Let* A_1, A_2, A_3 *and* A_4 *hold. Then there exist* $\delta>0$, $k_1>0$ *and* $k_2>0$ *such that for* ε *small enough and* $\|\alpha(\varepsilon)\|\leqslant\delta$, $T:B \to B$ *is a contraction on* B. *Further if* $m\neq 0$, *for any* α *there exist* $k_1>0$ *and* $k_2>0$ *such that for* ε *small enough* T *is a contraction on* B.

From this lemma, Banach's fixed point theorem [6] can be applied. This proves that T has one and only one solution in B.

4. REFERENCES

1. K.W.Chang, "Singular Perturbations of a General Boundary Value Problem", SIAM J. Math. Anal.3 (1972) 520-526.
2. K.W. Chang, "Diagonalization Method for a Vector Boundary Problem of Singular Perturbation Type" to appear.
3. W.A.Coppel, "Dichotomies and Reducibility", J. Diff. Eq. 3 (1967), 500-521.
4. A. Erdélyi, "On a Nonlinear Boundary Value Problem Involving a Small Parameter", J.Austral. Math. Soc. 2 (1962), 425-439.
5. P. Habets, "Singular Perturbations of a Nonlinear Boundary Value Problem", to appear.
6. J.K. Hale, "Ordinary Differential Equations", Wiley-Interscience, New York, 1969.

ASYMPTOTIC INTEGRATION OF LINEAR SYSTEMS

W. A. Harris, Jr[1]
D. S. Lutz[2]

1. Introduction.

The asymptotic behavior of solutions of an autonomous linear differential system $x' = Ax$ is determined essentially by the characteristic roots of the constant matrix A. If B(t) is a suitably small perturbation, then the asymptotic behavior of solutions of $x' = [A+B(t)]x$ is still determined by the limiting system $x' = Ax$. For example, if $P^{-1}AP = \Lambda$ is a diagonal matrix, B(t) is continuous for $t \geq t_0$, and $\int_{t_0}^{\infty} \|B(t)\| dt < \infty$, then $x' = [A + B(t)]x$ has a fundamental matrix X(t) satisfying, as $t \to \infty$, $X(t) = [P + o(1)] \exp(\Lambda t)$.

The classical theorems of Levinson [10] and Hartman-Wintner [9] are deeper results of this nature which describe the asymptotic behavior of solutions of the nonautonomous linear differential system $x' = A(t)x$ in terms of the characteristic roots of the matrix A(t).

In this note we present two preparatory lemmas which extend the validity of these fundamental results of Levinson and Hartman-Wintner to a wider class of systems $x' = A(t)x$. These extensions are in the same spirit as the fundamental results in that the asymptotic integration is given in terms of computable functions. In addition, we discuss the interrelation between the basic results and their extensions and give examples to delineate them.

Asymptotic integration, of interest in itself, is also a useful tool, for example, in the study of stability and boundary value problems. Our interest in the problem was motivated by recent results of Devinatz [3], [4] and Fedoryuk [5] in which extensions of Levinson's basic theorem are presented and utilized to determine the deficiency index of certain differential operators.

2. Preparatory Lemmas.

We are concerned with the linear differential system $y' = A(t)y$, in which the matrix A(t) has the form $\Lambda + V(t)$, where Λ is a constant diagonal matrix

[1]Supported in part by the United States Army under contract DAHCO4-74-G-0013
[2]Supported in part by the National Science Foundation under Grant GP-28149.

with distinct characteristic roots and $V(t) \to 0$ as $t \to \infty$. For t sufficiently large, $\Lambda + V(t)$ has distinct characteristic roots and there exists a matrix $S(t)$ for which

$$S^{-1}(t)\,[\Lambda + V(t)]\,S(t) = \Lambda(t),$$

where $\Lambda(t)$ is a diagonal matrix whose elements are the characteristic roots of $A(t)$ and $\Lambda(t) \to \Lambda$ as $t \to \infty$. Utilizing $S(t)$ as a transformation, $y = S(t)\,u$, the linear differential system $y' = A(t)\,y$ becomes $u' = B(t)\,u$, where

$$B(t) = S^{-1}(t)[\Lambda + V(t)]\,S(t) - S^{-1}(t)\,S'(t) = \Lambda(t) + \hat{V}(t).$$

If $\hat{V}(t)$ is "sufficiently regular" (in the sense of integrability[1]), then we can determine the asymptotic integration of $u' = B(t)u$ and hence, also that of the original system $y' = A(t)\,y$.

In order to determine the regularity of $\hat{V}(t)$, we must determine the regularity of $S(t)$, $S^{-1}(t)$, and $S'(t)$ in terms of the regularity of $V(t)$. Our basic preparatory lemma provides a means for studying this problem.

Lemma 1. *Let Λ be a constant diagonal matrix with distinct characteristc roots and $V(t)$ a continuous matrix for $t \geq t_0$ such that $V(t) \to 0$ as $t \to \infty$. Then there exists for $t \geq t_0' \geq t_0$ a matrix $Q(t)$, $Q(t) \to 0$ as $t \to \infty$, such that*

$$[I + Q(t)]^{-1}[\Lambda + V(t)]\,[I + Q(t)] = \Lambda(t),$$

where $\Lambda(t)$ is a diagonal matrix whose elements are the characteristic roots of $\Lambda + V(t)$. Furthermore, $Q(t)$ may be chosen so that diag $Q(t) \equiv 0$ and so that $Q(t)$ has the same regularity properties as $V(t)$, i.e.,

$$Q(t) = O(\|V(t)\|),\ Q'(t) = O(\|V'(t)\|),\ \textit{etc.},\ \text{as } t \to \infty .$$

---

1 $\|V(t)\| \in L_p(t_0, \infty)$ means that $\int_{t_0}^{\infty} \|V(t)\|^p\,dt < \infty$, where any convenient matrix norm may be used.

<u>Proof of Lemma 1.</u> We provide an elementary proof of this lemma which has other important ramifications, noting that aside from the normalization, diag $Q(t) \equiv 0$, this lemma was used by Levinson [10].

Write $\Lambda = \operatorname{diag}\{\lambda_1, \ldots, \lambda_n\}$, $\Lambda(t) = \Lambda + \operatorname{diag}\{d_{11}(t), d_{22}(t), \ldots, d_{nn}(t)\}$ $= \Lambda + D(t)$, $Q(t) = (q_{ij}(t))$, $1 \leq i, j \leq n$, $q_{ii}(t) \equiv 0$.

The existence of a matrix $Q(t) = o(1)$ as $t \to \infty$ for which

$$(2.1) \qquad [I + Q(t)]^{-1} [\Lambda + V(t)] [I + Q(t)] = \Lambda(t)$$

is equivalent to the existence of a solution $Q(t) = o(1)$ of the equation

$$\Lambda Q(t) - Q(t) \Lambda + V(t) Q(t) - Q(t) D(t) + V(t) - D(t) = 0,$$

or in component form,

$$(2.2) \qquad d_{ii} - v_{ii} - \sum_{i \neq k} v_{ik} q_{ki} = 0$$

$$(2.3) \qquad (\lambda_i - \lambda_k) q_{ik} + \sum_{\alpha \neq k} v_{i\alpha} q_{\alpha k} - q_{ik} d_{kk} + v_{ik} = 0,$$

where $d_{ii} = d_{ii}(t)$, $v_{ik} = v_{ik}(t)$, $q_{ik} = q_{ik}(t)$.

Solving equation (2.2) for d_{ii} and substituting the result into equation (2.3), we obtain the nonlinear system of equations

$$(2.4) \qquad (\lambda_i - \lambda_k) q_{ik} + \sum_{\alpha \neq k} v_{i\alpha} q_{\alpha k} - q_{ik}(v_{kk} + \sum_{\beta \neq k} v_{k\beta} q_{\beta k}) + v_{ik} = 0.$$

Conversely a solution to the nonlinear system of equations (2.4) together with equations (2.2) (for the definition of the d_{ii}) will yield a solution of the equation $[\Lambda + V(t)][I + Q(t)] = [I + Q(t)][\Lambda + D(t)]$. Hence if $Q(t) = o(1)$ as $t \to \infty$, $I + Q(t)$ will be nonsingular for t sufficiently large and equation (2.1) will be satisfied The nonlinear system of equations (2.4) is of the (vector) form $f(t, q) = 0$, where $f(\infty,0) = 0$ and $f_q(\infty,0)$ is nonsingular; hence, the standard implicit function theorem guarantees the existence of a unique solution to this equation, $q = o(1)$, which inherits the regularity properties of $V(t)$.

The proof of Lemma 1 provides a method for computing an approximation to $\Lambda(t)$ and $Q(t)$ of order $o(\|V(t)\|^2)$ $(t \to \infty)$ by solving the linear systems

$$\begin{cases} (\lambda_1-\lambda_k)\hat{q}_{ik} + \sum_{\alpha \neq k} v_{i\alpha}\hat{q}_{\alpha k} - v_{kk}\hat{q}_{ik} + v_{ik} = 0 \\ \hat{d}_{ii} = v_{ii} + \sum_{i \neq k} v_{ik}\hat{q}_{ki}. \end{cases}$$

In particular, an approximation of order $O(\|V(t)\|^2)$ is

$$\begin{cases} \tilde{q}_{ik} = (\lambda_k - \lambda_i)^{-1} v_{ik} + O(\|V(t)\|^2) \\ \tilde{d}_{ii} = v_{ii} + O(\|V(t)\|^2). \end{cases}$$

We formalize this remark as

Lemma 2. Let Λ be a constant diagonal matrix with distinct characteristic roots roots let $V(t)$ be a continuous matrix for $t \geq t_0$ such that $V(t) \to 0$ as $t \to \infty$. Then there exists for $t \geq t_0' \geq t_0$ a matrix $T(t)$, $\text{diag } T(t) \equiv 0$, $T(t) \to 0$ as $t \to \infty$ such that

$$[I + T(t)]^{-1} [\Lambda + V(t)] [I + T(t)] = \{\Lambda + \text{diag } V(t)\} + \hat{V}(t),$$

where $\hat{V}(t) = O(\|V(t)\|^2)$, $T(t) = O(\|V(t)\|)$, and $T(t)$ has the same regularity properties as $V(t)$.

This lemma shows that if Λ is a diagonal matrix with distinct characteristic roots and $V(t)$ is a continuous matrix such that $V(t) \to 0$ as $t \to \infty$, then the characteristic roots of $\Lambda + V(t)$ are the characteristics roots of $\Lambda + \text{diag } V(t) + O(\|V(t)\|^2)$ as $t \to \infty$.

If $\|V(t)\| \in L_2(t_0, \infty)$ and Λ is a diagonal matrix with distinct characteristic roots, then the characteristic roots of $\Lambda + V(t)$ are the diagonal entries of $\Lambda + V(t)$ to within integrable terms. This remark allows the results of Hartman-Wintner to be stated in terms of the characteristic roots of $\Lambda + V(t)$ instead of in the customary manner involving the diagonal entires of $\Lambda + V(t)$.

The proofs show that Lemmas 1 and 2 remain valid if the constant diagonal matrix Λ is replaced by the diagonal matrix $\Lambda(t) = \Lambda + \hat{\Lambda}(t)$, where $\hat{\Lambda}(t) \to 0$ as $t \to \infty$ and $\hat{\Lambda}(t)$ shares the same regularity properties as $V(t)$. This allows us to make repeated application of the lemmas.

3. Main Results and Applications.

In this section we show how the preparatory Lemmas 1 and 2 can be utilized to transform a given linear differential system so that the basic results of Levinson and Hartman-Wintner are applicable. For the convenience of the reader, we now state these basic results.

Theorem A. Levinson ([10], [1; pp. 92-95]) *Let* $\Lambda(t) = \mathrm{diag}\{\lambda_1(t), \lambda_2(t), \ldots, \lambda_n(t)\}$ *be continuous for* $t \geq t_0$ *and assume for each index pair* $j \neq k$ *that either*

(i) $$\int_{t_0}^{t} \mathrm{Re}(\lambda_k(s) - \lambda_j(s))\, ds \to \infty \text{ as } t \to \infty \text{ and}$$

$$\int_{s}^{t} \mathrm{Re}(\lambda_k(s) - \lambda_j(s))\, ds > -K \text{ for all } t_0 \leq s \leq t, \text{ or}$$

(ii) $$\int_{s}^{t} \mathrm{Re}(\lambda_k(s) - \lambda_j(s))\, ds < K \text{ for all } t_0 \leq s \leq t.$$

Furthermore, assume that $R(t)$ *is continuous for* $t \geq t_0$ *and* $\|R(t)\| \in L_1(t_0, \infty)$. *Then the linear differential system* $x' = (\Lambda(t) + R(t))x$ *has a fundamental matrix satisfying as* $t \to \infty$

$$X(t) = [I + o(1)] \exp\left(\int_{t_0}^{t} \Lambda(s)\, ds\right).$$

Theorem B. Hartman-Winter [9; pp. 71-72] *Let* $\Lambda(t) = \mathrm{diag}\{\lambda_1(t), \lambda_2(t), \ldots, \lambda_n(t)\}$ *be continuous for* $t \geq t_0$ *and assume that for each index pair* $j \neq k$, $|\mathrm{Re}(\lambda_j(t) - \lambda_k(t))| \geq \mu > 0$. *Furthermore, assume that* $V(t)$ *is continuous and* $\|V(t)\| \in L_2(t_0, \infty)$.

Then the linear differential system $x' = (\Lambda(t) + V(t))\,x$ *has a fundamental matrix satisfying as* $t \to \infty$

$$X(t) = [I + o(1)] \exp \left(\int_0^t [\Lambda(s) + \operatorname{diag} V(s)]\, ds\right).$$

The repeated application of the preparatory lemmas yields the following.

Theorem 1. Let A *be a constant matrix with distinct characteristic roots for some positive integer* k *let* $V^{(k)}(t)$ *and* $R(t)$ *be continuous for* $t \geq t_0$; $V^{(i)}(t) \to 0$ *as* $t \to \infty$ $0 \leq i \leq k-2$; $\| V^{(j)}(t) \| \varepsilon\, L_2(t_0, \infty)$ $1 \leq j \leq k-1$; $\| V(t) \|\ \| V'(t) \|$, $\| V^{(k)}(t) \|$, *and* $\| R(t) \| \varepsilon\, L_1(t_0, \infty)$; *and assume that the characteristic roots of the matrix* $A + V(t)$ *satisfy condition (i) or (ii) of Theorem A. Then the linear differential system* $x' = (A + V(t) + R(t))x$ *has a fundamental matrix satisfying as* $t \to \infty$

$$X(t) = [P + o(1)] \exp \left(\int_{t_1}^{t} \Lambda(s)\, ds\right),$$

$t_1 \geq t_0$, *where* $P^{-1}AP$ *is a diagonal matrix and* $\Lambda(t)$ *is a diagonal matrix whose components are the characteristic roots of* $A + V(t)$.

This result may be viewed not only as an extension of Levinson's Theorem A but also (under the additional assumption that $\| V(t) \| \varepsilon\, L_2(t_0, \infty)$) as an extension of Hartman-Wintner's Theorem B, in the sense that the restriction on the characteristic roots of $A + V(t)$ has been weakened at the expense of strengthening the assumptions on $V(t)$. The preparatory lemmas can also be used to obtain extensions of Hartman-Wintner's Theorem B by relaxation of the assumption that $\| V(t) \| \varepsilon\, L_2$. For example, if $V(t) \to 0$ as $t \to \infty$, $\| V'(t) \| \varepsilon\, L_2$, and the characteristic roots of A have distinct real parts, then by utilizing the transformation $y = P[I + Q(t)]\,u$ of Lemma 1, $P^{-1}AP$ diagonal, the system $y' = (A+V(t))\,y$ becomes $u' = (\hat{\Lambda}(t) + \hat{V}(t))\,u$, where $\hat{\Lambda}(t) = \Lambda(t) - \operatorname{diag} \{[I + Q(t)]^{-1} Q'(t)\}$ and $\hat{\Lambda}(t) + \hat{V}(t) = \Lambda(t) - [I + Q(t)]^{-1} Q'(t)$. Clearly, Theorem B applies to yield a fundamental matrix satisfying as $t \to \infty$

$$\hat{Y}(t) = [P + o(1)] \exp\left(\int_{t_1}^{t} \hat{\Lambda}(s)ds\right), \quad t_1 \geq t_0.$$

If, in addition, $\| V(t) \| \; \| V'(t) \| \in L_1$, then $\| \operatorname{diag} \{[I+Q(t)]^{-1} Q'(t)\} \| \in L_1$ and we have a fundamental matrix satisfying as $t \to \infty$

$$Y(t) = [P + o(1)] \exp \int_{t_1}^{t} \Lambda(s)ds.$$

We formalize this result as

<u>Theorem 2.</u> <u>Let A be a constant matrix with characteristic roots having distinct real parts, $V(t)$ a continuous matrix for $t \geq t_0$ such that $V(t) \to 0$ as $t \to \infty$, $V'(t)$ continuous for $t \geq t_0$ and $\| V'(t) \|^2$, $\| V(t) \| \; \| V'(t) \| \in L_1(t_0, \infty)$. Then the linear differential system $x' = (A + V(t))\,x$ has a fundamental matrix satisfying as $t \to \infty$</u>

$$X(t) = [P + o(1)] \exp\left(\int_{t_1}^{t} \Lambda(s)ds\right), \quad t_1 \geq t_0,$$

<u>where $P^{-1}AP$ is a diagonal matrix and $\Lambda(t)$ is a diagonal matrix whose components are the characteristic roots of $A + V(t)$.</u>

We emphasize that $\| V(t) \|^2$ is not necessarily integrable and hence, $\Lambda(t)$ is not necessarily $\operatorname{diag} \{ P^{-1}[A+V(t)]P \}$.

The assumption that $\| V(t) \| \; \| V'(t) \| \in L_1$ has allowed us to state the asymptotic integration in Theorems 1 and 2 in terms of the characteristic roots of $A + V(t)$. By modifying the characteristic roots of $A + V(t)$ by functions which are also computable, we can obtain the following results.

<u>Theorem 3.</u> <u>Let A be a constant matrix with distinct characteristic roots; for some positive integer k, let $V^{(k)}(t)$ and $R(t)$ be continuous for $t \geq t_0$; Let $V^{(i)}(t) \to 0$ as $t \to \infty$, $0 \leq i \leq k-2$, and $\| V^{(i)}(t) \| \in L_2$, $1 \leq i \leq k-1$; and let $\| V^{(k)}(t) \|$, $\| R(t) \| \in L_1(t_0, \infty)$. Let $P^{-1}AP$ be a diagonal matrix and</u>

let $Q(t)$ *be the unique matrix (of Lemma 1) for which*

$$[I + Q(t)]^{-1} P^{-1}[A + V(t)] P [I + Q(t)] = \Lambda(t),$$

$\operatorname{diag} Q(t) \equiv 0$, and let the characteristic roots of $\hat{\Lambda}(t) = \Lambda(t) - \operatorname{diag}\{[I + Q(t)]^{-1}Q'(t)\}$ *satisfy condition* (i) *or* (ii) *of Theorem* A. *Then the linear differential system* $x' = (A + V(t) + R(t))\,x$ *has a fundamental matrix satisfying as* $t \to \infty$

$$X(t) = [P + o(1)] \exp\left(\int_{t_1}^{t} \hat{\Lambda}(s)\, ds\right), \qquad t_1 \geq t_0.$$

We note that, for example, if $\|V(t)\|^2 \|V'(t)\| \in L_1(t_0, \infty)$, then $\operatorname{diag}\{[I + Q(t)]^{-1} Q'(t)\} = -\operatorname{diag}\{Q(t)\, Q'(t)\}$ to within integrable terms.

As a final example of the application of the preparatory lemmas, we present some sufficient conditions for the uniform stability of a linear differential system $x' = (A + V(t))\,x$, $V(t) \to 0$ as $t \to \infty$, which are extensions of results due to Conti and Cesari. (See, for example, [2 ; p. 114].)

Theorem 4. *Assume that the characteristic roots of* $A + V(t)$ *have non-positive real parts for* t *sufficiently large; that the characteristic roots of* A *with zero real part are simple; and that for some positive integer* k, $V^{(k)}(t)$ *is continuous for* $t \geq t_0$, $V^{(i)}(t) \to 0$ *as* $t \to \infty$, $0 \leq i \leq k-2$,
$\|V^{(i)}(t)\| \in L_2(t_0, \infty)$, $1 \leq i \leq k-1$, $\|V(t)\|\ \|V'(t)\| \in L_1(t_0, \infty)$, *and*
$\|V^{(k)}(t)\| \in L_1(t_0, \infty)$. *Then the zero solution of* $x' = (A + V(t))\,x$ *is uniformly stable for* $t \geq t_0$.

For a proof of these results see Harris and Lutz [6].

4. Examples.

In this section we give some example illustrating the type of results that can be obtained using Theorems 1-3, some examples which indicate the sharpness of the assumptions which are necessary for their applicability, as well as some examples which demonstrate the independence of the various hypotheses.

Example 1.

Let $A(t) = \Lambda + V(t) = \begin{pmatrix} 0 & 0 \\ 0 & 2 \end{pmatrix} + t^{-\alpha} \begin{pmatrix} 0 & 1 \\ 1 & 0 \end{pmatrix}$,

where $0 < \alpha \le 1$. The characteristic roots of $A(t)$ are $\lambda(t) = 1 \pm (1+t^{-2\alpha})^{1/2}$, i.e.,

$$\begin{cases} \lambda_1(t) = -\frac{1}{2} t^{-2\alpha} + \frac{1}{4} t^{-4\alpha} + \cdots , \\ \lambda_2(t) = 2 + \frac{1}{2} t^{-2\alpha} - \frac{1}{4} t^{-4\alpha} + \cdots , \end{cases}$$

and $Re(\lambda_2(t) - \lambda_1(t)) = 2 + O(t^{-2\alpha})$. Also $V'(t) = -\alpha t^{-\alpha-1}\begin{pmatrix} 0 & 1 \\ 1 & 0 \end{pmatrix}$ and $|V'(t)| \in L_1(1, \infty)$; hence, Levinson's Theorem applies. However, naively applying Hartman-Wintner's basic results indicates that the system $y' = A(t)y$ would have a fundamental matrix of the form

$$Y(t) = [I + o(1)] \begin{pmatrix} 1 & 0 \\ 0 & e^{2t} \end{pmatrix} \text{ as } t \to \infty ,$$

which is only valid if $\alpha > 1/2$ (i.e., $\|V(t)\| \in L_2$). Whereas, for example, the correct asymptotic integration for $\alpha = 1/2$ is

$$Y(t) = [I + o(1)] \begin{pmatrix} t^{-1/2} & 0 \\ 0 & t^{1/2} e^{2t} \end{pmatrix} \text{ as } t \to \infty .$$

Example 2.

To illustrate the independence of the various hypotheses on $V(t)$ and its derivatives, consider the (scalar) function $v(t) = t^{-\beta} \sin t^{1-\alpha}$.

(i) If $1/2 < \beta < 1$ and $(1-\beta)/(k+1) < \alpha \le (1-\beta)/k$, then

$|v^{(i)}(t)| \in L_2(1, \infty)$, $|v^{(i)}(t)| \notin L_1(1, \infty)$, $0 \le i \le k-1$, $|v^{(k)}(t)| \in L_1(1, \infty)$, and $|v(t)|\ |v'(t)| \in L_1(1,\infty)$. Hence for the system $y' = A(y)$, where

$$A(t) = \begin{pmatrix} i & 0 \\ 0 & -i \end{pmatrix} + v(t) \begin{pmatrix} 0 & 1 \\ 1 & 0 \end{pmatrix},$$

we may apply Lemma 2 k times to yield a system to which Levinson's basic result applies.

(ii) If $\beta = 1/2$ and $\alpha > 0$, then $|v(t)| \notin L_2(1,\infty)$; however, $|v'(t)| \in L_2(1,\infty)$ and $|v(t)||v'(t)| \in L_1(1,\infty)$. Hence, for the system $y' = A(t)y$, where

$$A(t) = \begin{pmatrix} 1 & 0 \\ 0 & -1 \end{pmatrix} + v(t) \begin{pmatrix} 0 & 1 \\ 1 & 0 \end{pmatrix},$$

we may apply Lemma 1 once to yield a system to which Hartman-Wintner's basic result applies. This case is covered by Theorem 2.

Example 3.

$$\text{Let } A(t) = \begin{pmatrix} i & 0 \\ 0 & -i \end{pmatrix} + \begin{pmatrix} 0 & -t^{1/4} & \sin t^{1/2} \\ t^{-1/4} \cos t^{1/4} & & 0 \end{pmatrix}.$$

The characteristic roots of $A(t)$ to within integrable terms are

$$\lambda_{1,2}(t) = \pm\, i(1 + t^{-1/2} \sin 2t^{1/2} - \frac{1}{4} t^{-1} \sin^2 2t^{1/2}).$$

Note that $\|V(t)\| \notin L_2(1,\infty)$ and $\|V(t)\|\ \|V'(t)\| \notin L_1(1,\infty)$. However, since $\|V''(t)\| \in L_1(1,\infty)$ and $\|V'(t)\| \in L_2(1,\infty)$, an application of Lemma 1 and Lemma 2 yields the linear differential system

$$v' = (\Lambda(t) + \operatorname{diag}\{Q(t)Q'(t)\} + R(t))\, v,$$

where $\|R(t)\| \in L_1$. A straightforward computation yields, for this example (to within integrable terms),

$$Q(t)Q'(t) = \begin{pmatrix} (8t)^{-1} \sin^2 t^{1/2} & 0 \\ 0 & -(8t)^{-1} \cos^2 t^{1/2} \end{pmatrix}.$$

Thus defining $\hat{\Lambda}(t) = \text{diag}\{\hat{\lambda}_1(t), \hat{\lambda}_2(t)\}$, where

$$\hat{\lambda}_1(t) = i(1 + t^{-1/2} \sin 2t^{1/2} - 1/4\, t^{-1} \sin^2 2t^{1/2}) + (8t)^{-1} \sin^2 t^{1/2}$$

and

$$\hat{\lambda}_2(t) = -i(1 + t^{-1/2} \sin 2t^{1/2} - 1/4\, t^{-1} \sin^2 2t^{1/2}) - (8t)^{-1} \cos^2 t^{1/2},$$

the resultant system has the form $v' = (\hat{\Lambda}(t) + \hat{R}(t))\, v$. Since $\text{Re}(\hat{\lambda}_1(t) - \hat{\lambda}_2(t)) = (4t)^{-1}$ and $\|\hat{R}(t)\| \in L_1$, Theorem A is applicable. We note that the modification is real although the original eigenvalues of $A(t)$ are purely imaginary and thus this modification is necessary to obtain the true character of the solutions. This example is covered by Theorem 3.

Example 4.

We close our discussion with the differential equation of an adiabatic oscillator, $w'' + (1 + at^{-1} \sin \lambda t)\, w = 0$, where a, λ are real parameters. This example has many interesting features:

(i) Levinson's basic result does not apply since $\|V'(t)\| \notin L_1$;

(ii) Hartman-Wintner's basic result does not apply since $\text{Re}(\lambda_1 - \lambda_2) = 0$;

(iii) when naively applied, Levinson and Hartman-Wintner yield the same result since $\|V(t)\| \in L_2$;

(iv) Lemma 1 and Lemma 2 may be applied as often as desired and yield systems to which neither Levinson's nor Hartman-Wintner's basic result applies; however, the asymptotic integration indicated by these theorems remains invariant;

(v) the asymptotic integration obtained by naively applying Levinson's or

Hartman - Wintner's Theorems can be either correct or incorrect, depending upon the value of the parameter λ.

Results on the asymptotic integration of adiabatic oscillators are contained in a forthcoming work of Harris and Lutz [7].

References.

1. E.A. Coddington and N. Levinson, Theory of Ordinary Differential Equations, New York, McGraw-Hill, 1955.

2. W. Coppel, Stability and Asymptotic Behavior of Differential Equations, Boston, Heath, 1965.

3. A. Devinatz, An asymptotic theorem for systems of linear differential equations, Trans. Amer. Math. Soc. 160(1971)353-363.

4. A. Devinatz, The deficiency index of a certain class of ordinary self-adjoint differential operators, Adv. in Math. 8(1972) 434-473.

5. M. Fedoryuk, Asymptotic methods in the theory of one-dimensional singular differential operators, Trans. Moskow Math. Soc., (1966) 333-386.

6. W. A. Harris, Jr. and D. A. Lutz, On the asymptotic integration of linear differential systems, J. Math. Anal. Appl. (to appear).

7. ______________, Asymptotic integration of adiabatic oscillators (to appear).

8. P. Hartman, Ordinary Differential Equations, New York Wiley, 1964.

9. P. Hartman and A. Wintner, Asymptotic integrations of linear differential equations, Amer. J. Math. 77(1955) 45-86 and 932.

10. N. Levinson, The asymptotic nature of solutions of linear differential equations, Duke Math. J. 15 (1948) 111-126.

Nonlinear perturbations of linear elliptic operators

Peter Hess

1. Let $\Omega \subset \mathbb{R}^N$ ($N \geq 1$) be a bounded domain with boundary $\partial\Omega$ in C^2. Let further

$$A = -\sum_{i,j=1}^{N} \frac{\partial}{\partial x_i}\left(a_{ij}\frac{\partial}{\partial x_j}\right) + \sum_{i=1}^{N} b_i \frac{\partial}{\partial x_i} + c,$$

where

$$\begin{cases} a_{ij} \in C^1(\overline{\Omega}) & (i,j = 1,\dots,N) \\ b_i,\ c \in L^\infty(\Omega) & (i = 1,\dots,N) \\ \sum_{i,j=1}^{N} a_{ij}(x)\,\xi_i\xi_j \geq \alpha|\xi|^2 \quad (\alpha>0) & \forall x \in \Omega,\ \xi \in \mathbb{R}^N, \end{cases}$$

be a uniformly elliptic linear differential expression of second order. It is assumed that all functions in consideration are real-valued. Moreover let $f\colon \Omega \times \mathbb{R} \to \mathbb{R}$ be a function satisfying the Caratheodory condition: $f(x,t)$ is measurable in $x \in \Omega$ for (fixed) $t \in \mathbb{R}$ and continuous in t for a.a. $x \in \Omega$. For given $g \in L^2(\Omega)$ the solvability of the nonlinear Dirichlet problem

$$\text{(I)} \qquad \begin{cases} (Au)(x) + f(x,u(x)) = g(x) & x \in \Omega \\ u(x) = 0 & x \in \partial\Omega \end{cases}$$

is investigated.

Theorem. Suppose for some constants $R_1 \leq 0 \leq R_2$,

(i) $$\int_\Omega \sup_{R_1 \leq t \leq R_2} |f(x,t)|^2\,dx < \infty$$

and

$$\text{(ii)}\qquad \left.\begin{aligned} c(x)R_1 + f(x,R_1) &\leq g(x)\\ c(x)R_2 + f(x,R_2) &\geq g(x)\end{aligned}\right\} \qquad \underline{\text{for}}\ \text{a.a. } x \in \Omega.$$

Then there exists $u \in H_0^1(\Omega) \cap H^2(\Omega)$, $R_1 \leq u(x) \leq R_2$ a.e. *on* Ω, *which is strong solution of problem* (I).

Unlike in numerous recent results, no assumption is made on the behaviour of $f(x,t)$ as $|t| \to \infty$. Noting that the functions $v(x) \equiv R_{1,2}$ $(x \in \Omega)$ are lower and upper solution of problem (I), respectively, and imposing in addition appropriate Hölder continuity on the functions involved, the solvability of (I) can be proved using a monotone iteration scheme ([2]). Our Theorem generalizes a related result by Martin [7] inasmuch as his L^∞-boundedness assumption on g and $f(.,t)$, $R_1 \leq t \leq R_2$, is weakened here to L^2-boundedness. The present method of proof seems to be new in this context; we show that the Theorem follows from a general result on unilateral elliptic problems, which is obtained by applying the theory of maximal monotone operators in Hilbert space. It is hoped that it will be possible to treat similarly also other boundary value problems, perhaps even assuming the knowledge of more general lower and upper solutions.

2. *Proof of the Theorem*. Let A be the closed linear operator in $L^2(\Omega)$ induced by $\mathcal{A}$:

$$D(A) = H_0^1(\Omega) \cap H^2(\Omega)$$

$$Au = \mathcal{A}\, u, \quad u \in D(A),$$

and let $\beta \subset \mathbb{R} \times \mathbb{R}$ be the maximal monotone graph given by

$$\beta(t) = \emptyset \text{ if } t < R_1,\ \beta(R_1) = (-\infty,0],\ \beta(t) = 0 \text{ if } R_1 < t < R_2,$$

$$\beta(R_2) = [0,+\infty),\ \beta(t) = \emptyset \text{ if } t > R_2.$$

By $\bar{\beta}$ we denote the maximal monotone extension of β to $L^2(\Omega) \times L^2(\Omega)$ defined as follows:

(1) $\quad v \in \bar{\beta}(u) \Longleftrightarrow u,\ v$ in $L^2(\Omega)$, $v(x) \in \beta(u(x))$, a.a. $x \in \Omega$.

Its effective domain is characterized by

$$D(\bar{\beta}) = \{u \in L^2(\Omega):\ \bar{\beta}(u) \neq \emptyset\}.$$

Moreover let

$$\tilde{f}(x,t) = \begin{cases} f(x,R_1) & t \leq R_1 \\ f(x,t) & R_1 \leq t \leq R_2 \\ f(x,R_2) & t \geq R_2, \end{cases}$$

$x \in \Omega$, and let $F\colon L^2(\Omega) \to L^2(\Omega)$ be the Nemytskii operator induced by $\tilde{f}$, i.e. $(Fu)(x) = \tilde{f}(x,u(x))$ for a.a. $x \in \Omega$. Note that the mapping F is uniformly bounded on $L^2(\Omega)$.

Without reference to the assumption (ii) of the Theorem we first prove the existence of $u \in D(A) \cap D(\bar{\beta})$ such that

(2) $$Au + \bar{\beta}(u) + F(u) \ni g.$$

To this purpose let $\beta_\lambda : \mathbb{R} \to \mathbb{R}$ be the Yosida approximation of β (e.g. [3]), which in this particular case can be computed as

$$\beta_\lambda(t) = \begin{cases} \lambda^{-1}(t - R_1) & t \leq R_1 \\ 0 & R_1 \leq t \leq R_2 \\ \lambda^{-1}(t - R_2) & t \geq R_2, \end{cases}$$

$\lambda > 0$. Let $\bar{\beta}_\lambda\colon L^2(\Omega) \to L^2(\Omega)$ be the corresponding extension, as defined in (1). By $A_0\colon H_0^1(\Omega) \to H^{-1}(\Omega)$ we denote the bounded linear operator implied by

$$(A_0 v, w) = \int_\Omega [\Sigma\ a_{ij} \frac{\partial v}{\partial x_j} \frac{\partial w}{\partial x_i} + \Sigma b_i \frac{\partial v}{\partial x_i} w + cvw]dx, \quad \forall v,\ w \text{ in } H_0^1(\Omega).$$

As a consequence of the uniform ellipticity of $\mathcal{A}$, there exist positive constants α_0, α_1 such that

$$(A_0 v,v) \geqq \alpha_0 \|v\|_1^2 - \alpha_1 \|v\|_0^2, \quad \forall v \in H_0^1(\Omega)$$

(here $\|\cdot\|_k$ is the norm in $H^k(\Omega)$). Let $R = \text{Max}\,\{-R_1, R_2\}$. Since $|\beta_\lambda(t)| \geqq \lambda^{-1}(|t| - R)$, we infer that

$$(\bar{\beta}_\lambda(v),v) \geqq \lambda^{-1}(\|v\|_0^2 - R|\Omega|^{\frac{1}{2}}\|v\|_0).$$

Thus

$$((A_0 + \bar{\beta}_\lambda + F)v,v) \geqq \alpha_0 \|v\|_1^2 + (\lambda^{-1} - \alpha_1)\|v\|_0^2 - (d_1 + d_2\lambda^{-1})\|v\|_0$$

$$\forall v \in H_0^1(\Omega),$$

with some uniform constants d_1, d_2. In the following let $0 < \lambda \leqq \alpha_1^{-1}$. Then (for fixed λ) the mapping $A_0 + \bar{\beta}_\lambda + F\colon H_0^1(\Omega) \to H^{-1}(\Omega)$ is coercive:

$$\|v\|_1^{-1}((A_0 + \bar{\beta}_\lambda + F)v,v) \to +\infty \quad (\|v\|_1 \to \infty).$$

A well-known existence theorem on pseudo-monotone mappings (e.g. [6, Chap. 2, Th. 2.7]) guarantees the existence of $u_\lambda \in H_0^1(\Omega)$ solving the equation

$$A_0 u_\lambda + \bar{\beta}_\lambda(u_\lambda) + Fu_\lambda = g.$$

Since $A_0 u_\lambda \in L^2(\Omega)$, a regularity result on linear elliptic equations (e.g. [1, Th. 9.8]) implies that $u_\lambda \in D(A)$. Hence

(3) $$Au_\lambda + \bar{\beta}_\lambda(u_\lambda) + Fu_\lambda = g.$$

Moreover, since for $\lambda < \lambda_0$, $(\bar{\beta}_\lambda(v),v) \geqq (\bar{\beta}_{\lambda_0}(v),v) \quad \forall v \in L^2(\Omega)$, the uniform boundedness in $H_0^1(\Omega)$ of the u_λ $(0 < \lambda \leqq \alpha_1^{-1})$ follows.

Let $\mathcal{A}' = -\Sigma \frac{\partial}{\partial x_i}(a_{ij} \frac{\partial}{\partial x_j})$ be the principal part of $\mathcal{A}$, and let A' be the operator in $L^2(\Omega)$ induced by $\mathcal{A}'$ on the domain $D(A') = D(A)$.

Let further $A'' = A - A'$ and A'' be the mapping in $L^2(\Omega)$ induced by A'', with $D(A'') = H_0^1(\Omega)$. Taking the inner product (in $L^2(\Omega)$) of (3) with $\bar\beta_\lambda(u_\lambda)$, we obtain

$$(A'u_\lambda, \bar\beta_\lambda(u_\lambda)) + \|\bar\beta_\lambda(u_\lambda)\|_0^2 = (g - Fu_\lambda - A''u_\lambda, \bar\beta_\lambda(u_\lambda)).$$

Since $(A'u_\lambda, \bar\beta_\lambda(u_\lambda)) \geqq 0$, the uniform boundedness in $L^2(\Omega)$ of $\bar\beta_\lambda(u_\lambda)$, $0 < \lambda \leqq \alpha_1^{-1}$, follows. By (3) we then conclude the uniform boundedness of Au_λ in $L^2(\Omega)$. For some sequence $\lambda_n \downarrow 0$ we now may assure that

$$\left.\begin{array}{l} u_{\lambda_n} \to u \\ Au_{\lambda_n} \rightharpoonup Au \\ \bar\beta_{\lambda_n}(u_{\lambda_n}) \rightharpoonup z \end{array}\right\} \quad \text{in } L^2(\Omega).$$

By a well-known result on Yosida approximations (e.g. [4, Lemma 4.5]), $u \in D(\bar\beta)$ and $z \in \bar\beta(u)$. Passage to the limit in (3) shows that u is an asserted solution of (2).

Obviously $R_1 \leqq u(x) \leqq R_2$ almost everywhere. We decompose the domain Ω into three parts $\Omega = \Omega_0 \cup \Omega_1 \cup \Omega_2$:

$$\Omega_0 = \{x \in \Omega: R_1 < u(x) < R_2\},$$
$$\Omega_1 = \{x \in \Omega: u(x) = R_1\},$$
$$\Omega_2 = \{x \in \Omega: u(x) = R_2\}.$$

By choice of β and (2),

$$(Au)(x) + f(x,u(x)) = g(x) \qquad \text{a.e. on } \Omega_0.$$

Next we note that the first and second partial derivatives of u vanish a.e. on Ω_1 (e.g. [5, Appendix to parts I and II]). Hence $(Au)(x) = c(x)R_1$ on Ω_1, and by (2),

$$c(x)R_1 + f(x,R_1) \geqq g(x), \qquad \text{a.a. } x \in \Omega_1.$$

Invoking assumption (ii) of the Theorem, we infer that

$$c(x)R_1 + f(x,R_1) = g(x)$$

and consequently

$$(4) \qquad (Au)(x) + f(x,u(x)) = g(x) \qquad \text{a.e. on } \Omega_1.$$

Similarly (4) holds on Ω_2. The Theorem is thus proved.

References.

[1] S. Agmon: Lectures on elliptic boundary value problems. Van Nostrand Math. Studies, No. 2, Van Nostrand, Princeton, N.J., 1965.

[2] H. Amann: On the existence of positive solutions of nonlinear elliptic boundary value problems. Indiana Univ. Math. J. 21 (1971), 125-146.

[3] H. Brézis: Opérateurs maximaux monotones. Mathematics Studies No. 5, North Holland, Amsterdam, 1973.

[4] T. Kato: Accretive operators and nonlinear evolution equations in Banach spaces. Proc. Sympos. Pure Math., Vol. 18, part 1, Amer. Math. Soc., Providence, R.I., 1970, 138-161.

[5] H. Lewy - G. Stampacchia: On the regularity of the solution of a variational inequality. Comm. Pure Appl. Math. 22 (1969), 153-188.

[6] J.-L. Lions: Quelques méthodes de résolutinn des problèmes aux limites non linéaires. Dunod, Gauthier-Villars, Paris, 1969.

[7] R.H. Martin: Nonlinear perturbations of second order elliptic operators. (to appear).

Limit point-limit circle criteria for $(py')' + qy = \lambda ky$

Don Hinton

1. Introduction. Let L be the differential operator defined on an open interval (a,b) by

$$L(y) = k^{-1}\{(py')' + qy\} , \quad a < x < b , \tag{1}$$

where the coefficients satisfy:

(2) (i) k is a positive continuous function on (a,b) ;

(ii) p and q are real functions on (a,b) , $p \geq 0$, and $1/p$ and q are $L_{loc}(a,b)$.

The notation $L_{loc}(a,b)$ indicates locally Lebesque integrable on (a,b); $AC_{loc}(a,b)$ indicates locally absolutely continuous. Thus the operator is singular only possibly at a or b . We suppose throughout only one endpoint is singular; if it is a , then we take $-\infty < a < b < \infty$, and if it is b , then we take $-\infty < a < b = \infty$. According to the Weyl classification L is said to be in the limit circle case (LC) if all solutions of $L(y) = 0$ satisfy $\int_a^b k|y|^2 < \infty$ (hence the same is true of all solutions of $L(y) = \lambda y$ for any complex λ); otherwise L is said to be in the limit point case (LP).

This research was supported in part by the NSF Grant GP-39113.

The use of the weight function k in (1) makes possible direct application to some of the classical operators. It also allows treatment of finite and infinite singularities from a single framework. Another use is in the derivation of limit point-limit circle criteria for the differential operator

$$(3) \qquad D(u) = K^{-1}\left\{\begin{pmatrix} 0 & 1 \\ -1 & 0 \end{pmatrix} u' + \begin{pmatrix} Q & R \\ R & P \end{pmatrix} u\right\}, \quad u = \begin{pmatrix} u_1 \\ u_2 \end{pmatrix}$$

since the transformation

$$u = \begin{pmatrix} 1/r & 0 \\ 0 & r \end{pmatrix}\begin{pmatrix} y \\ -py' \end{pmatrix} \text{ in } \begin{pmatrix} 0 & 1 \\ -1 & 0 \end{pmatrix}\begin{pmatrix} y \\ -py' \end{pmatrix}' = \begin{pmatrix} q & 0 \\ 0 & 1/p \end{pmatrix}\begin{pmatrix} y \\ -py' \end{pmatrix}$$

gives that u will satisfy

$$\begin{pmatrix} 0 & 1 \\ -1 & 0 \end{pmatrix} u' = \begin{pmatrix} qr^2 & r'/r \\ r'/r & 1/pr^2 \end{pmatrix} u .$$

With this transformation, Theorem 1 below yields results for (3) similar to the result of Titchmarsh given in [8].

We will not consider here limit point conditions which subject the coefficients to inequalities only on certain non-overlapping subintervals of (a,b) (cf. [1,3]). Instead we are interested in allowing perturbations to certain well behaved coefficients.

2. Limit point criteria. A large number of limit point theorems for (1) give effective criteria in only one of the cases $\int_a^b (k/p)^{1/2}$ finite or infinite. This dichotomy is illustrated by

$$L_1(y) = (x^\alpha y')' + q(x)y \ , \ 1 \le x < \infty \ ,$$

where for $\alpha \le 2$ it is known that L_1 is LP if for some $C > 0$, $q(x) \le Cx^{2-\alpha}$; on the other hand for $\alpha > 2$, L_1 is LP if $q(x) \le (2\alpha - 3)x^{\alpha-2}/4$, and moreover the constant $(2\alpha - 3)/4$ is sharp. Theorem 1 below applies to the case $\int_a^b (k/p)^{1/2} < \infty$. A recent paper by Everitt-Hinton-Wong [6] on $2n^{th}$ order equations gives results in the case $\int_a^b (k/p)^{1/2} = \infty$ and perturbation terms are present.

Theorem 1. *Let* L *be given by* (1) *where the coefficients satisfy* (2). *Then* L *is LP if there are real functions* $\eta > 0$ *and* h *on* (a,b) *such that*

(i) η *and* $p\eta'$ *are* $AC_{loc}(a,b)$; $h \in L_{loc}(ab)$.

(ii) $\int_a^b k\eta^2 = \infty$.

(iii) $q \le -(p\eta')'\eta^{-1} + h$, $a < x < b$.

(iv) *Either* $h \equiv 0$ *or the two integrals* $\int_a^b (p\eta^2)^{-1}$ *and* $\int_a^b \eta^2|h|\xi$ *are finite* ($\xi(x) = \int_a^x (p\eta^2)^{-1}$ *if a is singular and* $\xi(x) = \int_x^\infty (p\eta^2)^{-1}$ *if* b *is singular*).

Proof. First suppose $b = \infty$ is the singular endpoint. Let $y = \eta\phi$ in (1) where $L(y) = 0$. Then ϕ is a solution of

$$(p\eta^2\phi')' + \eta[(p\eta')' + q\eta]\phi = 0 \ . \tag{4}$$

Consider for comparison purposes the equation

$$(p\eta^2\psi')' + (h\eta^2)\psi = 0 \ . \tag{5}$$

For $h \neq 0$, define $s = [\int_x^\infty (pn^2)^{-1}]^{-1}$ and $z(s) = s\psi(x)$. Then a calculation shows that z satisfies

$$\ddot{z}(s) + s^{-4}[pn^4h](x) \quad z(s) = 0 \;, \; \cdot = \frac{d}{ds} \;. \tag{6}$$

Now in (6),

$$\int^\infty s|s^{-4}[pn^4h](x)|ds = \int^\infty n^2|h|\xi dx < \infty \;;$$

hence by a theorem of Ghizzetti, there is a solution z of (6) such that $s^{-1}z(s) \to 1$ as $s \to \infty$. Thus there is a solution ψ of (5) such that $\psi(x) \to 1$ as $x \to \infty$ (for $h = 0$, $\psi = 1$ is a solution of (5)); let $\psi(x) \geq 1/2$ for $x \geq x_0$. Let ϕ be the solution of (4) with initial conditions $\phi(x_0) = \psi(x_0)$ and $\phi'(x_0) = \psi'(x_0)$. Then condition (iii) implies $\phi(x) \geq \psi(x)$ for $x \geq x_0$ (cf. [11, p. 103]). Hence by condition (ii) the corresponding solution $y = n\phi$ of $L(y) = 0$ satisfies

$$\int_{x_0}^\infty k|y|^2 \, dx \geq \int_{x_0}^\infty (1/4)kn^2 = \infty \;,$$

and L is LP.

Suppose now a is the singular endpoint which we take to be zero. The change of variables $s = x^{-1}$, $Y(s) = y(x)$ transforms $L(y) = 0$ into $(\cdot = \frac{d}{ds})$

$$(p_1 Y^\cdot)^\cdot + q_1 Y = 0 \text{ and } \int_0^b k(x)|y(x)|^2 dx = \int_{b^{-1}}^\infty k_1(s)|Y(s)|^2 ds$$

where $p_1(s) = s^2 p(x)$, $q_1(s) = s^{-2} q(x)$, and $k_1(s) = s^{-2} k(x)$. The finite singularity case will now follow from the case above by choosing $\eta_1(s) = \eta(x)$ and $h_1(s) = s^{-2} h(x)$.

Theorem 1 with $h = 0$ and $k = 1$ was given by Kurss in [9]. Some related work by Wong and Zettl with $h = 0$ and $k = 1$ which uses change of variables methods is to be found in [12].

For $b = \infty$, $p(x) = x^{\alpha}$, and $k(x) = x^{\delta}$ $(\alpha > 2 + \delta)$, the choice $\eta(x) = x^{-(1+\delta)/2}$ gives that (1) is LP if

$$(7) \qquad q(x) \leq \frac{(1+\delta)(2\alpha-\delta-3)}{4} x^{\alpha-2} + h(x) , \int_a^b x^{1-\alpha}|h|dx < \infty .$$

For the singularity at $a = 0$, $p(x) = x^{\alpha}$, and $k(x) = x^{\delta}$ $(\alpha < 2 + \delta)$, the choice $\eta(x) = x^{-(1+\delta)/2}$ gives that (1) is LP if q satisfies (7). For $\delta = \alpha = 0$, this allows a perturbation term h to the criterion of Sears [10], i.e. $q(x) \leq -3/4x^2$, of size $\int_0^1 x|h|dx < \infty$.

The function p need not be differentiable. For example, with $b = \infty$, $p(x) = x^4 g(x)$ where $0 < c \leq g(x) \leq d$, and $k(x) = 1$, the choice $\eta(x) = \int_x^{\infty} (gx^{3/2})^{-1}$ gives that L is LP if $q(x) \leq 5cx^2/4 \leq 5x^{3/2}/2\eta$.

3. Limit circle criteria. One of the most powerful techniques to obtain limit circle criteria is to use asymptotic methods. However, these methods require considerable smoothness of the coefficients and do not allow for perturbations. An alternative approach to the problem is given by Everitt in [4], but the coefficients must still be sufficiently smooth. A third approach is that given by Halvorsen in [7] which allows rather large perturbations of LC equations which does not change the LC classification.

In Theorem 2, we have adapted Halvorsen's argument to the more general operator (1).

<u>Theorem</u> 2. <u>Let</u> L <u>be given by</u> (1) <u>where the coefficients satisfy</u> (2). <u>Then</u> L <u>is LC if there is a real function</u> $\eta > 0$ <u>on</u> (a,b) <u>and a number</u> $c > 0$ <u>such that</u>

(i) η <u>and</u> $p\eta'$ <u>are</u> $AC_{loc}(a,b)$.

(ii) $\int_a^b (p\eta^2)^{-1} = \infty$,

(iii) $\int_a^b k\eta^2 e^{g(x)}dx < \infty$ <u>where</u> $g(x) = c^{-1} \int_a^x h(\xi)d\xi$ <u>if</u> b <u>is singular</u>, $h = |q + \eta^{-1}(p\eta')' - c^2(p\eta^4)^{-1}|\eta^2$, <u>and</u> $g(x) = c^{-1} \int_x^b h(\xi)d\xi$ <u>if</u> a <u>is singular</u>.

<u>Proof.</u> Consider first the case with b singular. Let y be a solution of $L(y) = 0$,

$$z(t) = (\eta^{-1}y)(x) \ , \ t = \int_a^x (p\eta^2)^{-1} \ , \ \cdot = \frac{d}{dt} \ . \tag{8}$$

Then a calculation shows that

$$\ddot{z}(t) + f(t)z(t) = 0 \ , \ f(t) = [p\eta^4(\eta^{-1}(p\eta')' + q)](x) \ .$$

As shown by Halvorsen in [7], the function $E = (\dot{z}^2 + c^2z^2)/2$ satisfies an inequality

$$\begin{aligned} E(t) &\le K \exp c^{-1} \int_0^t |f(\gamma) - c^2|d\gamma \ , \ K = \text{constant} \\ &= K \exp g(x). \end{aligned} \tag{9}$$

Since $z^2 \le 2E/c^2$, we have that

$$\int_a^\infty ky^2dx \le \int_a^\infty k\eta^2(2K/c^2)e^{g(x)}dx < \infty \ ;$$

hence L is LC.

The proof for a singular follows a similar argument with $t = \int_x^b (p\eta^2)^{-1}$

If we express $q = q_1 + q_2$ where q_1 is a "smooth" part of q, then a reasonable choice of η with $b = \infty$ is $\eta = (pq_1)^{-1/4}$. If also $\int^\infty |\eta(p\eta')'|dx < \infty$ (a typical hypothesis for asymptotic methods, cf. [2, p. 120]), then the only other conditions required for L to be LC at ∞ are $\int^\infty (q_1/p)^{1/2} = \infty$ and

$$\int^\infty k(pq_1)^{-1/2}\{\exp \int_a^x |q_2|(pq_1)^{-1/2}\}dx < \infty .$$

For $p = 1$, $\int^\infty |\eta\eta''| < \infty$ is equivalent to $\int^\infty |q_1''q_1^{-3/2}| < \infty$ [2, p. 120], and we state this as a separate corollary.

Corollary 1. *Let* $L = k^{-1}\{y'' + (q_1 + q_2)y\}$ *on* $[a,\infty)$ *where* k *satisfies* 2 (i), q_1 *is* $C^{(2)}[a,\infty)$, *and* $q_2 \in L_{loc}(a,b)$. *Then* L *is* LC *at* ∞ *if* $\int_a^\infty |q_1''q_1^{-3/2}| < \infty$ *and*

$$\int_a^\infty kq_1^{-1/2}\{\exp \int_a^x |q_2|q_1^{-1/2}\}dx < \infty .$$

For example, if $k = 1$ and $q_1(x) = x^4$, then L is LC at ∞ if $|q_2(x)| \leq \alpha x$ for some $\alpha < 1$. To illustrate the scope of Halvorsen's approach, we now consider another perturbation of the equation

$$y'' + [x^4 + q_2(x)]y = 0 . \tag{10}$$

Let $y(x) = C(x)e^{H(x)}$ on $[1,\infty)$ where

$$C(x) = x^{-1} \cos(x^3/3) , \; H(x) = \int_1^x \xi^{-1} \cos^2(\xi^3/3)d\xi .$$

A calculation shows that as $x \to \infty$,

$$H(x) = (1/2)\ln x + k_0 + o(1) \ , \ k_0 = \text{constant},$$

from which we may calculate that $y \notin L_2(1,\infty)$. Another calculation shows that y satisfies an equation of the form (10) where as $x \to \infty$,

$$q_2(x) = 2x \sin(2x^3/3) + O(x^{-2}) \ .$$

Thus there is an LP equation of type (10) where $|q_2(x)| \leq (2+\varepsilon)\, x,\ \varepsilon > 0$.

Theorem 2 also applies to Euler type equations. For $p(x) = x^\alpha$ on $[1,\infty)(\alpha > 2)$ and $k = 1$, L is LC at ∞ if for some $c > 0$, $\delta > 0$,

$$(11) \qquad |q(x) - [c^2 + (1-\alpha)^2/4]x^{\alpha-2}| \leq c(|\alpha-2|-\delta)x^{\alpha-2} \ .$$

For $p(x) = x^\alpha$ on $(0,1]$ $(\alpha<2)$ and $k = 1$, L is LC at 0 if for some $c > 0$, $\delta > 0$, q satisfies (11) $(\eta(x) = x^{(1-\alpha)/2})$.

4. Strong limit point criteria. With some additional hypothesis to Theorem 1, it may be shown that (1) is strong limit point (SLP) at the singular point, i.e.

$$(12) \qquad \lim_{x\to X} pyz'(x) = 0 \ , \ (y,z \in \Delta)$$

where X is the singular point and Δ consists of all those y such that y and py' are $AC_{loc}(a,b)$, and y, $L(y)$ are Lebesque square integrable with respect to the weight function k . Equation (1) is said to satisfy the Dirichlet condition if $p^{1/2}y'$, $|q|^{1/2}y \in L_2(a,b)(y \in \Delta)$ and

$$(13)\qquad py\bar{z}\Big|_a^b + \int_a^b [-(py')\bar{z}' + qy\bar{z}] = \int_a^b kL(y)\bar{z}\ , \quad (y,\ z \in \Delta)\ .$$

<u>Theorem</u> 3. <u>Let</u> L <u>be given by</u> (1) <u>where the coefficients satisfy</u> (2). <u>Then</u> L <u>is</u> SLP <u>and satisfies the Dirichlet condition if there is a real function</u> $\eta > 0$ <u>on</u> (a,b) <u>such that</u>

(i) η <u>and</u> $p\eta'$ <u>are</u> $AC_{loc}(a,b)$.

(ii) $\int_a^b k\eta^2 = \infty$ <u>and</u> $\int_a^b (p\eta^2)^{-1} < \infty$.

(iii) $q \leq -(p\eta')'\eta^{-1}$, $a < x < b$.

(iv) $|p\eta'\eta| = O(\xi(x)^{-1})$ (<u>where</u> $\xi(x)$ <u>is as in Theorem</u> 1).

(v) $q_+ = \max\{0,q\} = O([p\eta^4]^{-1}\xi(x)^{-2})$.

<u>Proof</u>. First consider $b = \infty$. To establish (12) and (13) it is sufficient to consider y , z real,, and for (12) it is sufficient that $y = z$ [5]. Let $y \in \Delta$ and as in Theorem 1, let $y = \eta\phi$; then

$$(14)\qquad (p\eta^2\phi')' + \eta kL(\eta)\phi = \eta kL(y)\ .$$

Now $L(\eta) \leq 0$ by (iii); hence if we multiply (14) by ϕ and integrate, we have

$$(15)\qquad \int_a^X kyL(y) \leq p\eta^2\phi'\phi\Big|_a^X - \int_a^X p\eta^2(\phi')^2\ .$$

Now $\int_a^\infty k\eta^2\phi^2 < \infty$; thus by (ii) there is a sequence $\{X_n\}$, $X_n \to \infty$ as $n \to \infty$, such that $\phi(X_n)\phi'(X_n) \leq 0$. Using this in (15) it follows that $\int_a^\infty p\eta^2(\phi')^2 < \infty$. From $\int_a^\infty k\eta^2\phi^2 < \infty$ and

$$[\phi(x_2) - \phi(x_1)]^2 = [\int_{x_1}^{x_2} \phi']^2 \leq \int_{x_1}^{x_2} p\eta^2(\phi')^2 \int_{x_1}^{x_2} (p\eta^2)^{-1}\ ,$$

we may conclude that $\phi(x) \to 0$ as $x \to \infty$; hence also $\phi(x)^2\xi(x)^{-1} \to 0$ as

$x \to \infty$. Finally, from this limit, $\int_a^\infty p\eta^2(\phi')^2 < \infty$, and

$$\int_a^X (p\eta^2)^{-1}\xi(x)^{-2}\phi^2 = \xi(x)^{-1}\phi^2\Big|_a^X - 2\int_a^X \xi(x)^{-1}\phi\phi' ,$$

it follows from an application of Schwarz's inequality that

$$(16) \qquad \int_a^\infty (p\eta^2)^{-1}\xi(x)^{-2}\phi^2 < \infty .$$

Applying (iv), (16), and $\int_a^\infty p\eta^2(\phi')^2 < \infty$, we have that

$$(17) \qquad \int_a^\infty (p\eta^4)^{-1}\xi(x)^{-2}y^2 < \infty , \int_a^\infty p(y')^2 < \infty .$$

An integration of $kyL(y)$ by parts gives

$$\int_a^X kyL(y) = pyy'\Big|_a^X + \int_a^X [-p(y')^2 + qy^2]$$

from which we conclude by (17) and (v) that limit $(pyy')(X)$ as $X \to \infty$ exists either as a finite limit or ∞ . If $|pyy'| \geq \varepsilon > 0$, then

$$|p^{1/2}y'|\,|[p^{1/2}\eta^2\xi(x)]^{-1}y| \geq \varepsilon\ [p\eta^2\xi(x)]^{-1}$$

which is a contradiction since the left side is integrable and the right is not; thus L is SLP at ∞ . We may now also conclude from this argument that $\int_a^\infty [|q_-| + |q_+|]y^2 < \infty$. The Dirichlet condition follows immediately. The singularity at 0 case may be derived from the $b = \infty$ by the change of variable $s = 1/x$.

In the case of $p = k = 1$ and a finite singularity (take $\eta(x) = x^{-1/2}$) , this result agrees with that proved already by W. N. Everitt and M. Giertz (to appear in Math. Ann.).

References

1. F. V. Atkinson and W. D. Evans, "On solutions of a differential equation which are not of integrable square", Math. Z. 127(1972), 323-332.

2. W. A. Coppel, Stability and asymptotic behavior of differential equations (Heath, Boston, 1965).

3. M. S. P. Eastham, "On a limit-point method of Hartman", Bull. London Math. Soc. 4(1972), 340-344.

4. W. N. Everitt, "On the limit circle classification of second-order differential expressions", Quart. J. Math. (2), 23(1972), 193-196.

5. W. N. Everitt, M. Giertz, and J. B. McLeod, "On the strong and weak limit-point classification of second-order differential expressions", to appear.

6. W. N. Everitt, D. B. Hinton, and J. S. W. Wong, "On the strong limit-n classification of linear differential expressions of order 2n", J. London Math. Soc., to appear.

7. S. Halvorsen, "Bounds for solutions of second order linear differential equations, with applications to L^2-boundedness", D. K.N.V.S. Forhandlinger 36(1963), 36-40.

8. H. Kalf, "A limit-point criterion for separated Dirac operators and a little known result on Riccati's equation", Math. Z. 129(1972), 75-82.

9. H. Kurss, "A limit-point criterion for nonoscillatory Sturm-Liouville differential operators", Proc. Amer. Math. Soc. 18(1967), 445-449.

10. D. B. Sears, "On the solutions of a second order differential equation which are integrable square", J. London Math. Soc. 24(1949), 207-215.

11. F. G. Tricomi, Differential Equations (Hafner, New York, 1961).

12. J. S. W. Wong and A. Zettl, "On the limit point classification of second order differential equations", Math. Z. 132(1973), 297-304.

ON SECOND-ORDER DIFFERENTIAL OPERATORS OF LIMIT-CIRCLE TYPE

IAN KNOWLES

1. We will be concerned here with linear second-order differential operators of the form

$$\tau y(t) = y''(t) + q(t)y(t) , \quad 0 \leq t < \infty , \tag{1}$$

where $' \equiv \frac{d}{dt}$, and q is assumed to be real-valued and locally Lebesque integrable on the half-line $[0, \infty)$. The operator τ is said to be of limit-circle type at ∞ (LC) if every solution of the differential equation $\tau y = 0$ is of class $\mathcal{L}^2[0, \infty)$; otherwise τ is said to be of limit-point type at ∞ (LP).

In [7], I.A. Pavlyuk has shown that the condition

$$\int_0^\infty q^{-1/2}(t)\,dt < \infty$$

is both necessary and sufficient for τ to be LC, provided that q is positive and

$$\left| \tfrac{1}{4} q'' q^{-1} - \tfrac{5}{16} (q' q^{-1})^2 \right| \leq K < \infty$$

almost everywhere on the interval $[0, \infty)$. A somewhat similar condition also appears in [2] p.1414; both of these results have been generalized in [5].

One observes immediately that in all of these criteria we are placing rather severe differentiability restrictions directly onto the coefficient q. In this respect the following theorem is a considerable improvement :

THEOREM 1. *Let* ψ *be any non-vanishing locally absolutely continuous function defined on* $[0,\infty)$ *which satisfies either*

$$\left|\psi^{-4} - \psi^{-1}\psi'' - q\right| \leq K < \infty \tag{2}$$

almost everywhere on $[0, \infty)$, *or*

$$\int_0^\infty \psi^2(t) \left| \psi^{-4}(t) - \psi^{-1}(t)\psi''(t) - q(t) \right| dt < \infty . \tag{3}$$

Then τ *is* LC *if and only if*

$$\int_0^\infty \psi^2(t)\,dt < \infty \tag{4}$$

Proof. We reapply the method in [5] , in which the equation

$$y''(t) + q(t)y(t) = 0$$

is first transformed to the equation

$$\frac{d}{ds}\left(\psi^4(t)\frac{dz}{ds}\right) + \{q(t) + \phi(t)\}\, z(s) = 0 . \tag{5}$$

This is accomplished via the variable change $t = t(s)$ defined implicitly by (see [2] p. 1498)

$$s(t) = \int_0^t \psi^2(r)\,dr$$

where

$$z(s) = y(t)\ \{\tfrac{dt}{ds}\}^{\frac{1}{2}} \tag{6}$$

$$\phi(t) = \psi^{-1}(t)\ \psi''(t)$$

Furthermore, it follows from (6) that

$$\int_0^{s(\infty)} z^2(s)\,ds = \int_0^\infty y^2(t)\,dt \tag{7}$$

Consider now the equation

$$\frac{d}{ds}\left(\psi^4(t)\frac{dw}{ds}\right) + \psi^{-4}(t)w(s) = 0 \tag{8}$$

which has a fundamental set of solutions given by

$$u(s) = \sin\int_0^s \psi^{-4}(r)\,dr\ ,\quad v(s) = \cos\int_0^s \psi^{-4}(r)\,dr$$

It follows that

$$\int_0^{s(\infty)} \{u^2(s) + v^2(s)\} ds = \int_0^{\infty} \psi^2(t)\,dt \quad .$$

Consequently, every real solution w of (9) satisfies

$$\int_0^{s(\infty)} w^2(s)\,ds < \infty \tag{9}$$

if and only if (4) holds. Finally, given (7), the proof will be complete if we can show that (10) holds for every real solution w of (8) if and only if

$$\int_0^{s(\infty)} z^2(s)\,ds < \infty$$

for every real solution z of (5). This however follows from the comparison criteria of Bellman [1] and Halvorsen [4], used in conjunction with conditions (2) and (3) respectively (see [5]).

<u>Remarks.</u>

(i) The method readily extends to the more general equation considered in [5]. In this form Theorem 1 generalizes the result in [5].

(ii) In order to use Theorem 1, one must first find a suitable smooth function ψ which approximates q in the sense of (2) or (3). In some cases this could be difficult. As an example we have

COROLLARY. *Let P and h be real valued functions defined on $[0,\infty)$ and such that*

(i) $P > 0$ *in* $[0,\infty)$, *and* $P \in \mathcal{L}^2[0,\infty)$,

(ii) P' *locally absolutely continuous,*

(iii) $hP \in \mathcal{L}^1[0,\infty)$.

In $[0,\infty)$ *define*

$$q(t) = -h(t)P^{-1}(t) - P''(t)P^{-1}(t) + P^{-4}(t)$$

Then the corresponding operator τ *defined by* (1) *is LC.*

Proof: Substitute $\psi = P$, noting that (iii) ensures (3) is satisfied.

This result is comparable with the limit circle criterion in [3] (set $\phi = 0$).

2. In this final section I would like to draw attention to the limit-circle condition of Kupcov [6], which appears to have been largely overlooked in recent years.

THEOREM 2 (Kupcov [6]). *If there exist positive functions* ψ *and* Q *such that* Q' *and* ψ *are locally absolutely continuous, and*

$$\int_0^\infty \psi^2(t)\, \exp \int_0^t \left[\left| \frac{\psi'(s)}{\psi(s)} - \frac{Q'(s)}{Q(s)} \right| + \psi^2(s) \left| \psi^{-4}(s) - Q^{-1}(s) Q''(s) - q(s) \right| \right] ds\, dt < \infty$$

then τ *is* LC.

This result contains the limit-circle part of Theorem 1 (put $\psi = Q$). It also contains Theorem 1 in [3] (put $Q = P$ and $\psi = P(1+\phi)^{-1/4}$), as well as virtually every other limit-circle criterion. A question : can one generalize Theorem 1 in such a way as to include Theorem 2?

REFERENCES

[1] R. Bellman, "A stability property of solutions of linear differential equations", *Duke Math. J.*, 11(1944), 513-516.

[2] N. Dunford and J.T. Schwartz, *Linear operators; part II : Spectral theory* (Interscience, New York, 1963).

[3] M.S.P. Eastham, "Limit-circle differential expressions of the second order with an oscillating coefficient", *Quart. J. Math. Oxford (2)*, 24 (1973), 257-263.

[4] S. Halvorsen, "On the quadratic integrability of solutions of $d^2x/dt^2+f(t)x=0$" *Math. Scand.*, 14(1964), 111-119.

[5] I. Knowles, "On a limit-circle criterion for second-order differential operators", *Quart. J. Math. Oxford* (2), 24(1973), 451-455.

[6] N.P. Kupcov, "Conditions of non-selfadjointness of a second order linear differential operator", *Dokl. Akad. Nauk*, 138 (1961), 767-770.

[7] I.A. Pavlyuk, "Necessary and sufficient conditions for boundedness in the space $\mathcal{L}^2[0,\infty)$ for solutions of a class of linear differential equations of second-order", *Dopovidi Akad. Nauk Ukrain. RSR*, 1960, 156-158. MMR/26.2.74

A Dynamical Approach to Fourth Order Oscillations

Kurt Kreith

In [1] W. M. Whyburn shows how to represent the real selfadjoint fourth order differential equation

$$(1)\qquad (p_2(t)y'')'' - (p_1(t)y')' + p_0(t)y = 0 \qquad (p_2(t) > 0)$$

as a second order system of the form

$$(2)\qquad \begin{aligned} y'' &= a(t)y + b(t)x \qquad (b(t) > 0) \\ x'' &= c(t)y + a(t)x \end{aligned}$$

and uses the representation (2) to establish oscillation properties of solutions of (1). Subsequent studies of fourth order equations have developed new techniques for dealing with fourth order equations directly, most notable being the paper of Leighton and Nehari [2] which develops an extensive theory for the equation

$$(3)\qquad (p_2(t)y'')'' + p_0(t)y = 0 \qquad (p_2(t) > 0)$$

in the instances $p_0(t) < 0$ and $p_0(t) > 0$, respectively. Of interest is the fact that many of the techniques developed for the special equation (3) subsequent to the appearance of [1] allow a generalization to (2) or even to the more general second order system

$$(4)\qquad \begin{aligned} y'' &= a(t)y + b(t)x \qquad (b(t) > 0) \\ x'' &= c(t)y + d(t)x\ . \end{aligned}$$

Since it was shown in [3] that the general real fourth order differential equation can be reduced to the form (4), these generalizations allow one to study the oscillation properties of a much larger class of equations.

The system (4) has an obvious dynamical interpretation in terms of a particle of unit mass in the x,y-plane, subject to horizontal and vertical forces. Introducing polar coordinates

$$r^2 = x^2 + y^2 \quad ; \quad \theta = \arctan y/x$$

transforms (4) into

$$rr'' = r^2\theta'^2 + r^2 Q_1(\cos\theta, \sin\theta) \tag{4.1}$$

$$(r^2\theta')' = r^2 Q_2(\cos\theta, \sin\theta) \tag{4.2}$$

where Q_1 and Q_2 are quadratic forms defined by

$$Q_1(\xi,\eta) = a\eta^2 + (b+c)\xi\eta + d\xi^2$$

$$Q_2(\xi,\eta) = -c\eta^2 + (a-d)\xi\eta + b\xi^2.$$

While this representation appears to be singular at zeros of $r(t)$, such singularities can be avoided by a technique due to Taam [4] which calls for ordering such zeros in a sequence $\{t_k\}$ and defining

$$r(t) = \sqrt{x^2 + y^2} \quad \text{if} \quad t_{2k} \leq t \leq t_{2k+1}$$

$$r(t) = -\sqrt{x^2 + y^2} \quad \text{if} \quad t_{2k-1} \leq t \leq t_{2k}$$

and

$$\theta'(t_k) = 0.$$

In considering the systems (4), (4.1), and (4.2) we shall assume that the coefficients are continuous in $\mathcal{J} = [\alpha,\infty)$. The study of (4.1) and (4.2) is substantially simplified by the assumption that $Q_2(\xi,\eta)$ is positive definite (or negative definite) in $[\alpha,\infty)$. This assures [3] that $r^2\theta'$ can have at most one zero in $[\alpha,\infty)$ and that $\theta(t)$ is monotone for sufficiently large values of t. A solution $y(t)$ of a fourth order equation is then oscillatory at $t = \infty$ if and only if its polar representation satisfies

$$\lim_{t\to\infty} |\theta(t)| = \infty. \tag{5}$$

In [3] sufficient conditions for (5) are established by use of the theory of

H-oscillations. The following result, based on Riccati inequalities, strengthens Theorem 4.1 of [3].

Theorem 1. *If $Q_2(x,y)$ is positive definite on $[\alpha,\infty)$ and the equations*

$$u'' = a(t)u \qquad \text{and} \qquad u'' = d(t)u$$

are oscillatory at $t = \infty$, then

$$\lim_{t \to \infty} |\Theta(t)| = \infty. \tag{5}$$

Proof. Suppose (5) is not satisfied so that there exists a constant Θ_0 and a solution $r(t)$, $\Theta(t)$ of (4.1), (4.2) such that $\lim_{t \to \infty} \Theta(t) = \Theta_0$. Let $y(t)$, $x(t)$ be the corresponding solution of (4) and define

$$X(t) = \frac{x'(t)}{x(t)} \quad ; \quad Y(t) = \frac{y'(t)}{y(t)} .$$

If $\Theta(t)$ is eventually in the first or third quadrants, then

$$X' = c(t) \tan \Theta(t) + d(t) - X^2$$

and since $c(t) < 0$

$$X' \leq d(t) - X^2 \tag{6}$$

for all sufficiently large values of t. By means of the transformation $X = \frac{U'}{U}$, (6) implies that the differential inequality

$$UU'' \leq d(t)U^2 \tag{7}$$

has a solution $U(t)$ which is nonoscillatory at $t = \infty$. But if $u(t)$ is an oscillatory solution of $u'' + d(t)u = 0$ with zeros $t_1 < t_2 < \cdots$, then for sufficiently large t

$$\frac{d}{dt}\left[uu' - \frac{u^2}{U}U'\right] = \frac{u^2}{U^2}\left(dU^2 - UU''\right) + \left(u' - \frac{u}{U}U'\right)^2$$

and

$$0 = \left[uu' - \frac{u^2}{U}U^2\right]_{t_k}^{t_{k+1}} \geq \int_{t_k}^{t_{k+1}} \frac{u^2}{U^2}\left(dU^2 - UU''\right) d\tau$$

with equality if and only if $U(t)$ is a constant multiple of $u(t)$. But this contradicts (7). Similarly, if $\theta(t)$ is eventually in the second or fourth quadrants then

$$Y' = a(t) + b(t)\cot\theta(t) - Y^2,$$

and since $b(t) > 0$

$$Y' \leq a(t) - Y^2.$$

This last inequality contradicts the oscillatory behaviour of $u'' = a(t)u$ in an analogous way.

The hypotheses of Theorem 1 may be interpreted as requiring that the radial force in (4.1) is a sufficiently strong attractive force to assure that (5) is satisfied. The hypotheses of the following Theorem [6] relate more to the torque of (4.2) in assuring the same conclusion. We do, however, now impose the additional requirement that $a(x) \equiv 0$, $d(x) \equiv 0$ which effectively limits the application of Theorem 2 to selfadjoint equations of the form (3) with

$$p_2(t) = \frac{1}{b(t)} > 0 \quad \text{and} \quad p_0(t) = -c(t) > 0.$$

Theorem V of [1] asserts that if

$$\text{(8)} \qquad \int^{\infty} tb(t)\,dt = \infty \quad \text{and} \quad -\int^{\infty} tc(t)\,dt = \infty$$

then every solution $r(t)$, $\theta(t)$ of (4.1), (4.2) which satisfies $\theta'(\beta) > 0$ for some $\beta \geq \alpha$ also satisfies $\lim_{t\to\infty} \theta(t) = \infty$. In the following we eliminate the requirement $\theta'(\beta) > 0$.

Theorem 2. *If the coefficients of* (4) *satisfy* $a(t) \equiv 0$, $d(t) \equiv 0$, $b(t) > 0$, *and* $c(t) < 0$ *in* $[\alpha,\infty)$ *and if* (8) *is satisfied, then* $\lim_{t\to\infty} |\theta(t)| = \infty$ *for every nontrivial solution* $y(t)$, $x(t)$ *of* (4).

Proof. According to a theorem of Schneider [5], if no nontrivial solution of (1) has more than one double zero, then all nontrivial solutions oscillate or none oscillate. The hypotheses of Theorem 2 assure that $Q_2(x,y)$ is positive and that

$\Theta'(t)$ vanishes at most once. Since $\Theta'(t)$ vanishes at double zeros of $y(t)$, it follows that Schneider's theorem applies. Since by Theorem V of [1] some solutions of (4.1), (4.2) satisfy $\lim_{t \to \infty} \Theta(t) = \infty$, it now follows that all solutions satisfy $\lim_{t \to \infty} |\Theta(t)| = \infty$.

In a forthcoming paper [6] much of the separation theory of Part II of [2] is extended to more general systems of the form (4) and (2), and this naturally raises the question of how one generalizes Theorem 2 to such systems. The underlying problem is the lack of comparison theorems for the rate of rotation $\Theta(t)$ of such systems when the torque, given by $Q_2(x,y)$, is varied. Another question of interest is how one can establish (5) making use of both the radial forces and the torque simultaneously.

References

1. W. M. Whyburn, On self-adjoint ordinary differential equations of the fourth order, Amer. J. Math. 52(1930), 171-196.

2. W. Leighton and Z. Nehari, On the oscillation of solutions of self-adjoint linear differential equations of the fourth order, Trans. Amer. Math. Soc. 89(1958), 325-377.

3. K. Kreith, A nonselfadjoint dynamical system, Proc. Edinburgh Math. Soc., to appear.

4. C. T. Taam, Oscillation theorems, Amer. J. Math. 74(1952), 317-324.

5. L. J. Schneider, Oscillation properties of the 2-2 disconjugate fourth order selfadjoint differential equation, Proc. Amer. Math. Soc. 28(1971), 545-550.

6. K. Kreith, Rotation properties of a class of second order differential systems, to appear.

Over-Determined Systems and the Rectilinear Steady Flow of Simple Fluids

J.B. McLeod

1. Introduction

In a recent paper (1) Fosdick and Serrin have studied the possible forms of steady rectilinear flow for a simple fluid. The reader is referred to their paper for a detailed derivation of the governing equations of motion, and we content ourselves here with stating the equations.
They are

$$\mathrm{div}\,(\mu\, Du) = 2a, \tag{1.1}$$

$$Du\ \mathrm{div}\,(\phi\, Du) = Dg, \tag{1.2}$$

in which we are considering rectilinear flow in (say) the third space direction of an appropriate rectangular coordinate frame, with the velocity u a function only of the first two space coordinates x, y. The quantities μ and ϕ, which characterise shear and normal stress effects, are (as the derivation of the equations from the underlying constitutive relations shows) determinate functions of $|Du|^2 = (\partial u/\partial x) + (\partial u/\partial y)^2$, where Du is the gradient of u. The quantity a is a constant of integration, and g is a function (to be determined) which is closely related to the pressure. The equations thus represent three relations (since the second equation is a vector equation) between the two scalar functions u and g, and the system is over-determined.

In 1956 Ericksen (2) showed, essentially under the condition that μ and ϕ are not proportional, that the equations (1.1) and (1.2) are independent, and he did this by exhibiting a solution of (1.1) for which (1.2) was insoluble. This led him to conjecture that only quite special forms of steady rectilinear motion can exist, in particular that the only possible steady rectilinear motions are "either rigid or such that the curves of constant speed (in any plane orthogonal to the flow direction) are circles or straight lines"; and indeed explicit radially symmetric and plane symmetric solutions are relatively easy to obtain as, for example, in (1). Ericksen went on to argue that steady rectilinear flow through cylindrical pipes to which the fluid adheres will be "impossible except perhaps when the cylinder is made up of portions of planes and right circular cylinders".

Ericksen in fact derived the equations (1.1) - (1.2) from Rivlin's constitutive relations for a non-Newtonian fluid; Criminale, Ericksen and Filbey (3) derived them for a fluid of differential type, showing that steady rectilinear flow through a tube of elliptical cross-section is impossible with adherence at the wall; and Truesdell and Noll (4) derived the equations for a simple fluid, giving at the same time a slight variant of Ericksen's conjecture.

Fosdick and Serrin made it their objective to give a precise statement and rigorous proof of (a form of) Ericksen's conjecture, and they establish the following precise result:

Suppose an incompressible simple fluid moves rectilinearly and steadily in a fixed straight tube whose cross-section is a bounded and connected open set. If the fluid adheres to the tube wall (or alternatively if a certain slip condition is satisfied at the tube wall), if the functions μ _and_ ϕ _are analytic and not proportional, and if_ μ _satisfies a certain monotonicity condition, then the cross-section of the tube must be either circular or the annulus between two concentric circles._

Considering the conditions imposed in this result, we note first that the condition that μ and ϕ are not proportional (also imposed by Ericksen and by Truesdell and Noll) is clearly central; for if μ and ϕ are proportional, then (1.1) and (1.2) are immediately seen to be no longer independent and the whole nature of the problem changes. Also, Fosdick and Serrin show by means of specific examples that the result is no longer necessarily true if we allow the cross-section to become unbounded or the tube to move. The monotonicity condition on μ is that, for $K \geq 0$,

$$d(K\mu)/_{dK} > 0 \qquad (1.3)$$

where $K = |Du|$, and, with $'$ denoting differentiation with respect to K^2, we can rewrite (1.1) as

$$\mu\, \Delta u + 2\mu' (u_x{}^2 u_{xx} + 2u_x u_y u_{xy} + u_y{}^2 u_{yy}) = 2a$$

for which the characteristic quadratic form is

$$\mu(\xi_1{}^2 + \xi_2{}^2) + 2\mu' (u_x{}^2\, \xi_1{}^2 + 2u_x u_y\, \xi_1\xi_2 + u_y{}^2\xi_2{}^2)$$
$$= (\mu + 2\theta K^2\mu')(\xi_1{}^2 + \xi_2{}^2), \text{ say, where } 0 \leq \theta \leq 1,$$
$$= \{(1-\theta)\mu + \theta d(K\mu)/_{dK}\}\, (\xi_1{}^2 + \xi_2{}^2)$$
$$> 0 \qquad \text{by use of (1.3).}$$

Thus the condition (1.3), which is not unreasonable physically, and which is also used by Ericksen in (2), has the mathematical effect of rendering equation (1.1) elliptic. Since the assumption that an equation is elliptic is a very fundamental one for the behaviour of solutions, it seems reasonable to expect to retain this assumption and we shall in fact retain it throughout this paper.

The remaining condition is that μ and ϕ be analytic. (Fosdick and Serrin can ease this condition, but their argument becomes much more complicated, and they do not carry out the details.) In their proof they use the fact that the solutions of an elliptic equation with analytic coefficients are themselves analytic to expand u as a convergent power series, and it is by operations on this power series that they achieve their result. But essentially what they are using the power series to do is to establish a sort of uniqueness, by showing that the general solution of (1.1) and (1.2) contains only a finite number of parameters (in effect the early coefficients in the power series) and then showing that these parameters can all be accounted for by solutions which are either radially symmetric or plane symmetric. General association with differential equations suggests that a uniqueness result of this nature should not depend critically on whether or not the coefficients in the equations are analytic, and it is the principal aim of this paper to discuss another method of attack which makes no appeal to analyticity.

The Fosdick-Serrin result stated above is an almost immediate consequence of the following theorem which they prove.

Theorem 1. Let $u = u(x,y)$ be a solution of (1.1) and (1.2) in a domain Ω, where a is a non-zero constant and the coefficients μ and ϕ are analytic functions of K^2 with μ satisfying (1.3). If there is a point in Ω where $Du = 0$, and if μ and ϕ are not proportional, then u must be either radially symmetric or plane symmetric, i.e. in an appropriate rectangular coordinate frame we must have either

$$u = u(r) \quad \text{or} \quad u = u(x)$$

where

$$r^2 = x^2 + y^2.$$

(The condition $a \neq 0$ does not figure in the result stated earlier, but if $a = 0$ and Ω is bounded and connected, then the maximum principle assures us that the only possible solution to (1.1) and the boundary condition $u = 0$ on $\partial\Omega$ is $u = 0$. Similarly, the condition that $Du = 0$ somewhere in Ω is trivially satisfied when we have $u = 0$ on $\partial\Omega$.)

Our object is to obtain the following theorem.

Theorem 2. Let $u = u(x,y)$ be a solution of (1.1) and (1.2) in a domain Ω, where the coefficients μ and ϕ are three times differentiable as functions of K^2, the third derivatives satisfying a Lipschitz condition, and where μ satisfies (1.3). We suppose also that the values of K^2 for which $\mu\phi' - \mu'\phi \neq 0$ are dense. Then one or other of two mutually exclusive possibilities must arise. The first is that the solution is identically of the form

$$u(x,y) = Ax + By + C \tag{1.4}$$

where A, B, C are constants with either $A=B=0$ or A^2+B^2 equal to one of the roots of $\mu\phi' - \mu'\phi = 0$, and this possibility arises if and only if

$a = 0$, <u>since trivially a solution of the form (1.4) satisfies (1.1) if and only if</u> $a = 0$.

<u>The second possibility is that there exists a point</u> (x_o, y_o) <u>such that at</u> (x_o, y_o) <u>neither</u> $(u_x, u_y) = (0,0)$ <u>nor is</u> $u_x^2 + u_y^2$ <u>a root of</u> $\mu\phi' - \mu'\phi = 0$, <u>and</u> u <u>is then determined by at most 6 arbitrary constants, these being specifically the values of, say,</u> u, u_x u_y, u_{xx}, u_{xy} <u>and one of the third derivatives of</u> u <u>at</u> (x_o, y_o).

It is clear that the two possibilities in Theorem 2 are mutually exclusive. It is not immediately clear that one or other must arise, since there would seem no reason why there should not exist a solution for which $u_x^2 + u_y^2$ is constant at a root of $\mu\phi' - \mu'\phi = 0$ without u_x and u_y being separately constant; that this cannot happen is part of the theorem.

Comparison of Theorem 2 with Theorem 1 is not entirely straightforward. There is no longer any restriction on a, nor on the vanishing of Du, although it is clear that the case $a = 0$ and points for which $Du = 0$ are in some way exceptional; and the assumption of analyticity has been replaced by a differentiability assumption which is probably the least that allows any progress analytically with the problem. The non-proportionality of μ and ϕ is now expressed slightly differently, but the two conditions are of course equivalent for analytic functions.

On the degree of generality possible in the solution Theorem 2 says that, if we exclude the linear solutions and the solutions for which $Du = 0$ at some base point (x_o, y_o), then the remaining solutions have at most 6 degrees of freedom , and we can actually specify a set of constants which will generate all those remaining solutions. Intuitively, we can explain some of these degrees of freedom in terms of invariants of the equations. Thus addition of an arbitrary constant to u does not alter the equations, nor does translation and rotation of the coordinate axes. This in effect explains 4 of the 6 degrees of freedom, and there are circumstances inwhich we can explain more or even all of them in terms of such invariants. For example, if $a = 0$, and if $u = u_o(x,y)$ provides a solution of (1.1) and (1.2), then so also does $u^* = A^{-1}u_o(Ax,Ay)$. Alternatively, for any a, if μ, ϕ are powers of $|Du|$, with, say, $\mu = |Du|^{2p}$, and if $u_o(x,y)$ provides a solution of (1.1) and (1.2), then so also does $u^* = A^{-(2p+2)}u_o(A^{2p+1}x, A^{2p+1}y)$. If we combine both cases, so that $a = 0$ and μ,ϕ are powers of $|Du|$, then $u^* = Bu_o(Ax,Ay)$ provides a solution for arbitrary constants A,B, so that in this case we have found the full 6 invariants. It should however be remarked that, because of the essential non-linearity of the problem, even if we produce a solution with 6 arbitrary constants, and even if we are prepared to exclude the linear solutions and the solutions for which $Du = 0$ at some base point, we may still not have the general solution, since it may well happen that there is not a one-to-one correspondence between the 6 constants we have and the 6 that are specified in Theorem 2, and this point can be specifically exemplified.

In the case of Theorem 1, under the conditions which prevail there, the solution must be either radially symmetric or plane symmetric. If it is radially symmetric, then we have 2 degrees of freedom in the choice of the centre of symmetry, and as is shown in (1), the solution is then reduced to the solution of a second-order ordinary differential equation, the integration of which introduces 2 further arbitrary constants. There are thus 4 degrees of freedom in all. A similar argument with plane symmetric solutions shows that in these there are 3 degrees of freedom, one lying in the choice of the variable of which the solution is a function, and 2 again being constants of integration. In all, therefore, the system has 4 degrees of freedom, compared with the 6 which are possible (although not of course in all cases necessarily attained) according to Theorem 2.

We prove Theorem 2 in the next section, and then close with some brief remarks on other approaches to the problem.

2. Proof of Theorem 2

Since μ satisfies (1.3), equation (1.1) is, as we have already remarked, elliptic. Since μ is $C^{3+\alpha}$ for any $\alpha<1$, we know from standard results for elliptic equations that the solution u must be $C^{4+\alpha}$, and we are therefore justified in differentiating (1.1) twice.

So far as (1.2) is concerned, we can eliminate g by taking the curl of both sides, obtaining

$$\frac{\partial}{\partial x}\left\{\frac{\partial u}{\partial y} \operatorname{div} (\phi\ Du)\right\} - \frac{\partial}{\partial y}\left\{\frac{\partial u}{\partial x} \operatorname{div} (\phi\ Du)\right\} = 0, \qquad (2.1)$$

and this we are justified in differentiating once. If therefore we denote (1.1) by $F_1 = 0$ and (2.1) by $F_2 = 0$, we can form the equations

$$\frac{\partial F_1}{\partial x} = 0, \qquad \frac{\partial F_1}{\partial y} = 0, \qquad (2.2)$$

$$\frac{\partial^2 F_1}{\partial x^2} = 0, \quad \frac{\partial^2 F_1}{\partial x \partial y} = 0, \quad \frac{\partial^2 F_1}{\partial y^2} = 0, \quad \frac{\partial F_2}{\partial x} = 0, \quad \frac{\partial F_2}{\partial y} = 0, \qquad (2.3)$$

and the five equations (2.3) provide us with equations for the five fourth-order derivatives $\partial^4 u/\partial x^i \partial y^{4-i}$ $(i = 0,1,2,3,4)$ in terms of the derivatives of lower order. Further, the equations (2.3) are linear in the fourth-order derivatives, and so the equations are uniquely soluble provided the usual coefficient determinant does not vanish, and we now therefore have to investigate this determinant, Δ say, proving in fact that

$$\Delta = 4\,(\mu'\phi - \mu\phi')^2\,(u_x{}^2 + u_y{}^2)^3\,\{\mu + 2\mu'\,(u_x{}^2 + u_y{}^2)\}. \qquad (2.4)$$

The proof of (2.4) is a straightforward but cumbersome piece of algebra which we omit.

Assuming now that (2.4) holds, we observe first that the factor $\mu + 2\mu'\,(u_x{}^2 + u_y{}^2)$ cannot vanish because of (1.3). Two cases therefore arise : Case I, in which $\Delta = 0$ everywhere in Ω; and Case II, in which there exists some (x_0, y_0) for which $\Delta \neq 0$, and where therefore $u_x^2 + u_y^2$ is neither zero nor a root of $\mu'\phi - \mu\phi' = 0$. We now consider these cases in turn.

Case I. One possibility here is that $u_x = u_y = 0$ everywhere in Ω; this is one of the alternatives envisaged in the theorem, and nothing more then needs to be said.

If this is not the case, then there must exist some (x_1, y_1) for which $u_x^2 + u_y^2 \neq 0$ and so (necessarily) (x_1, y_1) is a root of $\mu'\phi - \mu\phi' = 0$. In view of the density of the non-zeros of $\mu'\phi - \mu\phi'$, $u_x^2 + u_y^2$ must remain constant at its value at (x_1, y_1). What we have to show is that if $u_x^2 + u_y^2$ is constant, then necessarily u_x and u_y are separately constant. We thus have

$$u_x^2 + u_y^2 = k^2, \tag{2.5}$$

say, and from (1.1), since μ is now constant,

$$u_{xx} + u_{yy} = c, \qquad c = 2a/\mu. \tag{2.6}$$

Differentiating (2.5) with respect to x and y, we obtain

$$u_x u_{xx} + u_y u_{xy} = 0, \tag{2.7}$$

$$u_x u_{xy} + u_y u_{yy} = 0, \tag{2.8}$$

and (2.6), (2.7) and (2.8) are now 3 linear equations for u_{xx}, u_{xy}, u_{yy}, the determinant of the coefficients being readily verified to be $u_x^2 + u_y^2 \neq 0$, since we have already dealt with the case $u_x = u_y = 0$. We can therefore solve the equations, obtaining explicitly

$$u_{xx} = cu_y^2/(u_x^2 + u_y^2), \quad u_{xy} = -cu_x u_y/(u_x^2 + u_y^2),$$
$$u_{yy} = c\,u_x^2/(u_x^2 + u_y^2). \tag{2.9}$$

We may assume $c \neq 0$, since $c = 0$ leads at once to $u_{xx} = u_{xy} = u_{yy} = 0$ and so to u_x and u_y separately constant. Since we must have $\partial u_{xx}/\partial_y = \partial u_{xy}/\partial x$, we can carry out the requisite differentiations, remembering that $u_x^2 + u_y^2$ is constant, to obtain

$$2u_y u_{yy} = -(u_y u_{xx} + u_x u_{xy}),$$

which from (2.9) leads to

$$2u_y u_x^2 = -u_y^3 + u_x^2 u_y,$$

or

$$u_y\,(u_x^{\,2} + u_y^{\,2}) = 0.$$

Since $u_x = u_y = 0$ is excluded, we deduce that $u_y = 0$, and this implies from (2.9) that $u_{xx} = u_{xy} = 0$, so that u_x is constant, as required.

<u>Case II</u>. In this case, at (x_o,y_o) we have

$$u_x^{\,2} + u_y^{\,2} \neq 0, \qquad \mu'\phi - \mu\phi' \neq 0,$$

and so from continuity these inequalities persist in some neighbourhood of (x_o,y_o). Until further notice, we shall restrict ourselves to such a neighbourhood.

Since $\Delta \neq 0$, we can solve the equations (2.3) to give the five fourth-order derivatives in terms of the derivatives of lower order. If for convenience of notation we set

$$u_o = u,\ u_1 = u_x,\ u_2 = u_y,\ u_3 = u_{xx},\ u_4 = u_{xy},\ u_5 = u_{yy},\ u_6 = u_{xxx},$$

$$u_7 = u_{xxy},\ u_8 = u_{xyy},\ u_9 = u_{yyy},$$

then in particular we have, say,

$$u_{xxxx} = f\,(u_o, \ldots, u_9),\quad u_{xxxy} = g\,(u_o, \ldots, u_9),$$

$$u_{xxyy} = h\,(u_o, \ldots, u_9),\quad u_{xyyy} = k\,(u_o, \ldots, u_9).$$

This means that $u_o, \ldots, u_9$, regarded as functions of x, with y merely a parameter, satisfy the 10 simultaneous first-order ordinary differential equations given by

$$u_o' = u_1,\ u_1' = u_3,\ u_2' = u_4,\ u_3' = u_6,\ u_4' = u_7,\ u_5' = u_8,\ u_6' = f,$$

$$u_7' = g,\ u_8' = h,\ u_9' = k,$$

and the conditions of Theorem 2 are sufficient to ensure that the right-hand sides are Lipschitzian in $u_o, \ldots, u_9$. Consequently, u is uniquely determined as a function of x, with $y = y_o$, if we are given the values of $u_o, \ldots, u_9$ at (x_o,y_o).

In fact, it is sufficient to give the values of $u_o,\ldots,u_4$ and one of $u_6,\ldots, u_9$. It is unnecessary to give u_5, i.e. u_{yy}, because it is given in terms of u_x, u_y, u_{xx}, u_{xy} by (1.1), which we can rewrite in the form

$$(\mu + 2\mu' u_x^2) u_{xx} + 4\mu' u_x u_y u_{xy} + (\mu + 2\mu' u_y^2) u_{yy} = 2a. \quad (2.10)$$

And we can also show that the equation (2.1) and the two equations (2.2) can be solved to give some three of $u_6, \ldots, u_9$ in terms of the fourth and $u_o, \ldots, u_5$. For since these three equations are linear in $u_6, \ldots, u_9$, i.e. in u_{xxx}, u_{xxy}, u_{xyy}, u_{yyy}, the solvability for any three depends upon the non-vanishing of the corresponding coefficient determinant. If we rewrite (2.1) in the form

$$u_y \{(\phi + 2\phi' u_x^2) u_{xxx} + 4\phi' u_x u_y u_{xxy} + (\phi + 2\phi' u_y^2) u_{xyy}\} -$$
$$- u_x \{(\phi + 2\phi' u_x^2) u_{xxy} + 4\phi' u_x u_y u_{xyy} + (\phi + 2\phi' u_y^2) u_{yyy} = R, \quad (2.11)$$

where none of the terms in R contains derivatives of u to order higher than the second, then it is easy to see that (2.11) and (2.2) will be solvable for some trio of u_{xxx}, u_{xxy}, u_{xyy}, u_{yyy}, provided that the matrix M given by

$$\begin{matrix} \mu + 2\mu' \mu_x^2 & 4\mu' \mu_x u_y & \mu + 2\mu' u_y^2 & 0 \\ 0 & \mu + 2\mu' u_x^2 & 4\mu' u_x u_y & \mu + 2\mu' u_y^2 \\ u_y (\phi + 2\phi' u_x^2) & 4\phi' u_x u_y^2 - & u_y (\phi + 2\phi' u_y^2) - & -u_x (\phi + 2\phi' u_y^2) \\ & - u_x (\phi + 2\phi' u_x^2) & - 4\phi' u_x^2 u_y & \end{matrix}$$

has rank 3, and this again is a straightforward piece of algebra.

Given the requisite 6 arbitrary constants, therefore, $u(x,y_o)$ is determined, along with all its first, second, and third order derivatives with respect to both x and y at (x, y_o), for x sufficiently close to x_o. Further, at any such point (x^*, y_o), we can set up 10 simultaneous first-order ordinary differential equations, regarding $u_o, \ldots, u_9$ now as functions of y, with $x = x^*$ as a parameter, and solve, with the initial conditions now known at (x^*, y_o), to give u and all its first, second and third order derivatives at (x^*, y), for y sufficiently close to y_o. The end result is that u is determined locally near (x_o, y_o), and in view of the unique continuation property for solutions of an elliptic equation (as, for example, in (5)), it is determined globally. This completes the proof of the theorem.

3. A case soluble in elementary terms: $\mu = 1$, $a = 0$

If $\mu = 1$ and $a = 0$, (1.1) and (1.2) can be solved explicitly, the solution involving nothing worse than integrals of elementary functions. Only certain functions ϕ are possible, and for each such ϕ the general solution u can be obtained. The solution depends upon consideration of a number of cases, and we will not attempt to go through these here, but perhaps the most useful aspect is that it provides a ready source of examples or counter-examples. Although Fosdick and Serrin do not claim in (1) to have carried through this solution, it is a reasonable assumption that they must have in order to produce the examples given there.

4. The hodograph transformation

The hodograph transformation is to take $u_x = p$ and $u_y = q$ as independent variables, and U as the dependent variable, where

$$U = px + qy - u.$$

This has the obvious attraction that μ, ϕ become functions of the independent variables, indeed, of only one of the independent variables if we make the further transformation to polar coordinates

$$\rho^2 = p^2 + q^2, \qquad \theta = \tan^{-1} (q/p).$$

(The existence of this transformation demands, of course, that at least locally p and q must be independent.) While work on this approach is not yet complete, it is certainly possible to make progress towards the ultimate aim of unifying Theorems 1, 2 and so providing a complete discussion of the solutions of the over-determined system.

References

1. R.L. Fosdick and J. Serrin, "Rectilinear steady flow of simple fluids", Proc. Royal Soc. London A 332 (1973), 311-333.

2. J.L. Ericksen, "Over-determination of the speed in rectilinear motion of non-Newtonian fluids", Quart. Appl. Math. 14 (1956), 318-321.

3. W.O. Criminale, Jr., J.L. Ericksen and G.L. Filbey, Jr., "Steady shear flow of non-Newtonian fluids", Arch. Rat. Mech. Anal. 1 (1958), 410-417.

4. C. Truesdell and W. Noll, "The non-linear field theories of mechanics", Flügge's Handbuch der Physik III/3 (Berlin-Heidelberg-New York, Springer, 1965).

5. N. Aronszajn, "A unique continuation theorem for solutions of elliptic partial differential equations or inequalities of second order", J.de Math. Pures et Appl. 36 (1957), 235-249.

A NECESSARY AND SUFFICIENT LIMIT-CIRCLE CRITERION FOR LEFT-DEFINITE EIGENVALUE PROBLEMS

H.- D. NIESSEN

ABSTRACT: For left - definite singular eigenvalue problems arrising from canonical systems of differential equations a limit-circle criterion is deduced which is at the same time necessary and sufficient. This criterion is applied to differential equations of the form $M(\eta)= \lambda N(\eta)$ with M and N differential operators. In case of the Sturm-Liouville-equation the criterion also implies a necessary and sufficient limit-point criterion.

1. BASIC DEFINITIONS AND RESULTS

Let us consider the differential system

$$(1.1) \qquad y' + \begin{pmatrix} -D_{12}^* & -D_{22} \\ D_{11} & D_{12} \end{pmatrix} y = \lambda \begin{pmatrix} 0 & 0 \\ E_{11} & 0 \end{pmatrix} y \qquad (\lambda \in \mathbb{C})$$

on an interval I with endpoints a and b $(-\infty \leq a < b \leq \infty)$. The coefficients D_{ij}, E_{11} shall be (k,k)- matrix-valued continuous functions on I such that $D_{11}(x)$, $D_{22}(x)$ and $E_{11}(x)$ are hermitian and

$$(1.2) \qquad D_{11}(x) \geq 0 \geq D_{22}(x) \qquad (x \in I).$$

Furthermore, with a positive constant ρ the inequality

$$-\rho\, D_{11}(x) \leq E_{11}(x) \leq \rho\, D_{11}(x) \qquad (x \in I)$$

is assumed to hold. Finally, for any nonzero solution y of (1.1) to $\lambda=0$ the "Dirichlet-integral" shall be positive, i.e.,

$$0 < \int_I y^* \begin{pmatrix} D_{11} & 0 \\ 0 & -D_{22} \end{pmatrix} y \ (\leq \infty).$$

With these assumptions the canonical system (1.1) is said to be "left-definite". Such left-definite differential systems, the corresponding boundary-value problems arrising from suitable boundary conditions, and the corresponding expansion theorems have been treated in [2]. A few definitions and results of [2] will be needed:

For any subinterval J of I denote by $E_0(J)$ the set of all locally absolutely continuous solutions y of (1.1) to $\lambda=0$, for which the integral $\int_J y^* \begin{pmatrix} D_{11} & 0 \\ 0 & -D_{22} \end{pmatrix} y$

is finite. Fix an arbitrary point $x_o \in I$ and define

$$I_a: = I \cap [a,x_o],\ I_b: = I \cap [x_o,b],$$

$$r_a: = \dim E_o(I_a), r_b: = \dim E_o(I_b), r: = \dim E_o(I).$$

Then r is the number of boundary conditions which have to be imposed ([2],5.) and

(1.3) $k \le r_a,\ r_b \le 2k$ ([2], 3.),

(1.4) $r = r_a + r_b - 2k$ ([2],(5.8)).

In view of (1.4), r may be calculated, if r_a and r_b are known. Therefore, the following considerations will be restricted to I_b (Similar results hold for I_a).

(1.5) Definition: (1.1) is said to be in the limit-circle case at b iff $r_b = 2k$.

This means that all solutions of (1.1) to $\lambda=0$ lie in $E_o(I_b)$.

Let Y be a fundamental matrix of (1.1) to $\lambda=0$ with initial condition $Y(x_o) = E_{2k}$ and define

$$D: = - Y^* \begin{pmatrix} 0 & E_k \\ E_k & 0 \end{pmatrix} Y.$$

Then [2], chapter 3 implies

(1.6) Theorem: (1.1) is limit-circle iff D is bounded in I_b.

Let H denote the matrix $\begin{pmatrix} 0 & E_k \\ -E_k & 0 \end{pmatrix}$. Then ([2],(1.6))

(1.7) $Y^* H Y = H.$

2. THE LIMIT-CIRCLE CRITERION

Let the fundamental matrix Y (with $Y(x_o)= E_{2k}$) be decomposed into (k,k)- matrices Y_j:

$$Y=\begin{pmatrix} Y_1 & Y_3 \\ Y_2 & Y_4 \end{pmatrix}.$$

Then (1.7) is equivalent to

(2.1) $Y_1^* Y_2= Y_2^* Y_1, Y_3^* Y_4= Y_4^* Y_3, Y_1^* Y_4= E_k + Y_2^* Y_3.$

Furthermore, (1.1) and (1.2) imply

(2.2) $(Y_1^* Y_2)' = Y_2^* D_{22} Y_2 - Y_1^* D_{11} Y_1 \le 0.$

<u>(2.3) Lemma</u>: $Y_1(x)$ is regular for $x \in I$.

Proof: Assume, e.g., that $Y_1(x)$ is singular for some $x > x_o$ and define $x_1 := \min\{x > x_o / Y_1(x) \text{ singular}\} (> x_o)$. Let $Y_1(x_1)c=0$ with $c \neq 0$. Then $c^* Y_1^*(x_1) Y_2(x_1) c = 0$. Since $Y_2(x_o)=0$ also $Y_1^*(x_o)Y_2(x_o)=0$ and (2.2) implies that $Y_1^* Y_2$ is hermitian and monotonously decreasing. Therefore $c^* Y_1^*(x) Y_2(x) c = 0$ for all $x \in [x_o, x_1)$, which implies $Y_1^*(x) Y_2(x) c = 0$ since $Y_1^* Y_2$ is hermitian. Now $Y_1(x)$ is regular for $x \in [x_o, x_1)$, and therefore $Y_2(x)c=0$. Since Y_2 is continuous, we get $Y_2(x_1)c=0$. Multiplying the third equation of (2.1) for $x=x_1$ by c^* from the left gives the contradiction $c=0$.

Now it is possible to define

$$\text{(2.4)} \qquad Z := \begin{pmatrix} Y_1^{-1} & 0 \\ 0 & Y_1^* \end{pmatrix} Y, \begin{pmatrix} Z_1 & Z_3 \\ Z_2 & Z_4 \end{pmatrix} := Z.$$

Then (1.7) and (2.1) are valid for Z and Z_i instead of Y and Y_i, resp.. Furthermore, D is invariant with respect to the transformation (2.4). Using (2.1) we get

$$\text{(2.5)} \qquad D = -Z^* \begin{pmatrix} 0 & E_k \\ E_k & 0 \end{pmatrix} Z = -\begin{pmatrix} 2\, Z_1^* Z_2 & 2\, Z_2^* Z_3 + E_k \\ 2\, Z_3^* Z_2 + E_k & 2\, Z_3^* Z_4 \end{pmatrix}.$$

Now it is possible to prove

<u>(2.6) Lemma</u>: Z is bounded on I_b if (1.1) is limit-circle at b.

Proof: (1.6) and (2.5) imply that $Z_1^* Z_2$, $Z_2^* Z_3$ and $Z_3^* Z_4$ are bounded. Since $Z_1 = E_{2k}$, it follows that Z_1, Z_2 and $Z_4 = E_k + Z_2^* Z_3$ are bounded. To prove the boundedness of Z_3, remark that $Z_2 = Y_1^* Y_2$ is hermitian and negative-semidefinite in I_b. Furthermore it may be shown by the aid of (1.1) and (2.1) that $Z_3' = Y_1^{-1} D_{22} Y_1^{*-1} \le 0$. Especially Z_3 is hermitian and negative-semidefinite in I_b. Multiplying the equation $Z_4 = E_k + Z_2^* Z_3$ from the left by Z_3 we get $Z_3 Z_4 = Z_3 + Z_3 Z_2^* Z_3 \le Z_3 \le 0$. Since $Z_3 Z_4$ is bounded, this implies the boundedness of Z_3.

(2.7) Corollary: If (1.1) is limit-circle at b, then $Y_1^*D_{11}Y_1$, $Y_1^{-1}D_{22}\,Y_1^{*-1}$, $Y_1^{-1}Y_1' - Y_1^{-1}D_{12}^*Y_1$ are integrable over I_b.

Proof: By lemma 2.6 Z is bounded on I_b. Therefore Z' is integrable over I_b and by (1.7) $Z^{-1}=H^{-1}Z^*H$ is bounded on I_b. This implies by (1.1) and (2.4) that

$$\begin{pmatrix} -(Y_1^{-1}Y_1' - Y_1^{-1}D_{12}^*Y_1) & Y_1^{-1}D_{22}Y_1^{*-1} \\ -Y_1^*D_{11}Y_1 & (Y_1^{-1}Y_1' - Y_1^{-1}D_{12}^*Y_1)^* \end{pmatrix} = Z'Z^{-1}$$

is integrable over I_b.

Specialisation of a known theorem (e.g.[1],p.88) on asymptotic behavior of solutions of differential systems gives the following

(2.8) Lemma: Let C be a continuous matrix-valued function which is integrable over I_b. Then there exists a fundamental matrix W of $W'= CW$ such that $W(x)\longrightarrow E(x\to b)$.

Denote by U a fundamental matrix of $U'= D_{12}^*\,U$. Then we can prove the

(2.9) Limit-circle criterion: (1.1) is limit-circle at b iff

$U^*D_{11}U$ and $U^{-1}D_{22}U^{*-1}$ are integrable over I_b.

Proof: 1) Let (1.1) be limit-circle at b. Then $C:=(Y_1^{-1}Y_1' - Y_1^{-1}D_{12}^*Y_1)^*$ is integrable over I_b by Corollary 2.7. According to lemma 2.8 there exists a fundamental matrix W of $W'=CW$ converging to E_k for $x\to b$. Define $K: = U^{-1}Y_1W^{*-1}$. Then a simple calculation using the differential equations for U and W shows that K is constant - and obviously regular. Therefore, Corollary 2.7 implies that $WK^*U^*D_{11}UKW^*$ and $W^{*-1}K^{-1}U^{-1}D_{22}U^{*-1}K^{*-1}W^{-1}$ are integrable over I_b. Since W converges to E_k and K is regular, it follows that $U^*D_{11}U$ and $U^{-1}D_{22}U^{*-1}$ are integrable.

2) If $U^*D_{11}U$ and $U^{-1}D_{22}U^{*-1}$ are integrable, there exists a fundamental matrix W of

$$W'=\begin{pmatrix} 0 & U^{-1}D_{22}U^{*-1} \\ -U^*D_{11}U & 0 \end{pmatrix}W$$

converging to E_{2k} $(x\to b)$. Then

$$Y:=\begin{pmatrix} U & 0 \\ 0 & U^{*-1} \end{pmatrix}W \tag{2.10}$$

is a fundamental matrix of (1.1) to $\lambda=0$. Since D is invariant with respect to the transformation (2.10),

$$D = - W^{*}\begin{pmatrix} 0 & E_k \\ E_k & 0 \end{pmatrix} W$$

is bounded. In view of theorem 1.6 (1.1) is limit-circle.

<u>(2.11) Corollary</u>: Let D_{12} be integrable over I_b. Then (1.1) is limit-circle iff D_{11} and D_{22} are integrable over I_b.

Proof: By lemma 2.8 in this case U itself may be choosen to converge to E_k for $x \to b$. Therefore, the conclusion follows from (2.9).

3. APPLICATION TO DIFFERENTIAL EQUATIONS

The Sturm-Liouville equation

(3.1) $-(p\eta')' + q\eta = \lambda k\eta$

may be transformed to the canonical system

(3.2) $y' + \begin{pmatrix} 0 & \frac{1}{p} \\ q & 0 \end{pmatrix} y = \lambda \begin{pmatrix} 0 & 0 \\ k & 0 \end{pmatrix} y$

by the transformation $y = \begin{pmatrix} \eta \\ -p\eta' \end{pmatrix}$. This is left-definite if p,q and k are continuous, if $p > 0$, $q \geq 0$ and not identically zero and if there exists a positive constant ρ such that $|k| \leq \rho q$. Since $D_{12}=0$, Corollary 2.11 gives the following

<u>(3.3) Theorem</u>: (3.1) is limit-circle at b (considered as left-definite eigenvalue problem!) iff $\frac{1}{p}$ and q are integrable over I_b. (3.1) is limit-point at b, i.e., $r_b=1$ iff $\frac{1}{p}$ or q is not integrable over I_b.

The last assertion follows from the first since $r_b \geq 1$ by (1.3). Especially in the left-definite case the equation $-\eta'' + q\eta = \lambda k\eta$ is always limit-point if I_b is unbounded.

Now consider the equation

(3.4) $M(\eta) = \lambda N(\eta)$

with $M(\eta) = \sum_{i=0}^{k} (-1)^i (m_i \eta^{(i)})^{(i)}$, $N(\eta) = \sum_{i=0}^{k-1} (-1)^i (n_i \eta^{(i)})^{(i)}$.

By a suitable transformation (3.4) my be transformed into a canonical system (1.1) with

$$D_{12}= (\delta_{i,j+1}),\ D_{11}= \operatorname{diag}(m_o,m_1 \ldots,m_{k-1}),$$

$$D_{22}= \operatorname{diag}(0,\ldots,0,-m_k^{-1}),\ E_{11}= \operatorname{diag}\ (n_o,n_1,\ldots,n_{k-1}).$$

In this case U may be choosen to be

$$U = \sum_{i=o}^{k-1} \frac{(x-x_o)^i}{i!} \quad (D_{12}^*)^i$$

and (2.9) implies

(3.5) Theorem: Let (3.4) be left-definite. Then (3.4) is limit-circle at b iff

$(x-x_o)^{2k-2}\ m_k^{-1}$ and all $(x-x_o)^{2(k-i-1)} m_i\,(i=o,\ldots,k-1)$

are integrable over I_b.

REFERENCES

[1] COPPEL,W.A., Stability and Asymptotic Behavior of Differential Equations, Boston: Heath and Company 1965

[2] SCHNEIDER,A. and H.D. NIESSEN, Linksdefinite singuläre kanonische Eigenwertprobleme I. To appear in Journ. f.d. Reine und Angew.Math.

GENERALIZED WEYL CIRCLES

Åke Pleijel

SUMMARY. Weyl circles are studied for second order, formally selfadjoint ordinary differential equations $Su = \lambda\, r(x)u$ on a half-open interval, first for equations with $r(x) \geq 0$. It is then indicated how the study carries over to equations of polar type, provided S possesses a non-negative Dirichlet integral. Finally the method is applied to such equations but in an interesting situation recently introduced by Atkinson, Everitt and Ong. The theory is then based upon a not necessarily non-negative form. Hilbert space theories are referred to in the next, but only for motivation, and conditions for them are not deliberated. The paper is the result of stimulating discussions with W.N. Everitt during the author's visit to Dundee, September-November 1973.

1. NON-NEGATIVE LOWER ORDER OPERATOR.

Let

$$S = Da_{11}D + Da_{10} + a_{01}D + a_{00} \tag{1.1}$$

with $D = id/dx$. Here a_{11} and a_{00} shall be real while $\bar{a}_{10} = a_{01}$, all these functions being defined on $[ab) = \{x: a \leq x < b\}$. It is assumed that

$$a_{11}(x) > 0 \quad \text{on} \quad [ab). \tag{1.2}$$

The operator S is considered together with an operator T of the form $Tu = r(x)u$, first under the condition

$$r(x) \geq 0, \quad \text{not identically } 0 \text{ on } [ab). \tag{1.3}$$

The function r as well as the functions in (1.1) must satisfy certain regularity conditions. We shall suppose that they are continuous and that a_{11}, a_{10} have continuous derivatives. These conditions are sufficient but can be relaxed.

If $r(x) > 0$ on $[ab)$ the equality $Su = T\dot{u}$ determines $\dot{u} = r^{-1}Su$ but is otherwise a linear relation which is here defined as the set of ordered pairs

$$E[ab) = \{U = (u,\dot{u}) \in C^{(2)} \times C^{(1)} : Su = T\dot{u} \ \text{ a.e.}\} . \tag{1.4}$$

In this definition $C^{(\ell)}$ or $C^{(\ell)}[ab)$ denotes the set of all complex functions u with continuous derivatives $u, Du, \ldots, D^{\ell-1}u$ of which $D^{\ell-1}u$ is absolutely continuous and has an integrable square derivative $D^{\ell}u$ on compact subintervals of $[ab)$. (In the present case $C^{(1)}$ can be replaced by L^2_{loc} in (1.4).)

If $U = (u,\dot{u})$, $V = (v,\dot{v})$ are pairs in $E[ab)$, partial integrations over any compact subinterval of $[ab)$ lead to the Green's formula

$$i^{-1}\left((\dot{u},v)_T\Big|_{\alpha}^{\beta} - (u,\dot{v})_T\Big|_{\alpha}^{\beta}\right) = [q_x(U,V)]_{\alpha}^{\beta} \tag{1.5}$$

in which

$$(u,v)_T\Big|_{\alpha}^{\beta} = \int_{\alpha}^{\beta} r u \bar{v} . \tag{1.6}$$

The expression q_x is given by

$$q_x(U,V) = (a_{11}Du + a_{10}u)\bar{v} + u(\overline{a_{11}Dv + a_{10}v}) . \tag{1.7}$$

Because of (1.3) the form (1.6) is non-negative, $(u,u)_T\Big|_{\alpha}^{\beta} \geq 0$. The operator T is therefore called non-negative.

The identity

$$q_x(U,U) = \frac{1}{2}\,|a_{11}Du + a_{10}u + u|^2 - \frac{1}{2}\,|a_{11}Du + a_{10}u - u|^2 \tag{1.8}$$

shows that the signature of the hermitean form q_x, i.e. the pair of its positive and negative inertia indices, satisfies the inequality

$$\text{sig } q_x \leq (1,1) \tag{1.9}$$

on every finite dimensional subspace of $E[ab)$. For any complex number λ let the solution space $E_\lambda[ab)$ be defined by

$$E_\lambda[ab) = \{U = (u,\lambda u) \in E[ab)\} . \tag{1.10}$$

The elements of $E_\lambda[ab)$ are related to the solutions of $Su = \lambda Tu$ and $E_\lambda[ab)$ is 2-dimensional because of (1.2). If $U = (u,\lambda u)$ is in $E_\lambda[ab)$, the expressions

$$a_{11}Du + a_{10}u \pm u \tag{1.11}$$

can be given arbitrary values at $x \in [ab)$ by solving for $u(x)$ and $u'(x)$ and integrating $Su = \lambda Tu$ with these initial values. As a consequence the expressions (1.11) are linearly independent on $E_\lambda[ab)$ and

$$\operatorname{sig} q_x = (1,1) \quad \text{on} \quad E_\lambda[ab), \quad x \in [ab) . \tag{1.12}$$

2. <u>GENERALIZED WEYL CIRCLES</u>. Let $U = (u,\dot{u}) \in E[ab)$ and put $U = (u, \lambda u + f)$. From (1.5) one deduces

$$q_{x'}(U,U) - q_x(U,U) = c(u,u)_T\Big|_x^{x'} - i(f,u)_T\Big|_x^{x'} + i(u,f)_T\Big|_x^{x'} , \tag{2.1}$$

where $c = c(\lambda) = i^{-1}(\lambda - \bar{\lambda})$. Since $(\cdot,\cdot)_T\Big|_x^{x'}$ is non-negative, Cauchy-Schwarz's inequality can be applied and yields

$$cq_{x'}(U,U) - cq_x(U,U) \geq (|c|\,\|u\|_T\Big|_x^{x'} - \|f\|_T\Big|_x^{x'})^2 - (f,f)_T\Big|_x^{x'}$$

with obvious meaning of $\|\cdot\|_T\Big|_x^{x'}$. If $V = (v,\lambda v)$ belongs to $E_\lambda[ab)$, the sum $U + V$ can be written in the same way as $U = (u, \lambda u + f)$ and with the same f. It follows that

$$\begin{aligned} &cq_{x'}(U+V,\ U+V) + (f,f)_T\Big|_a^{x'} \geq \\ &\geq cq_x(U+V,\ U+V) + (f,f)_T\Big|_a^{x} + (|c|\,\|u+v\|_T\Big|_x^{x'} - \|f\|_T\Big|_x^{x'})^2 \end{aligned} \tag{2.2}$$

holds true for all V in $E_\lambda[ab)$. The value λ shall be non-real so that $c = c(\lambda) \neq 0$.

Consider a 1-dimensional subspace of $E_\lambda[ab)$ on which $c(\lambda)q_x$ is positive definite for all x near b, i.e. consider a $\Phi = (\varphi, \lambda\varphi)$ in $E_\lambda[ab)$ such that

$$cq_x(\Phi,\Phi) > 0 \quad \text{for} \quad x \quad \text{near} \quad b . \tag{2.3}$$

This is possible, for according to (1.12) one can take $\Phi \neq 0$ such that $cq_a(\Phi,\Phi) \geq 0$. With $U = V = \Phi$ $(\dot{u} = \lambda\varphi)$ and with $\alpha = a$, $\beta = x$ the identity (1.5) reduces to

$$cq_x(\Phi,\Phi) = cq_a(\Phi,\Phi) + c^2 \int_a^x (\varphi,\varphi)_T \tag{2.4}$$

and (2.3) is satisfied for $x > x_1$ if $r(x_1) \neq 0$. If $U = (u, \lambda u + f)$ fulfils the condition

$$\int_a^b (f,f)_T < +\infty\,, \tag{2.5}$$

let $\Sigma_x = \Sigma_x(U,\Phi)$ be the set of all complex numbers ℓ for which

$$\Sigma_x(U,\Phi):\ cq_x(U + \ell\Phi,\ U + \ell\Phi) + \int_a^x (f,f)_T \leq \int_a^b (f,f)_T\ . \tag{2.6}$$

From (2.2) with $V = \ell\Phi$ it follows that $\Sigma_x \supset \Sigma_{x'}$ when $a \leq x < x' < b$. Because of (2.3) the set (2.6) is a circular disc, a point or the empty set when x is near b. Its center ℓ^* is determined by $q_x(U + \ell^*\Phi,\ \Phi) = 0$ so that

$$\ell^* = -q_x(U,\Phi)/q_x(\Phi,\Phi)\ . \tag{2.7}$$

It is asserted that

$$cq_x(U + \ell^*\ \Phi,\ U + \ell^*\Phi) \leq 0\,. \tag{2.8}$$

The opposite inequality would imply that cq_x is positive definite on the linear hull $[U,\Phi]$ which is contradicted by (1.9). Having proved (2.8) we conclude that Σ_x is non-empty.

Because of the compactness of Σ_x when x is near b there is at least one ℓ contained in all Σ_x. According to (2.2) the left hand side of (2.6) tends non-decreasingly to a finite limit for such a value ℓ as x tends to b. The last term of (2.2) tends to 0 when x and x' tend to b and $V = \ell\Phi$. As a consequence the integral

$$\int_a^b (u + \ell\varphi,\ u + \ell\varphi)_T$$

is finite. It follows that $U + \ell\Phi = (u + \ell\varphi,\ \lambda(u + \ell\varphi) + f)$ belongs to

$$E[ab] = \{(u,\dot u) \in E[ab) : \int_a^b (u,u)_T < +\infty, \quad \int_a^b (\dot u,\dot u)_T < +\infty\}, \qquad (2.9)$$

where $]$ indicates the finiteness of the integrals $\int_a^b (.,.)_T$. For any $U = (u,\dot u)$, $V = (v,\dot v)$ in $E[ab]$ the left hand side of (1.5) has a limit when β tends to b. It follows that

$$q_b(U,V) = \lim_{x\to b} q_x(U,V)$$

exists on $E[ab]$. Since $U + \ell\Phi \in E[ab]$ if ℓ is contained in all Σ_x, a transition to the limit in (2.6) shows that

$$c(\lambda)q_b(U + \ell\Phi,\ U + \ell\Phi) \leq 0. \qquad (2.10)$$

On the other hand, if $U + \ell\Phi \in E[ab]$ and (2.10) is fulfilled it follows from (2.2) that ℓ is in all Σ_x.

<u>3. LIMIT-POINT AND LIMIT-CIRCLE.</u> For $U = \Theta = (\theta,\lambda\theta)$ in $E_\lambda[ab)$ the function $f = \dot u - \lambda u$ is 0, and (2.6) becomes

$$\Sigma_x(\Theta,\Phi) : cq_x(\Theta + \ell\Phi,\ \Theta + \ell\Phi) \leq 0. \qquad (3.1)$$

If $E_\lambda[ab) = \{\Theta,\Phi\}$ and $\ell \in \Sigma_x(\Theta,\Phi)$ for all x, the element $\Theta + \ell\Phi$ is $\neq 0$ and belongs to

$$E_\lambda[ab] = E_\lambda[ab) \cap E[ab].$$

Thus $E_\lambda[ab]$ is at least 1-dimensional.

If $\dim E_\lambda[ab] = 1$, take Φ in $E_\lambda[ab)$ but outside $E_\lambda[ab]$. Then (2.3) is fulfilled because of $\int_a^b (\varphi,\varphi)_T = +\infty$ and (2.4). For U in $E[ab]$ the function $f = \dot u - \lambda u$ satisfies (2.5) and there is a $U + \ell\Phi$ in $E[ab]$ satisfying (2.10). Since Φ is outside $E[ab]$, the value of ℓ must be 0 and (2.10) reduces to $c(\lambda)\ q_b(U,U) \leq 0$ for all U in $E[ab]$.

On the other side, if q_b does not take both signs on $E[ab]$, the space $E_\lambda[ab]$ must be 1-dimensional for any λ in the half-plane $c(\lambda) > 0$ or < 0, where $c(\lambda)\ q_b(U,U) \leq 0$. Otherwise last inequality violates the existence in

$E_\lambda[ab)$ of a Φ satisfying (2.3), for $E_\lambda[ab) = E_\lambda[ab]$ if $\dim E_\lambda[ab] = 2$.

It follows that one of the equalities

$$(\dim E_\lambda[ab],\ \dim E_{\bar\lambda}[ab]) = (2,2),\ (2,1),\ (1,2),\ (1,1) \tag{3.2}$$

is valid for all λ in $\mathrm{Im}(\lambda) > 0$. If S is real, only (2,2) and (1,1) are possible. Easy computations show that this is also true for S non-real. The pair S,T is in Weyl's limit-circle case when (3.2) equals (2,2), and is in the limit-point case when the value is (1,1). The limit-point case occurs if and only if $q_b(U,V) = 0$ on $E[ab]$.

It is easy to see that in the limit-point case only one ℓ is contained in all $\Sigma_x(\Theta,\Phi)$, $a \le x < b$. This ℓ, denoted by m, is Weyl's m-coefficient in a wide sense of Θ, Φ. Since $\Theta + m\,\Phi \in E[ab]$, the equality

$$q_b(\Theta + m\Phi,\ \Theta + m\Phi) = 0 \tag{3.3}$$

is valid. In the limit-circle case an m-coefficient is any point $\ell = m$ on the boundary of the limit-circle $c\, q_b(\Theta + \ell\Phi\ ,\ \Theta + \ell\,\Phi) \le 0$. The equality (3.3) still holds true. The proper m-coefficients require specialization of Θ, Φ .

4. ABOUT SYMMETRIC BOUNDARY CONDITIONS.

A Hilbert space theory can be obtained by considering symmetric boundary conditions i.e. linear subspaces of $E[ab]$ on which

$$\overset{b}{\underset{a}{(\dot u, v)_T}} = \overset{b}{\underset{a}{(u, \dot v)_T}}\ , \tag{4.1}$$

see Pleijel [11], [12]. If U,V are in $E[ab]$ the value $q_b(U,V)$ exists as a limit and

$$i^{-1}\Big(\overset{b}{\underset{a}{(\dot u, v)_T}} - \overset{b}{\underset{a}{(u, \dot v)_T}}\Big) = q_b(U,V) - q_a(U,V). \tag{4.2}$$

The symmetry (4.1) requires the vanishing of the hermitean form $Q = q_b - q_a$ or, because of linearity, the vanishing of

$$Q(U,U) = q_b(U,U) - q_a(U,U) \tag{4.3}$$

when U satisfies a symmetric boundary condition.

It is essential to consider subspaces $F \subset E[ab]$ of (maximal) dimension $\dim E_\lambda[ab] + \dim E_{\bar\lambda}[ab]$ on which Q is non-degenerate. The Q-projection U' of $U \in E[ab]$ on F is determined by $Q(U - U', F) = 0$, $U' \in F$. It can be shown that $Q(U - U', E[ab]) = 0$ from which it follows that $Q(U,V) = Q(U',V')$ under Q-projection on F. All spaces F are isomorphic and Q-true to each other under Q-projection. Any symmetric boundary condition is determined by a linear subspace Z of a space F if $Q(Z,Z) = 0$ and $\dim Z = \min(\dim E_\lambda[ab], \dim E_{\bar\lambda}[ab])$. An element U in $E[ab]$ satisfies the symmetric boundary condition if its Q-projection on F belongs to Z. The direct sum $E_\lambda[ab] \dotplus E_{\bar\lambda}[ab]$ is a space F when λ is non-real.

Previous statements remain valid in $E[a\cdot b]$ which is similarly defined as $E[ab]$ but contains pairs $U = (u,\dot u)$ only defined and regular in the union $a\cdot b$ of neighbourhoods $a\cdot$ and $\cdot b$ of a and b. This leads to new F-spaces. For S in (1.1) and $Tu = r(x)u$, $\dim F = 2 \dim Z = 4$ or 2 since this is so for $F = E_\lambda[ab] \dotplus E_{\bar\lambda}[ab]$. In the limit-circle case the 4-dimensional space of all elements

$$U = (u,\lambda u) : Su = \lambda Tu \quad \text{in} \quad a\cdot \quad \text{and in} \quad \cdot b \tag{4.4}$$

is a space of type F.

5. SYMMETRIC BOUNDARY CONDITIONS IN WEYL'S CASE.

If the pair of S,T in section 1 is in the limit-point case, the form Q equals $-q_a$ on $E[ab]$. Due to (1.7) the vanishing of $q_a(U,U)$ requires that

$$(a_{11}Du + a_{10}u)/(iu) \tag{5.1}$$

takes a real value $\cot\alpha$ at $x = a$. The value $\cot\alpha = \infty$ corresponds to $u(a) = 0$. It follows that a symmetric boundary condition has the form

$$(a_{11}Du + a_{10}u)\sin\alpha - iu\cos\alpha = 0, \quad x = a, \tag{5.2}$$

in the limit-point case and in a theory based upon $\overset{b}{\underset{a}{(\cdot,\cdot)}}_T$.

In the limit-circle case let F be the space (4.4) for a non-real λ. Let Φ in $E_\lambda[ab)$ satisfy (5.2) for a certain α and let $\Psi = \Theta + m\Phi$ with m on the boundary of the limit-circle so that $q_b(\Psi,\Psi) = 0$ according to (3.3). The space Z spanned by Φ in $a\cdot$, $Z = \{\Phi\}$ in $a\cdot$, and by Ψ in $\cdot b$, $Z = \{\Psi\}$ in $\cdot b$, is a 2-dimensional nullspace for $Q = q_b - q_a$, and determines a symmetric boundary condition. Considering Q-projections on F it follows that $U \in E[ab]$ satisfies this symmetric boundary condition if U satisfies (5.2) at $x = a$ and if

$$q_b(U, \Theta + m\,\Phi) = 0. \tag{5.3}$$

6. WEYL'S CHOICE. In Weyl's theory it is essential to take a Φ in $E_\lambda[ab)$ fulfilling a prescribed condition (5.2), α given. This together with a (less important) choice of Θ and certain normalizations of Θ and Φ leads to the definition of the proper Weyl m-coefficients.

Because of Green's formula the inequality (3.1) can also be written

$$c\, q_a(\Theta + \ell\Phi\,,\, \Theta + \ell\Phi) + c^2(\theta + \ell\varphi,\, \theta + \ell\varphi)_T{}_a^x \leq 0. \tag{6.1}$$

Due to Weyl's choice of Φ the term $|\ell|^2\cdot c\, q_a(\Phi,\Phi)$ vanishes, and if Θ is taken so as to satisfy (5.2) with a different α, also $c\, q_a(\Theta,\Theta)$ vanishes. For Θ Weyl takes (5.2) with α replaced by $\alpha + \pi/2$. The elements Θ, Φ are further normalized so as to satisfy

$$a_{11}D\varphi + a_{10}\varphi = i\cos\alpha, \tag{6.2}$$

$$\varphi = \sin\alpha, \tag{6.3}$$

$$a_{11}D\theta + a_{10}\theta = -i\sin\alpha, \tag{6.4}$$

$$\theta = \cos\alpha, \tag{6.5}$$

for $x = a$. Then $q_a(\Phi,\Theta) = i$ and (6.1) reduces to Weyl's well-known inequality

$$(\theta + \ell\varphi,\, \theta + \ell\varphi)_T{}_a^x \leq \operatorname{Im}(\ell)/\operatorname{Im}(\lambda) \tag{6.6}$$

for Σ_x. The radius $\rho(x)$ of Σ_x equals $|\ell - \ell^*|$, where ℓ^* is the center (2.7) and ℓ lies on the boundary of (6.6) or (3.1) so that $q_x(U,U) = 0$ for $U = \Theta + \ell\Phi$. This is true if (5.1) is real i.e. if $u = \theta + \ell\varphi$ satisfies a condition (5.2) at x. Taking $\alpha = 0$ in this condition one obtains $\ell = -\theta(x)/\varphi(x)$ and computations as in [14] give

$$(\rho(x))^{-1} = 2|\mathrm{Im}(\lambda)| \int_a^x (\varphi,\varphi)_T . \tag{6.7}$$

Replacing Φ,Θ by Weyl's choice (3.1) is transformed into (6.6). Due to (6.7) the set (6.6), then also (3.1), is a finite disc in the limit circle case.

According to section 5 a prescribed condition (5.2) can be completed to a symmetric boundary condition in the limit-circle case by the choice of $\ell = m(\lambda,\alpha)$ on $\int_a^b (\Theta + \ell\varphi,\ \Theta + \ell\varphi)_T = \mathrm{Im}(\ell)/\mathrm{Im}(\lambda)$. To coordinate points $m(\lambda,\alpha)$ belonging to different values λ but to the same boundary condition, one takes $m(\lambda,\alpha)$ as the limit when x tends to b of points $\ell(x,\lambda,\alpha)$ on the boundaries of (6.6). This leads, also in the limit-point case, to a function $m(\lambda,\alpha)$ which is analytic in $\mathrm{Im}(\lambda) > 0$ and in $\mathrm{Im}(\lambda) < 0$. The analyticity is due to the boundedness of the circles (3.1) when λ is bounded and bounded away from $\mathrm{Im}(\lambda) = 0$. A full discussion of the limit circle case is only recently given by Fulton in an as yet unpublished paper based on his thesis [6].

Weyl's m-coefficient $m(\lambda,\alpha)$ determines the weight measure in Fourier integral formulas corresponding to the spectral expansion in a Hilbert space theory when $r(x) > 0$, see [2].

7. NON-NEGATIVE HIGHER ORDER OPERATOR.

A second Green's formula

$$i^{-1}\left(\int_\alpha^\beta (\dot u,v)_S - \int_\alpha^\beta (u,\dot v)_S\right) = \left[q_x(U,V)\right]_\alpha^\beta \tag{7.1}$$

is valid when $U = (u,\dot u)$ and $V = (v,\dot v)$ belong to $E[ab)$. In (7.1)

$$\int_\alpha^\beta (u,v)_S = \int_\alpha^\beta a_{11}\, Du\, \overline{Dv} + a_{10} u\, \overline{Dv} + a_{01}\, Du\, \bar v + a_{00} u \bar v \tag{7.2}$$

and q_x is the new hermitean form

$$q_x(U,V) = (a_{11}Du + a_{10}u)\bar{\dot{v}} + \dot{u}(\overline{a_{11}Dv} + \overline{a_{10}v}). \qquad (7.3)$$

The integrand of (7.2) can be written

$$a_{11}^{-1}((a_{11}Du + a_{10}u)\,(\overline{a_{11}Dv} + \overline{a_{10}v}) + \begin{vmatrix} a_{00} & a_{01} \\ a_{10} & a_{11} \end{vmatrix} u\bar{v}). \qquad (7.4)$$

In addition to (1.2) it is assumed that the determinant here is ≥ 0, $\not\equiv 0$. Then $\overset{\beta}{\underset{\alpha}{(u,u)_S}} \geq 0$ which is expressed by calling S non-negative. Such an operator is now considered together with T, $Tu = r(x)u$, where $r(x)$ is allowed to take both positive and negative values.

The theory in the sections 1-3 can be verbally carried over to the present situation if $(.,.)_T$ is replaced by $(.,.)_S$. Also for the new q_x the value $q_x(U,U)$ is the difference of two squares and (1.9) remains valid. For $U = (u,\lambda u)$, $a_{11}Du + a_{10}u \pm \dot{u} = a_{11}Du + a_{10}u \pm \lambda u$, and the corresponding expressions are linearly independent on $E_\lambda[ab)$ defined as in (1.10), provided $\lambda \neq 0$. Thus $\text{sig}\, q_x = (1,1)$ on $E_\lambda[ab)$ when λ is non-real. As a consequence of (7.1) an identity (2.4) holds true for the new q_x and with $\overset{x}{\underset{a}{(.,.)_S}}$ instead of $\overset{x}{\underset{a}{(.,.)_T}}$. It follows that there are elements $\Phi = (\varphi,\lambda\varphi)$ in $E_\lambda[ab)$ such that $c\, q_x(\Phi,\Phi) > 0$ near b which holds if $cq_a(\Phi,\Phi) \geq 0$. If $U = (u,\lambda u + f) \in E[ab)$, λ non-real and $\overset{b}{\underset{a}{(f,f)_S}} < +\infty$, the sets

$$\Sigma_x(U,\Phi) : c\, q_x(U + \ell\Phi,\, U + \ell\Phi) + \overset{x}{\underset{a}{(f,f)_S}} \leq \overset{b}{\underset{a}{(f,f)_S}} \qquad (7.5)$$

have the same properties as the sets (2.6). Also the reasoning in section 3 can be repeated. In the identities which correspond to (3.2) the values (2,1), (1,2) are excluded (compare the recent generalizations by Bert Karlsson [7], [8] of the theorem in Everitt [3]). Statements of the type in section 4 remain valid when $(.,.)_T$ is replaced by $(.,.)_S$ and a Hilbert space theory can be established also for the present operators, Pleijel [13].

A deviation from the theory occurs by the explicit definition of symmetric boundary conditions for S,T on [ab). Let E[ab] be defined as in (2.9), $\overset{b}{\underset{a}{(.,.)}}_S$ instead of $\overset{b}{\underset{a}{(.,.)}}_T$. Then $q_b = 0$ on E[ab] is charcteristic for the limit-point case also in the present theory, but leads to the form

$$(a_{11}Du + a_{10}u)\sin\alpha - i\,\dot{u}\cos\alpha = 0, \quad x = a, \tag{7.6}$$

of a symmetric boundary condition in the new theory. In the limit-circle case a condition (7.6) can be completed to a symmetric boundary condition by the choice of a point m on the boundary of a limit circle (x = b in (7.5), $U = \Theta$) $c(\lambda)\, q_b(\Theta + \ell\Phi\ ,\ \Theta + \ell\Phi) \leq 0$ constructed for a non-real λ and with Θ and Φ in $E_\lambda[ab)$ of which $\Phi = (\varphi,\lambda\varphi)$ satisfies (7.6). The direct generalization of Weyl's theory including his m-coefficients is obtained by letting Θ and Φ satisfy

$$a_{11}D\varphi + a_{10}\varphi = i\cos\alpha, \quad \lambda\varphi = \sin\alpha, \tag{7.7}$$

$$a_{11}D\theta + a_{10}\theta = -i\sin\alpha, \quad \lambda\theta = \cos\alpha \tag{7.8}$$

at x = a. This leads to formulas (6.6) and (6.7) but with $\overset{x}{\underset{a}{(.,.)}}_S$ instead of $\overset{x}{\underset{a}{(.,.)}}_T$. Coordinated by a symmetric boundary condition as in the classical case the m-coefficient $m(\lambda,\alpha)$ corresponding to (7.7), (7.8) is analytic in $\mathrm{Im}(\lambda) > 0$ and in $\mathrm{Im}(\lambda) < 0$. The determination by $m(\lambda,\alpha)$ of a weight measure in a Fourier integral theorem encounters certain complications. Friedrichs extensions as presented for a polar case in Pleijel [10] define symmetric boundary conditions for which the development in for instance [2] can be copied.

The preceeding theory is valid with adequate modifications also when $T = Db_{10} + b_{01}D + b_{00}$, b_{00} real and $\overline{b}_{10} = b_{01}$. See Karlsson [7], [8].

8. ON A PAPER BY ATKINSON, EVERITT AND ONG.

In the paper [1] by Atkinson, Everitt, Ong the operators of section 7 are considered together with a condition

$$(a_{11}Du + a_{10}u)\sin\gamma - i\,u\cos\gamma = 0, \quad x = a, \tag{8.1}$$

in which $\cot\gamma$ is real and finite. It is assumed that the pair S,T is in the limit-point case with respect to $\overset{b}{\underset{a}{(\cdot,\cdot)}}_S$. The condition (8.1) is <u>not</u> a symmetric boundary condition with respect to $\overset{b}{\underset{a}{(\cdot,\cdot)}}_S$ if $\gamma \neq \pi/2$, compare (7.6). But a spectral study of S,T can be founded also on other forms, namely

$$\overset{x}{\underset{a}{(u,v)}}{}^A_S = \overset{x}{\underset{a}{(u,v)}}_S + A\, u(a)\,\overline{v(a)} \tag{8.2}$$

with $x = b$ and with A real. If (8.2) is introduced into (7.1) with $\alpha = a$, $\beta = x$ one finds that the identity

$$i^{-1}\left(\overset{x}{\underset{a}{(\dot u,v)}}{}^A_S - \overset{x}{\underset{a}{(u,\dot v)}}{}^A_S\right) = q_x(U,V) - q_a^A(U,V) \tag{8.3}$$

is valid when $U = (u,\dot u)$, $V = (v,\dot v)$ are in $E[ab)$ and

$$q_a^A(U,V) = (a_{11}Du + a_{10}u - i\,Au)\bar{\dot v} + \dot u(\overline{a_{11}Dv} + \overline{a_{10}v} - \overline{i\,Av}). \tag{8.4}$$

Because of the limit-point assumption q_b vanishes on $E[ab]$. On account of (8.3) with $x = b$ and due to (8.4) the symmetry $\overset{b}{\underset{a}{(\dot u,v)}}{}^A_S = \overset{b}{\underset{a}{(u,\dot v)}}{}^A_S$ leads to the condition

$$(a_{11}Du + a_{10}u - i\,Au)\sin\alpha - i\,\dot u\cos\alpha = 0, \quad x = a, \tag{8.5}$$

which is a symmetric boundary condition with respect to $\overset{b}{\underset{a}{(\cdot,\cdot)}}{}^A_S$. This condition is reduced to (8.1) when $\alpha = \pi/2$ and $A = \cot\gamma$.

For a generalization of Weyl's proper m-coefficient it is essential to consider a $\Phi = (\varphi,\lambda\varphi)$ in $E_\lambda[ab)$ which satisfies a prescribed boundary condition (8.5) which might be (8.1) if $\alpha = \pi/2$. The set $\Sigma_x(\Theta,\Phi)$ is defined by (7.5) with $\Theta = (\theta,\lambda\theta)$ in $E_\lambda[ab)$ so that $f = 0$. It is a circular disc if $c(\lambda)\, q_x(\Phi,\Phi) > 0$. For $V = U = \Phi$ in (8.3) one obtains

$$c\, q_x(\Phi,\Phi) = c^2 \overset{x}{\underset{a}{(\varphi,\varphi)}}_S + c^2A|\varphi(a)|^2, \tag{8.6}$$

$c = c(\lambda) = i^{-1}(\lambda-\bar\lambda)$, since (8.5) implies that $q_a^A(\Phi,\Phi) = 0$. It follows that $c\, q_x(\Phi,\Phi) > 0$ when x is near b provided $A \geq 0$. Then all relevant statements of section 7 remain valid if (7.6) is replaced by (8.5), the form

$\overset{b}{\underset{a}{(\cdot,\cdot)}}_S$ by $\overset{b}{\underset{a}{(\cdot,\cdot)}}{}_S^A$. This includes the case when the pair S,T is in the limit-circle case and Hilbert space theories can be established as in section 7. Weyl's m-coefficient can be defined similarly as in section 7 and is analytic for non-real values of λ.

The case when $A < 0$ is different. The inequality $c(\lambda)\, q_x(\Phi,\Phi) > 0$ is valid near b if $\overset{b}{\underset{a}{(\varphi,\varphi)}}_S = +\infty$ but may not be fulfilled otherwise. If Φ satisfies (8.5) with $u = \varphi$ and $\dot{u} = \lambda\varphi$, and if $\overset{b}{\underset{a}{(\varphi,\varphi)}}_S < +\infty$, the value λ is an eigenvalue. In [1] it is proved that there are at most two non-real eigenvalues $\lambda_o,\bar{\lambda}_o$. For other non-real values of λ Weyl's generalized m-coefficient can be conceived. By a choice of Φ,Θ corresponding to the one in section 7, i.e. by taking $a_{11}D\varphi + a_{10}\varphi - i\,A\varphi = i\cos\alpha$, $\lambda\varphi = \sin\alpha$ instead of (7.7) etc., one obtains (6.6) and (6.7) with $\overset{x}{\underset{a}{(\cdot,\cdot)}}{}_S^A$ instead of $\overset{x}{\underset{a}{(\cdot,\cdot)}}_T$. The choice of Θ in [1] gives transformed formulas. The analyticity of the obtained m-coefficient $m(\lambda,\alpha)$ follows as in the classical case when $\lambda \neq \lambda_o,\bar{\lambda}_o$. The limitation to the condition (8.1) in [1] is motivated by the wish to use $m(\lambda,\alpha)$ as a weight measure in cases similar to those at the end of section 7.

As indicated in Everitt [5] the theory exemplifies Iohvidov-Krein's study of symmetric operators in a Hilbert space with a non-definite inner product containing a dominant definite part, [9].

The proof by Atkinson, Everitt and Ong of the theorem about non-real eigenvalues runs as follows. Let u and v in $U = (u,\lambda u)$, $V = (u,\mu v)$ be eigensolutions and assume that

$$\lambda \neq \bar{\mu}, \quad \lambda \neq \bar{\lambda}, \quad \mu \neq \bar{\mu}\,. \tag{8.7}$$

Because of (8.7) and since U and V satisfy the same symmetric boundary condition, the symmetry $\overset{b}{\underset{a}{(\dot{u},v)}}{}_S^A = \overset{b}{\underset{a}{(u,\dot{v})}}{}_S^A$ leads to $\overset{b}{\underset{a}{(u,v)}}_S = -A\,u(a)\,\overline{v(a)}$ and to $\overset{b}{\underset{a}{(u,u)}}_S = -A|u(a)|^2$, $\overset{b}{\underset{a}{(v,v)}}_S = -A|v(a)|^2$. Thus $|\overset{b}{\underset{a}{(u,v)}}_S|^2 = \overset{b}{\underset{a}{(u,u)}}_S\,\overset{b}{\underset{a}{(v,v)}}_S$ which means equality in a Cauchy-Schwarz inequality in $C^{(1)}[ab)$, $\overset{b}{\underset{a}{(\cdot,\cdot)}}_S$

finite. But $\int_a^b (w,w)_S = 0$ in this space implies $a_{11}Dw + a_{10}w = 0$, $w = 0$ on sub-interval, so that w is identically 0. Hence our solutions of $Su = \lambda Tu$, $Sv = \mu Tv$ are linearly dependent which gives a contradiction if $\lambda \neq \mu$.

On the other hand any non-real λ_o determines a constant $A < 0$ and a symmetric boundary condition (8.5) such that λ_o becomes an eigenvalue. For a non-trivial solution of $S\psi = \lambda_o T\psi$ with $\int_a^b (\psi,\psi)_S < +\infty$ is essentially unique in the limit-point case. The reality of the spectrum of any problem concerned in section 7 excludes the possibility that $\psi(a,\lambda_o) = 0$. If one takes the value $A = -\int_a^b (\psi,\psi)_S / |\psi(a,\lambda_o)|^2$ it follows that $\int_a^b (\psi,\psi)_S^A = 0$. Then $q_a^A(\Psi,\Psi) = 0$ because of (8.3) with $x = b$. Then (8.5) with $u = \psi$, $\dot{u} = \lambda\psi$ is fulfilled for some α.

The last remark in section 7 is also valid for the present section if a certain condition is fulfilled, see Karlsson [7], [8].

References

[1] F.V. Atkinson, W.N. Everitt and K.S. Ong, On the m-coefficient of Weyl for a differential equation with an indefinite weight function, to appear in Proc. London Math. Soc.

[2] E.A. Coddington and N. Levinson, Theory of Ordinary Differential Equations, McGraw-Hill Book Company, Inc., New York - Toronto - London, 1955.

[3] W.N. Everitt, Singular differential equations I: the even order case, Math. Ann. 156, 9-24, 1964.

[4] W.N. Everitt, Integrable-square, analytic solutions of odd-order, formally symmetric, ordinary differential equations, Proc. London Math. Soc. 25, Series III, 156-182, 1972.

[5] W.N. Everitt, Some remarks on a differential expression with an in-

definite weight function, Conference in Spectral Theory and Asymptotics of Differential Equations, Scheveningen, The Netherlands, September 1973, to appear in Lecture Notes in Mathematics, Springer-Verlag.

[6] Charles Fulton, Parametrizations of Titchmarsh's $m(\lambda)$-function in the limit circle case, doctoral dissertation, Thein.-Westf. Hochschule, Aachen. A paper based upon this dissertation will appear soon.

[7] Bert Karlsson, Generalization of a theorem of Everitt, to appear in Proc. London Math. Soc.

[8] Bert Karlsson, unpublished paper containing a continuation of the preceeding one, to appear in the preprint series Uppsala University, Department of Mathematics Reports.

[9] I.S. Iohvidov and M.G. Krein, Spectral theory of operators in spaces with an indefinite metric, Parts I and II, Amer. Math. Soc. Transl. 13, 105-175, and 34, 283-373.

[10] Åke Pleijel, Le problème spectral de certain équations aux dérivées partielles, Arkiv för Matematik, Astronomi och Fysik 30 A, N:o 21, 1-47, 1944.

[11] Åke Pleijel, Spectral theory for pairs of ordinary formally self-adjoint differential operators, Journal Indian Math. Soc. 34, 259-268, 1970.

[12] Åke Pleijel, Green's functions for pairs of formally selfadjoint ordinary differential operators, Conference on the Theory of Ordinary and Partial Differential Equations, Dundee, Scotland, March 1972, Lecture Notes in Mathematics 280, 131-146, 1972, Springer-Verlag.

[13] Åke Pleijel, A positive symmetric ordinary differential operator combined with one of lower order, Conference in Spectral Theory and Asymptotics of Differential Equations, Scheveningen, The Netherlands, September 1973, to appear in Lecture Notes in Mathematics, Springer-Verlag.

[14] E.C. Titchmarsh, Eigenfunction Expansions Associated with Second-order Differential Equations, Part I, Second Edition, Clarendon Press, Oxford 1962.

Existence and Multiplicity of Solutions of Some Nonlinear Equations

by

Klaus Schmitt

University of Utah and Universität Würzburg

1. Introduction. In studying some properties (such as isolatedness, muliplicity, bifurcation) of solutions of nonlinear equations, the variational equation plays an important role. In this paper we show that to certain classes of nonlinear equations set-valued variational equations may be associated which may be used in much the same way as linear variational equations. We use the concept of a set-valued derivative as used by Chow and Lasota [4]; first we establish some elementary properties of solutions and then apply these results to some boundary value problems for nonlinear second order differential equations. Similar results may be established for certain boundary value problems for elliptic partial differential equations by using the ideas in [1], [10] together with those presented here.

2. Some Properties of Solutions of Operator Equations.

Let E be a real Banach space with norm $||\cdot||$ and denote by $c(E)$ the collection of all nonempty closed and bounded convex subsets of E. A mapping $F:E \to c(E)$ is called upper-semicontinuous in case the graph $\{(x, y) : x \in F(x)\}$ is closed in $E \times E$, it is called compact, in case for every bounded set V in E the set $F(V) = \{y : y \in F(x), x \in V\}$ is precompact in E; it is called completely continuous, in case it is both upper-semicontinuous and compact. We call F positive homogeneous, if or every scalar $\lambda, \lambda \geqq 0$, $F(\lambda x) = \lambda F(x)$, $x \in E$, and homogeneous if the latter equality holds for all $\lambda\in R=(-\infty,\infty)$.

The concept of a set-valued derivative to be introduced now is motivated by the paper of Chow and Lasota. A mapping $f:E\to E$ is called <u>c-differentiable</u> at a point $x_o\in E$ if there exists a mapping $F:E\to c(E)$ which is positive homogeneous and a function $r:E\to E$ such that

(1) $$f(x) - f(x_o) \in F(x-x_o) + r(x-x_o) ,$$

where $$\|r(x-x_o)\| = o(\|x-x_o\|) \quad \text{as } x\to x_o.$$

Let us now consider the operator equation

(2) $$x = f(x) ,$$

where $f:E\to E$ is completely continuous and x_o is a solution of (2) with f having a completely continuous c-derivative F at x_o. Using these assumptions we establish some properties of x_o and F.

<u>Proposition 1</u>. Let $\{x_n\}_{n=1}^{\infty}$ be a sequence of solutions of (2) with $x_n-x_o \neq 0$, $n=1,2,\ldots$, and $\lim_{n\to\infty} x_n=x_o$. Then there exists $y(\neq 0)\in E$ such that

(3) $$y\in F(y).$$

<u>Proof.</u> $x_n-x_o = f(x_n) - f(x_o) \in F(x_n-x_o)+r(x_n-x_o)$,

thus $y_n = \frac{x_n-x_o}{\|x_n-x_o\|}$ satisfies

$$y_n \in F(y_n) + \frac{r(x_n-x_o)}{\|x_n-x_o\|} \quad ;$$

now use the complete continuity of F and the o-property of r to conclude that (3) has a nontrivial solution.

<u>Corollary 2.</u> If in proposition 1. $x_n - x_o \in K$, where K is a closed subset of E with the property that $\alpha K\subseteq K$, for all $\alpha\geq 0$, then (3) has a solution $y\in K-\{0\}$.

Corollary 3. Let f be of the special form f=h+g, where g is completely continuous linear and let K be as in corollary 2. Further let $\{x_n\}_{n=1}^{\infty}$ be a sequence of solutions of (2) such that $\lim_{n\to\infty} g(x_n) = g(x_o)$ and that $\lim_{n\to\infty} x_n = \bar{x}$ exists. Further let $x_n - x_o \in K-\{0\}, n=1,2,\ldots,$ and let (3) have no nontrivial solutions in K. Then $\bar{x}-x_o \in K-\{0\}$. (And, of course, $\bar{x}$ is a solution of (2)).

Corollary 4. Let the relation (3) imply y=0, then x_o is an isolated solution of (2).

Proposition 5. Let relation (3) imply y=0 and let F be homogeneous. Then the Leray-Schauder index of x_o is an odd integer.

Proof. Since x_o is an isolated solution of (2), there exists $\varepsilon_o>0$ such that x_o is the only solution of (2) in the spheres $B_\varepsilon = \{x: \|x-x_o\|<\varepsilon\}$, $0<\varepsilon\leq\varepsilon_o$. Thus the Leray-Schauder degree $d(I-f,B_\varepsilon,0)$ is defined and constant for $0<\varepsilon\leq\varepsilon_o$; this constant is the index of x_o (see [11]). Define $g:\bar{U}_\varepsilon = \bar{B}_\varepsilon - x_o \to E$ by $g(y) = f(x) - f(x_o)$, where $y = x-x_o$. Then $d(I-g, U_\varepsilon,0) = d(I-f, B_\varepsilon,0)$, $0<\varepsilon\leq\varepsilon_o$. We now show that for $\varepsilon>0$ sufficiently small I-g is homotopic to an odd vector field on ∂U_ε and hence since U_ε is a symmetric convex neighborhood of 0 $d(I-g, U_\varepsilon,0)$ will be an odd integer. To this end consider the family of vector fields $y-h(\lambda,y) = y - \frac{1}{1+\lambda} g(y) + \frac{\lambda}{1+\lambda} g(-y)$, $0\leq\lambda\leq1$. If this family is not zero-free on ∂U_ε, for all $\varepsilon>0$ sufficiently small, one may find sequences, $\{\varepsilon_n\}$, $\lim_{n\to\infty} \varepsilon_n=0$, $\{y_n\}$, $\|y_n\|=\varepsilon_n$, and $\{\lambda_n\}\subset(0,1]$ such that

$$y_n = \frac{1}{1+\lambda_n} g(y_n) - \frac{\lambda_n}{1+\lambda_n} g(-y_n) \in F(y_n) + r_1(y_n) ,$$

where $\|r_1(y_n)\| = o(\|y_n\|)$, as $y_n\to0$. Now put $z_n = \frac{y_n}{\|y_n\|}$ and use an argument similar to the one in proposition 1. to conclude that (3) has a nontrivial solution. From this contradiction, we obtain that for

$\varepsilon>0$ sufficiently small $d(I-g,U_\varepsilon,0) = d(I-h(1,0), U_\varepsilon,0)$. On the other hand $y-h(1,y)$ is an odd vector field, and hence by Borsuk's theorem (see [11]) the required degree is nonzero.

Remark. The argument used in the above proof was suggested by and is similar to an argument in [4].

Corollary 6. Assume the conditions of proposition 5. and let $g:E\times E\to E$ be completely continuous with $\|q(x,v)\|=O(\|v\|)$ as $v\to 0$ uniformly on bounded x-sets. Then there exists $\varepsilon>0$ and $\sigma>0$ such that for all $v, \|v\|\leq\sigma$ the perturbed equation

(4) $$x = f(x)+q(x,v)$$

has a solution x_v with $\|x_v-x_o\|\leq\varepsilon$. Further $\|x_v-x_o\|\to 0$ as $\|v\|\to 0$

Proof Choose $\varepsilon>0$ such that x_o is the only solution of (2) in B_ε. Then choose $\sigma>0$ such that $\|v\|\leq\sigma$ implies that $x-f(x)-\lambda q(x,v)$ is zero free on ∂B_ε, $0\leq\lambda\leq 1$. Thus by homotopy invariance of Leray-Schauder degree we obtain that $0 \neq d(I-f,B_\varepsilon,0) = d(I-f-q(\cdot,v),B_\varepsilon,0)$. Hence (4) has a solution with the desired properties.

Remark. Corollary 6 is the implicit function theorem of [4]. This proof here shows that this theorem actually is a special case of the more general result: If x_o is the only solution of (2) in a bounded open set U and if $d(I-f,U,0)\neq 0$, then there exists $\sigma>0$ such tnat the perturbed equation (4) has a solution $x_v\in U$ for all $\|v\|\leq\sigma$, further $\|x_v-x_o\|\to 0$ as $v\to 0$. Since $d(I-f,B_\varepsilon,0)\neq 0$ for $\varepsilon>0$ sufficiently small it follows from the definition and the properties of Leray-Schauder degree that the solution x_o may be computed as a limit of a sequence of finite dimensional problems and in particular if E has a basis (see [12]) then Galerkin's method (see[8],[13 may be used to compute x_o. Letting $\{\varphi_i\}_{i=1}^{\infty}$, be a basis for E and E_n be the finite dimensional subspace of E spanned by $\{\varphi_i\}_{i=1}^{n}$ and P_n the usual projection of E onto E_n (P_n - the truncation operator), then it is known that the projections are uniformly bounded, i.e. there exists $K>0$ such that $\|P_n\|\leq K$, $n=1,2,\ldots$ (see e.g.[12]). Thus for n sufficiently large each of the finite-dimensional problems

(5) $$x = P_n f(x)$$

has a solution $x_n \in B_\varepsilon$ $0 \neq d(I-f,B_\varepsilon,0) = d(I-P_nf,B_\varepsilon,0) = d(I-P_nf,B_\varepsilon \cap E_n,0)$, for n sufficiently large), and $\lim_{n\to\infty} x_n = x_o$. If f is Fréchet-differentiable at x_o, then it is known that the error estimate

(6) $$\|x_n - x_o\| \leq Q(1+\varepsilon_n)\|P_n x_o - x_o\|$$

holds [13], where $\lim_{n\to\infty} \varepsilon_n = 0$. If instead, we assume the less restrictive c-differentiability of f at x_o considered above, one may verify the following.

<u>Corollary 7.</u> Let the conditions of proposition 5 hold. Then there exists a constant $Q>0$ and a sequence $\{\alpha_n\}$, $\lim_{n\to\infty} \alpha_n = 0$ such that the Galerkin approximations satisfy

(7) $$\|x_n - x_o\| \leq Q(1+\alpha_n)\|P_n x_o - x_o\| .$$

The proof is somewhat lengthy, hence will be omitted here, it should be pointed out that the compactness and the positive homogeneity of F plays an important role in establishing (7).

3. Boundary Value Problems for Ordinary Differential Equations.

We next consider some applications of the results in § 2 to boundary value problems for second order ordinary scalar differential equations of the form

(8) $$x''+f(t,x,x')=0, 0\leq t\leq 1$$

(9) $$ax(0)-bx'(0)=A, \quad cx(1)+dx'(1)=B,$$

where a,b,c,d are nonnegative constants and $a+b>0$, $c+d>0$, $a+c>0$, and

$f:[0,1]\times R\times R\to R$ is continuous.

The boundary conditions (9) may, in some of the considerations to follow, without much additional detail, be replaced by the more general nonlinear constraints considered in [3] (see also [5],[7]). Also similar results will hold for coupled systems of the form (8) we refer the interested reader to [10] where such problems are studied for systems of equations where f is continuously differentiable.

A function $\alpha\in C([0,1],R)$ is called a <u>lower solution</u> of (8)-(9) in case

$$\alpha'' + f(t,\alpha,\alpha') \geq 0 \tag{10}$$
$$a\alpha(0)-b\alpha'(0)\leq A, \quad c\alpha(1)+d\alpha'(1)\leq B.$$

An <u>upper solution</u> β is defined by replacing α by β in (10) and reversing inequalities.

Let there exist α and β, lower and upper solutions of (8)-(9), respectively such that $\alpha(t)\leq\beta(t), 0\leq t\leq 1$. f is said to satisfy a <u>Nagumo condition</u> on the set $w = \{(t,x):\alpha(t)\leq x\leq\beta(t),\ 0\leq t\leq 1\}$, if there exists a positive continuous function $\varphi(s)$ such that $|f(t,x,y)|\leq\varphi(|y|)$, $(t,x)\in w$, and $\int^{\infty}\frac{s\,ds}{\varphi(s)} = +\infty$.

<u>Lemma 8.</u> Let there exist lower and upper solutions α and β, respectively of (8)-(9), such that $\alpha(t)\leq\beta(t)$, $0\leq t\leq 1$, and let f satisfy a Nagumo condition with respect to the pair α,β. Then the problem (8)-(9) has a solution $x_0(t)$ with $\alpha(t)\leq x_0(t)\leq\beta(t)$.

For proofs of this result and applications thereof we refer to [3], [5], [9], [10].

<u>Lemma 9.</u> Let the hypotheses of Lemma 8 hold and let $x_0(t)$ be a solution of (8)-(9) with $\alpha(t)\leq x_0(t)\leq\beta(t)$. Then the following hold:

(a) If β does not satisfy the boundary conditions (9), there exists a sequence of solutions $\{x_n\}$ of (8) different from x_o converging uniformly (together with the sequence of derivatives) to a solution of (8)-(9) and $x_n(t)-x_o(t)\geq 0$, $0\leq t\leq 1$.

(b) If α does not satisfy the boundary conditions (8), then the same conclusion as in (a) holds except that $x_n(t)-x_o(t)\leq 0, 0\leq t\leq 1$.

<u>Proof.</u> We verify a particular case of part (a). Assume $a\beta(0)-b\beta'(0)>A$. Pick a strictly monotone decreasing sequence $\{\varepsilon_n\}$, $\lim_{n\to\infty}\varepsilon_n=0$, such that $a\beta(0)-b\beta'(0)\geq A+\varepsilon_n$. Let x_n be a solution of (8) satisfying $ax_n(0)-bx_n'(0) = A+\varepsilon_n$, $cx_n(1)+dx_n'(1) = B$ and $x_o(t)\leq x_n(t)\leq\beta(t)$ (apply Lemma 8 with x_o as lower solution and β as upper solution). From this sequence one may select a subsequence having the above properties.

Assume that $x_o(t)$ is a solution of (8)-(9) and let there exist a neighborhood N of $\{(t,x_o(t),x_o'(t)) : 0\leq t\leq 1\}$ such that for all $(t,y,y')\in N$ and $z=y-x_o(t)$, $z'=y'-x_o'(t)$

$$\text{(11)}\qquad g_1(t,z,z')+G_1(t,z,z')\leq f(t,y,y') - f(t,x_o(t),x_o'(t))\leq G_2(t,z,z')+g_2(t,z,z'),$$

where $|g_i(t,z,z')| = o(|z|+|z'|)$, $i = 1,2$, uniformly in t and G_1 and G_2 are continuous functions with the properties:

(A) $G_i(t,\lambda z,\lambda z') = \lambda G_i(t,z,z')$, $\lambda\geq 0$, $0\leq t\leq 1$, $-\infty<z,z'<\infty, i=1,2$.

(B) Initial value problems for $z''= G_i(t,z,z')$ are uniquely solvable and solutions are extendable to [0,1].

(The Lipschitz condition (11) is essentially that used throughout [2], except that here it is only assumed in a neighborhood of the solution trajectory.)

Let E be the Banach space $C^1([0,1],R)$ where for $x \in E$, $\|x\| = \max_{[0,1]}|x(t)| + \max_{[0,1]}|x'(t)|$. Let G(t,s) be the Green's function associated with the differential operator $\frac{d^2}{dt^2}$ and the boundary conditions (9), when A=0=B. Further let φ be the unique linear function (of t) satisfying the boundary conditions (9). Define $f: E \to E$ by

$$f(x)(t) = \int_0^1 G(t,s)f(s,x(s),x'(s))ds + \varphi(t), \tag{12}$$

then the solution x_0 of (8)-(9) is a fixed point of f in E. Furthermore condition (11) guarantees that f is c-differentiable at x_0 with c-derivative $F: E \to c(E)$ defined by (here one needs the fact that $G(t,s) \geq 0$, $0 \leq t, s \leq 1$)

$$F(y) = \{z \in E: z(t) = \int_0^1 G(t,s)u(s)ds,\ u \in C([0,1]R), \tag{13}$$

$$G_1(t,y(t),y'(t)) \leq u(t) \leq G_2(t,y(t),y'(t)), 0 \leq t \leq 1\}.$$

That F is completely continuous follows easily from the definition.

<u>Theorem 10.</u> Let G_1 and G_2 have the additional property that the only solution y of

$$y'' + G_1(t,y,y') \leq 0 \leq y'' + G_2(t,y,y') \tag{14}$$

which satisfies the boundary conditions (9) (with A=B=0) is the zero solution. Then x_0 is an isolated solution of (8)-(9).

<u>Proof.</u> That (14) together with the homogeneous boundary conditions (9) have only the trivial solution implies that when $y \in F(y)$, then $y = 0$, thus corollary 4 is applicable.

<u>Remark.</u> If in the boundary conditions (9) a=1=c, b=0=d, then a simple condition which would imply the additional requirement imposed by Theorem 10 upon G_1 and G_2 is the following: The equation $y'' + G_2(t,y,y')$ should be disconjugate on [0,1].

In what is to follow G(t,s) shall be the Green's function associated with the differential operator $\frac{d^2}{dt^2}$ and the boundary conditions $x(0)=0=x(1)$; we define $h:E\to E$ and $g:E\to E$ by setting

$$g(x)(t) = (1-t)x(0)+tx(1)$$
$$h(x)(t) = \int_0^1 G(t,s)f(s,x(s),x'(s))ds ,$$

then h is completely continuous and g is completely continuous linear, further under the assumptions imposed on f(t,x,y) ((11)) we obtain that the operator f = h+g is completely continuous c-differentiable at x_o, a c-derivative being given by

$$(14) \qquad F(y) = \{z\in E: z(t)= \int_0^1 G(t,s)u(s)ds+(1-t)y(0)+ty(1),$$
$$u\in C([0,1],R), G_1(t,y(t),y'(t))\le u(t)\le G_2(t,y(t),y'(t))\}.$$

Theorem 11. Let the conditions of Lemma 9 hold and assume that the solution v(t) of $v''+G_1(t,v,v')=0$, $v(0)=0, v'(0)=1$ has a zero in [0,1). Then under the conditions of part a. of lemma 9 the sequence $\{x_n\}$ must converge to a solution of (8)-(9) different from x_o.

Proof. Let $\{x_n\}$ be the sequence in part a. of lemma 9. Let $K = \{x\in E: x(t)\ge 0,\ 0\le t\le 1\}$, then $x_n-x_o\in K-\{0\}$ and $\lim_{n\to\infty} g(x_n) = g(x_o)$. Thus in order to establish the theorem we must show that (3) has no nontrivial solutions in K. This we argue indirectly. If $y\in K-\{0\}$ satisfies $y\in F(y)$, then $y''+u(t)=0$, where $G_1(t,y(t),y'(t))\le u(t)\le G_2(t,y(t),y'(t))$, hence one may easily conclude that $y(t)>0, 0<t<1$ (one uses that $y''+G_1(t,y,y')\le 0$ and the fact that initial value problems for $y''+G_1(t,y,y') = 0$ are uniquely solvable). Let now t_1 be the first zero of v, then $y(t)>0, 0<t\le t_1$,

thus will be an upper solution for the boundary value problem $z'' + G_1(t, z, z') = 0$, $z(0) = 0$, $z(t_1) = y(t_1)$, a suitable positive multiple of $u(t)$, on the other hand, say $\lambda u(t)$ will be a lower solution, with $\lambda u(t) \leqq y(t)$, $0 \leqq t \leqq t_1$. Using lemma 8 one obtains the existence of a solution $z(t)$ of this problem with $\lambda u(t) \leqq z(t) \leqq y(t)$. The assumptions about G_1, however, imply that $z(t)$ must be a multiple of $u(t)$, contradicting that $z(t_1) \neq 0$.

Example. Let $k(x)$ be a continuous odd function such that $xk(x) > 0$, $x \neq 0$ and let $k(1) \leqq 1$. Further assume that there exist positive constants a and b such that for small positive x, $ax \leqq k(x) \leqq bx$. Further let $a > \pi^2$, then the boundary value problem

$$x'' + k(x) - x^3 = 0\ , \quad x(0) = 0 = x(1)$$

has a nontrivial solution $x(t)$ with $x(t) > 0$, $0 < t < 1$.

3. References.

1. H. Amann, Existence of multiple solutions for nonlinear elliptic boundary value problems, Ind. Univ. Math. J. 21 (1972), 925-935.
2. P. Bailey, L. Shampine and P. Waltman, Nonlinear Two Point Boundary Value Problems, Academic Press, New York, 1968.
3. J. Bebernes and R. Wilhelmsen, A remark concerning a boundary value problem, J. Diff. Equations 10 (1971), 389-391.
4. S-N. Chow and A. Lasota, An implicit function theorem for nondifferentiable mappings, Proc. Amer. Math. Soc. 84 (1972), 141-146.
5. L. Erbe, Nonlinear boundary value problems for second order differential equations, J. Diff. Equations 7 (1970), 459-472.
6. L. Jackson, Subfunctions and boundary value problems for second order ordinary differential equations, Advances in Math. 2 (1968), 307-363.
7. H. Knobloch, Second order differential inequalities and a nonlinear boundary value problem, J. Diff. Equations 5 (1969), 55-71.
8. M. Krasnosel' skii, Toplogical Methods in The Theory of Nonlinear Integral Equations, Pergamon, New York, 1964.
9. K. Schmitt, A nonlinear boundary value problem, J. Diff. Equations 7 (1970), 527-537.
10. K. Schmitt, Applications of variational equations to ordinary and partial differntial equations - multiple solutions of boundary value problems, J. Differential Equations, to appear.

11. J. Schwartz, Nonlinear Functional Analysis, Gordon and Breach, New York, 1969.

12. I. Singer, Bases in Banach Spaces I, Springer Verlag, New York 1970.

13. S. Shirali, A note on Galerkin's method for nonlinear equations, Aequat. Math. 4 (1970), 198-200.

4. Acknowledgement. This work was supported by U. S. Army research grant no. ARO-D-31-124-72-G56 and by the Deutsche Forschungsgemeinschaft while the author was Visiting Professor at the University of Würzburg, Germany.

The Floquet Problem for Almost Periodic Linear Differential Equations

George R. Sell

I. Introduction.

The Floquet theory for linear periodic differential equations is a fundamental tool for the study of both linear and non-linear periodic phenomena arising from differential equations. We shall present a summary of this theory in a moment, but recall that this theory has two aspects. First, there is a representation theorem whereby one can express the matrix solution of a linear periodic differential equation as the product of periodic matrix and an exponential e^{tR}. Secondly, by using this representation, together with the spectrum of the matrix R, one is able to give an essentially complete description of the asymptotic behavior of the solutions of the original differential equation.

The Floquet problem for linear almost periodic differential equations is simply to "extend" the Floquet theory to the almost periodic case. My objective in this lecture is to examine the meaning of such an extension. We shall see that, even for first order almost periodic equations, one cannot always derive a representation theorem whereby the solution is expressed as the product of an almost periodic function and an exponential. Thus the "representation theorem" part of the Floquet theory cannot be extended to the almost periodic case without some modification.

There is however some justification for expecting that the second part of the Floquet theory, which describes the asymptotic behavior of solutions, can be extended to the almost periodic case. It is this aspect of the theory that we wish to study in more detail in this lecture. I would also like to point out how recent work by V.M. Millionščikov [7-9], R. J Sacker and G.R. Sell [12-14] and J.F. Selgrade [15] has led to further insight into this rather old problem.

Before going on, I would like to point out that many of the results described in this lecture come from joint work with Robert Sacker. I would like to express my gratitude to him for his contribution to my understanding of the subject.

II. Floquet Theory; Periodic Linear Equations.

Consider the nth-order periodic linear differential equation

$$x' = A(t)x \tag{1}$$

where $x \in X$ (and $X = R^n$ or $X = C^n$) and $A(t)$ is ω-periodic in t. Let $\Phi(t)$ be the $n \times n$ matrix solution of Eqn (1) with $\Phi(0) = I$, the identity. Then $\Phi(t)$ is nonsingular for all t. Now define F by $\Phi(\omega) = e^{\omega F}$, and define $\Psi(t)$ by $\Psi(t) = \Phi(t)e^{-tF}$. One then shows that $\Psi(t)$ is ω-periodic in t, i.e.

$$\Phi(t) = \Psi(t)e^{tF} \tag{2}$$

is the Floquet representation of Φ as the product of a periodic matrix and an exponential matrix.

Let $\{\lambda_1,\ldots,\lambda_k\}$ denote the spectrum of F, i.e. the eigenvalues of F. Let V_i denote the generalized eigenspace associated with λ_i, i.e.

$$V_i\{x \in X: (F-\lambda_i I)^p x = 0 \quad \text{for some} \quad p = 0,1,\ldots,n\} .$$

Then $V_i \cap V_j = \{0\}$ if $i \neq j$ and

$$X = V_1 + \ldots + V_k .$$

For each $\lambda \in R$ with $\lambda \neq \mathrm{Re}\lambda_i$ for any i, define S_λ to be the sum of the V_i with $\mathrm{Re}\,\lambda_i < \lambda$ and U_λ to be the sum of the V_i with $\mathrm{Re}\,\lambda_i > \lambda$. Then $S_\lambda \cap U_\lambda = \{0\}$ and $X = S_\lambda + U_\lambda$. Furthermore, one has

$$e^{-\lambda t}\Phi(t)x \to 0 \quad \text{as} \quad t \to +\infty$$

if and only if $x \in S_\lambda$, and

$$e^{-\lambda t}\Phi(t)x \to 0 \quad \text{as} \quad t \to -\infty$$

if and only if $x \in U_\lambda$. Also

$$\sup_{t \in R} \|e^{-\lambda t}\,\Phi(t)x\| < +\infty$$

if and only if $x = 0$.

The set of numbers $\{\mathrm{Re}\,\lambda_1,\ldots,\mathrm{Re}\,\lambda_k\}$ we shall call the <u>spectrum</u> of Eqn. (1). This concept does generalize to almost periodic equation and as such leads to the heart of the Floquet problem, as we view this problem.

Before going on let us also note that the spectrum of Eqn. (1) is precisely the collection of Lyapunov-type members for Eqn. (1), cf. L. Cesari [2] and P. Hartman [5].

III. First Order Almost Periodic Equations.

Consider the first order equation

$$x' = a(t)x \tag{3}$$

where $x \in R^1$ and $a(t)$ is a real-valued almost periodic function of t. Let $\mathcal{G}$ denote the hull of a, i.e.

$$\mathcal{G} = \mathrm{Cl}\{a_\tau : \tau \in R\}$$

where $a_\tau(t) = a(\tau + t)$ and the closure is taken in the topology of uniform convergence on compact sets, cf. [16]. For each $a \in \mathcal{G}$ the fundamental solution $\Phi(a,t)$ of Eqn. (3) that satisfies $\Phi(a,0) = 1$ is

$$\Phi(a,t) = e^{\int_0^t a(s)ds} .$$

If one were to try to find a Floquet representation of Φ of the form

$$\Phi(a,t) = \Psi(a,t)e^{\lambda t}$$

where $\Psi(a,t)$ is almost periodic in t and λ is constant, then it is not hard to see that λ must be the mean value

$$m(a) = \lim_{T\to+\infty} \frac{1}{T}\int_0^T a(s)ds = \lim_{T\to-\infty} \frac{1}{T}\int_0^T a(s)ds .$$

Therefore if one assumes that $m(a) = 0$ in Eqn. (1), then it is well known that $\Phi(a,t)$ is almost periodic in t if and only if the integral

$$\int_0^t a(s)ds$$

is bounded in t. Since there do exist almost periodic functions $a(t)$

with $m(a) = 0$ and $\int_0^t a(s)ds$ unbounded, we see, then, that there cannot exist a Floquet representation of $\Phi(a,t)$ as the product of an almost periodic function and an exponential. Therefore the Floquet representation theorem cannot be extended to the almost periodic case without some modification of the form of the representation.

Let us now turn to the second part of the Floquet theory, viz. the asymptotic behavior of the solutions of Eqn. (3). First note that $m(a) =$ constant for $a \in \mathcal{A}$. Next, if $\lambda > m(a)$ then

$$e^{-\lambda t}\,\Phi(a,t) \to 0 \quad \text{as} \quad t \to +\infty$$

for all $a \in \mathcal{A}$. Also if $\lambda < m(a)$ then

$$e^{-\lambda t}\ \ (a,t) \to 0 \quad \text{as} \quad t \to -\infty$$

for all $a \in \mathcal{A}$. Define the bounded set by

$$\mathcal{B}_\lambda = \{(x,a): \|e^{-\lambda t}\,\Phi(a,t)x\| \text{ is uniformly bounded in } t\}.$$

Then it follows that if $\lambda \neq m(a)$ one has

$$\mathcal{B}_\lambda = \{0\} \times \mathcal{A}$$

i.e. $\mathcal{B}_\lambda$ is trivial.

The spectrum of Eqn. (3), which we shall define in the next section is the set $\{m(a)\}$, which is also the set of Lyapunov-type numbers of Eqn. (3).

IV. The Spectrum of an Almost Periodic Linear Differential Equation.

Consider now an n-th order equation

$$x' = A(t)x$$

where $x \in X$ $(X = R^n$ or $X = C^n)$ and $A(t)$ is almost periodic in t. Let $\mathfrak{A}$ denote the hull of A and let $\varphi(x,A,t)$ denote the solution of the initial value problem

$$x' = A(t)x \ , \quad x(0) = x$$

where $x \in X$ and $A \in \mathfrak{A}$. Let A_τ be defined by $A_\tau(t) = A(\tau + t)$. Then the mapping $\pi \colon X \times \mathfrak{A} \times R \to X \times \mathfrak{A}$ given by

$$\pi(x,A,\tau) = (\varphi(x,A,\tau), A_\tau)$$

defines a linear skew product flow on $X \times \mathfrak{A}$, cf. [12]. Since $\varphi(x,A,\tau)$ is linear in x, the equation

$$\Phi(A,\tau)x = \varphi(x,A,\tau)$$

defines a linear transformation $\Phi(A,\tau)$ on X.

For each real number λ let

$$\pi_\lambda(x,A,\tau) = (e^{-\lambda\tau}\ \varphi(x,A,\tau), A_\tau) \ .$$

This too is a linear skew product flow on $X \times \mathfrak{A}$. Next define the <u>bounded set</u>

$$\mathfrak{B}_\lambda = \{ (x,A) \colon \|e^{-\lambda t}\ \varphi(x,A,t)\| \text{ is uniformly bounded in } t \}$$

the stable set

$$\mathcal{S}_\lambda = \{(x,A): \|e^{-\lambda t}\,\varphi(x,A,t)\| \to 0 \quad \text{as} \quad t \to +\infty\} ,$$

and the unstable set

$$\mathcal{U}_\lambda = \{(x,A): \|e^{-\lambda t}\,\varphi(x,A,t)\| \to 0 \quad \text{as} \quad t \to -\infty\} .$$

Also define the fibers

$$\mathcal{B}_\lambda(A) = \{x \in X: (x,A) \in \mathcal{B}_\lambda\}$$

$$\mathcal{S}_\lambda(A) = \{x \in X: (x,A) \in \mathcal{S}_\lambda\}$$

$$\mathcal{U}_\lambda(A) = \{x \in X: (x,A) \in \mathcal{U}_\lambda\} .$$

The fibers $\mathcal{B}_\lambda(A)$, $\mathcal{S}_\lambda(A)$ and $\mathcal{U}_\lambda(A)$ are linear subspaces of X . The following theorem is proved in Sacker and Sell [12, 14].

Theorem 1. *A necessary and sufficient condition for*

(4) $$X \times G = \mathcal{S}_\lambda + \mathcal{U}_\lambda \quad \text{(Whitney sum)}$$

is that $\mathcal{B}_\lambda$ *is trivial, i.e.* $\mathcal{B}_\lambda = \{0\} \times G$. *Furthermore, in this case there exists for each* $A \in G$ *a projection* $P_\lambda(A): X \to X$ *such that* $P_\lambda(A)x$ *is jointly continuous* x *and* A *and*

(5)
$$|\Phi_\lambda(A,t)\, P_\lambda(A)\, \Phi_\lambda^{-1}(A,s)| \le K\, e^{-\alpha(t-s)} , \quad s \le t$$
$$|\Phi_\lambda(A,t)[I - P_\lambda(A)]\, \Phi_\lambda^{-1}(A,s)| \le K\, e^{-\alpha(s-t)} , \quad t \le s$$

for positive constants K *and* α . (Here $\Phi_\lambda(A,t) = e^{-\lambda t}\,\Phi(A,t)$.)

The meaning of Equation (4) is that $\dim \mathcal{S}_\lambda(A)$ and $\dim \mathcal{U}_\lambda(A)$ are constant over A, and that $\mathcal{S}_\lambda(A)$ and $\mathcal{U}_\lambda(A)$ vary continuously in A, and that

$$X = \mathcal{S}_\lambda(A) + \mathcal{U}_\lambda(A)$$

for all $A \in \mathcal{Q}$. That is $\mathcal{S}_\lambda$ and $\mathcal{U}_\lambda$ are invariant subbundles for π, cf. [17].

The Inequality (5) says that every equation $A \in \mathcal{Q}$ admits exponential dichotomy. It is not hard to see that the range of $P_\lambda(A)$ is precisely $\mathcal{S}_\lambda(A)$ and the null space is $\mathcal{U}_\lambda(A)$. In particular we see that the rate of decay in $\mathcal{S}_\lambda$ and $\mathcal{U}_\lambda$ is exponential, provided $\mathcal{B}_\lambda$ is trivial.

We now define the <u>resolvent set</u> $\rho(\mathcal{Q})$ to be the collection of all $\lambda \in R$ for which $\mathcal{B}_\lambda = \{0\} \times \mathcal{Q}$. The complement $\sigma(\mathcal{Q}) = R - \rho(\mathcal{Q})$ is called the <u>spectrum</u> of A. This includes the concept of spectrum discussed in Sections II and III as special cases.

The following theorem is proved in Sacker and Sell [14].

<u>Theorem</u> 2. <u>Assume that</u> $\dim X = n \geq 1$. <u>Then the spectrum</u> $\sigma(\mathcal{Q})$ <u>is nonempty and compact and consists of the union of</u> k <u>non-overlapping closed intervals</u> $[a_i, b_i]$, $i = 1, \ldots, k$ <u>where</u> $k \leq n$. <u>Furthermore, associated with each interval</u> $[a_i, b_i]$ <u>there is a nontrivial invariant subbundle</u> $\mathcal{V}_i$ <u>with the property that for every</u> $\epsilon > 0$ <u>there exist positive numbers</u> K <u>and</u> α <u>such that for all</u> $(x, A) \in \mathcal{V}_i$ <u>one has</u>

$$\|e^{-(b_i + \epsilon)t} \varphi(x, A, t)\| \leq K|x|e^{-\alpha t}, \quad t \geq 0 \tag{6}$$

and

$$(7) \qquad \|e^{-(a_i - \epsilon)t} \varphi(x,A,t)\| \leq K|x|e^{+\alpha t} , \quad t \leq 0 .$$

Finally one has

$$X \times G = \mathcal{V}_1 + \ldots + \mathcal{V}_k \quad \text{(Whitney sum)} .$$

Now let us turn to the Lyapunov-type numbers or, as they are sometimes called, the Lyapunov characteristic exponents. For each $(x,A) \in X \times G$ with $x \neq 0$ define the four Lyapunov-type numbers by

$$\lambda_s^+(x,A) = \limsup_{T \to +\infty} \frac{1}{T} \log \|\varphi(x,A,T)\|$$

$$\lambda_i^+(x,A) = \liminf_{T \to +\infty} \frac{1}{T} \log \|\varphi(x,A,T)\|$$

$$\lambda_s^-(x,A) = \limsup_{T \to -\infty} \frac{1}{T} \log \|\varphi(x,A,T)\|$$

$$\lambda_i^-(s,A) = \liminf_{T \to -\infty} \frac{1}{T} \log \|\varphi(x,A,T)\| .$$

The theory of Lyapunov-type numbers has played an important role in study of both linear and nonlinear differential equations. The reader may see [1 - 11] for example. For our purposes here we have the following result.

Theorem 3. For all $(x,A) \in X \times G$, $x \neq 0$, the Lyapunov type numbers lie in the spectrum $\sigma(G)$.

Let us now consider $(x,A) \in \mathcal{V}_i$ $(x \neq 0)$ where $\mathcal{V}_i$ is given by Theorem 2. Let $\lambda(x,A)$ designate any of the four Lyapunov-type numbers. Then it follows from Theorem 2 that

$$a_i \leq \lambda(x,A) \leq b_i \ .$$

In particular, if the spectral interval $[a_i,b_i]$ degenerates to a single point $\{a_i\}$, i.e. $a_i = b_i$, then we see that all four Lyapunov-type numbers are the same, and consequently for all $(x,A) \in \mathcal{U}_i$ $(x \neq 0)$ one has

$$\lim_{T\to+\infty} \frac{1}{T} \log \|\varphi(x,A,T)\| = \lim_{T\to-\infty} \frac{1}{T} \log \|\varphi(x,A,T)\| = a_i$$

This is the same asymptotic behavior we observed in Section III for the first order equation.

We can now give a precise formulation of the Floquet problem for almost periodic equations, viz.

<u>Do the spectral intervals</u> $[a_i,b_i]$ <u>degenerate to points</u>?

We might say the spectrum $\sigma(\mathcal{G})$ contains continuous spectrum if it contains a nondegenerate interval. The Floquet problem is then equivalent to asking whether there exists an almost periodic equation with continuous spectrum.

Before turning to an application let us make note of an interesting concept due to Millionščikov, viz. the concept of the probable spectrum, cf. [8]. Assume that $\dim X = n$. For any nonsingular $n \times n$ matrix B, the matrix B^*B is nonsingular and self adjoint. Furthermore B^*B has precisely n-eigenvalues (counting multiplicity) and all of these are positive real numbers. Let $d_1(B),\ldots,d_n(B)$ denote the positive square roots of these eigenvalues and order them by $d_1(B) \geq d_2(B) \geq \ldots \geq d_n(B)$.

Now consider $x' = A(t)x$ where $A \in \mathcal{G}$. For $i = 1,\ldots,n$ define

$$\nu_i(A) = \lim_{\tau\to+\infty} \limsup_{s\to+\infty} \frac{1}{s\tau} \Sigma_{j=0}^{s-1} \log \; d_i(\Phi(A_{j\tau},\tau)) \; .$$

It is not difficult to verify that for each i, $\nu_i(A)$ is a bounded measurable function of $A \in \mathcal{G}$ where the measure on $\mathcal{G}$ is the Haar measure μ. Therefore the Birkhoff Ergodic Theorem can be applied and we have that

$$\nu_i(A) = \int_{\mathcal{G}} \nu_i(A)\, \mu(dA) = \bar{\nu}_i$$

for almost every $A \in \mathcal{G}$. The collection of real numbers $\{\bar{\nu}_1,\ldots,\bar{\nu}_n\}$ is the probable spectrum of A.

Let us now turn to an application.

V. An Application.

Consider the almost periodic equation

$$x' = A(t)x$$

where $x \in X$ and $X = R^n$ or $X = C^n$. Assume that $A(t)$ is continuous and for all $t \in R$ one has the commutivity relationship

$$A(t) \int_0^t A(s)ds = \int_0^t A(s)ds \;\; A(t) \; . \tag{8}$$

It then follows that

$$\Phi(A,t) = \exp(\int_0^t A(s)ds) \; .$$

Next let

$$M(A) = \lim_{T\to+\infty} \frac{1}{T} \int_0^T A(s)ds = \lim_{T\to-\infty} \frac{1}{T} \int_0^T A(s)ds$$

be the mean value matrix. We then have the following result.

Theorem 4. *Assume that the commutivity relationsip* (8) *is satisfied. Then the spectrum* $\sigma(G)$ *of* $x' = A(t)x$ *agrees with the spectrum of the constant coefficient equation*

$$x' = M(A)x \ ,$$

i.e. $\sigma(G) = \{\text{Re } \lambda_1, \ldots, \text{Re } \lambda_k\}$ *where* $\{\lambda_1, \ldots, \lambda_k\}$ *denote the eigenvalues of* $M(A)$. *In particular* $\sigma(G)$ *consists of point spectrum only.*

References

1. B.V. Bylov, et al. *Theory of Lyapunov Exponents*. Izdat. "Nauka", Moscow, 1966.

2. L. Cesari. *Asymptotic Behavior and Stability Problems in Ordinary Differential Equations*. Springer-Verlag, Berlin, 1959.

3. S.P. Dilberto. On systems of ordinary differential equations. *Contrib. to the Theory of Nonlinear Oscillations*. Ann. Math. Studies, no. 20, 1-38, Princeton Univ. Press, 1950.

4. A. Halanay. On a Linear Differential Equation with an Almost Periodic Coefficient. Dokl. Akad. Nauk SSSR (N.S.) 88 (1953), 419-422.

5. P. Hartman. *Ordinary Differential Equations*. Wiley, New York, 1964.

6. J.C. Lillo. A Note on the Continuity of Characteristic Exponents. Proc. Nat. Acad. Sci. U.S.A. 46 (1960), 247-250.

7. V.M. Millionščikov. A Stability Criterion... Almost Periodic Coefficients. Mat. Sbornik. 78 (1969), 179-201.

8. V.M. Millionščikov. On the Theory of Lyapunov Exponents. Mat. Zametki 7 (1970), 503-513.

9. V.M. Millionščikov. Linear Systems of Ordinary Differential Equations. *Actes Congr. Internat. Mathématiciens* (Nice, 1970), Tome 2, pp. 915-919, Gauthier-Villars, Paris, 1971. Also in Differencial'nye Uravnenija 7 (1971), 387-390.

10. O. Perron. Uber ein Matrixtransformation. Math. Zeit. 32 (1930), 465-473.

11. R.J. Sacker. A New Approach to the Perturbation Theory of Invariant Surfaces. Communications Pure Applied Math. 18 (1965), 717-732.

12. R.J. Sacker and G.R. Sell. Existence of Dichotomies and Invariant Splittings for Linear Differential Systems. I. J. Differential Equations (to appear).

13. R.J. Sacker and G.R. Sell. Existence of Dichotomies and Invariant Splittings for Linear Differential Systems. II. (to appear).

14. R.J. Sacker and G.R. Sell. A Spectral Theory for Linear Time-Varying Equations (to appear).

15. J.F. Selgrade. Isolated Invariant Sets for Flows on Vector Bundles. (to appear).

16. G.R. Sell. Topological Dynamics and Differential Equations. Van Nostrand-Reinhold, London, 1971.

17. G.R. Sell. Linear Differential Systems. Lecture Notes. Univ. of Minnesota, 1974.

This research was supported in part by NSF Grant No. 38955.

GLOBAL ESTIMATES FOR NONLINEAR REACTION AND DIFFUSION

Ivar Stakgold

1. INTRODUCTION

We shall obtain gradient bounds and some global estimates for the solution u(x) of the nonlinear problem of combined diffusion and reaction

$$(1) \qquad -\Delta u = f(u) \quad , \quad x \in D \quad ; \quad \frac{\partial u}{\partial \nu} + hu = 0 \quad , \quad x \in \partial D \quad .$$

Here D is a bounded domain in R_n with boundary ∂D and outward normal ν, Δ is the n-dimensional Laplacian, h is a positive constant ($h = \infty$ corresponds to vanishing Dirichlet data), and f is such that (1) has a unique positive solution u(x). We assume throughout that f(z) is continuous for $z \geq 0$ and that $f(0) = 0$; the forced case $f(0) > 0$ and the nonlinear boundary condition $\partial u/\partial \nu + hu = p(u)$ will be treated elsewhere.

Problem (1) arises in a variety of applied contexts such as: a) steady operation of a homogeneous, monoenergetic nuclear reactor with feedback - here u is the neutron density; b) nonlinear heating such as Joule heating in a homogeneous medium (with u being the temperature); c) nonlinear chemical reaction combined with diffusion in a biochemical setting - here u is the concentration of a reactant. It should perhaps be noted that the equation $-\operatorname{div}(k(v) \operatorname{grad} v) = q(v)$ corresponding to a nonlinear diffusion coefficient can be transformed to equation (1) by the change of variable $u = \int_0^v k(z)dz$.

In the analysis of (1) an important role is played by the linear problem

$$(2) \qquad -\Delta\phi = \lambda\phi \quad , \quad x \in D \quad ; \quad \frac{\partial\phi}{\partial\nu} + h\phi = 0 \quad , \quad x \in \partial D \quad .$$

The fundamental eigenvalue $\lambda_1 = \lambda_1(h)$ of (2) is simple, positive, and increases with h, while the corresponding eigenfunction does not vanish in D. We shall let $\phi_1(x)$ be the <u>positive</u> fundamental eigenfunction whose <u>maximum value</u> is 1. If $h \nless \infty$, ϕ_1 is positive on the closure of D.

By setting $f(z) = \lambda g(z)$, we can regard (1) as a branching problem. Under suitable conditions on g, (1) will then have a branch of positive solutions emanating from the trivial solution at $\lambda = \lambda_1$. Preliminary results on existence and uniqueness of positive solutions will be obtained by using monotone methods (see KELLER [2], SIMPSON and COHEN [7], SATTINGER [6], STAKGOLD and PAYNE [8]). There is little that is new here apart from slight improvements in some of the proofs. Next we derive gradient bounds by an indirect use of Hopf's second maximum principle[1]. We find that $J(x) = |\text{grad } u|^2 + \int_0^u f(z)dz$ obeys an elliptic inequality except where grad u vanishes. A calculation shows that $\frac{\partial J}{\partial\nu} \leq 0$ on the boundary so that the maximum of J must occur at an exceptional point, leading to the desired bound. We then introduce level surface coordinates for u, enabling us to derive a number of isoperimetric norm estimates. Finally, by using the volume enclosed by a level surface as a new independent variable, we obtain an upper bound for the total flux of u through the boundary, this quantity having special importance in applications. In general outline the approach is similar to that in some of our previous papers (PAYNE and STAKGOLD [4,5,8]) but additional difficulties - both technical and conceptual - arise in virtue of the nonlinearity of the equation and the nature of the boundary condition. At the same time we are able to extend and deepen some of our earlier results, at the sacrifice of restricting ourselves to convex domains.

2. EXISTENCE AND UNIQUENESS OF POSITIVE SOLUTIONS

A <u>positive</u> solution of (1) is understood to be a solution $u(x)$ which is nonnegative but does not vanish identically in D. Let us set

$$f(z) = \lambda g(z) \quad , \quad \lambda > 0 \quad , \tag{3a}$$

where $g(z)$ is continuous for $z \geq 0$ and

$$g'(z) \text{ is strictly decreasing for } z > 0 \quad , \tag{3b}$$

$$g'(0) = 1 \quad , \; g(0) = 0 \quad . \tag{3c}$$

We then define

$$\mu = \begin{cases} \lim\limits_{z \to \infty} z/g(z) & \text{if this limit is positive} \\ + \infty & \text{otherwise} \quad . \end{cases} \tag{4}$$

The boundary value problem (1) takes the form

$$-\Delta u = f(u) \text{ or, equivalently } -\Delta u = \lambda g(u) \quad , \quad x \in D, \tag{5a}$$

$$\frac{\partial u}{\partial \nu} + hu = 0 \quad , \quad x \in \partial D \quad . \tag{5b}$$

<u>Remarks</u>. 1. We do not exclude the possibility that g might become negative as z increases.

2. For each λ, $\lambda_1 < \lambda < \mu\lambda_1$, the curves $\lambda_1 z$ and $f(z)$ intersect at exactly one positive value of z. For any other λ, these curves do not intersect for $z > 0$.

Theorem 1. If f satisfies conditions (3), the boundary value problem (5) has one and only one positive solution for $\lambda_1 < \lambda < \mu\lambda_1$ and no positive solution for any other λ; moreover, for $\lambda \leq \lambda_1$ a solution of (5) can not be positive anywhere in D.

Proof A. Nonexistence. For (5) to have a positive solution, $f(z)-\lambda_1 z$ must change sign for $z > 0$. Indeed, adding $-\lambda_1 u$ to both sides of (5a) and using the Fredholm alternative we find that $\int_D [f(u)-\lambda_1 u]\phi_1 dx = 0$ which implies that either $f(z)-\lambda_1 z$ changes sign for $z > 0$ or that u is a constant - a positive zero of $f(z)-\lambda_1 z$. This latter possibility is eliminated because a nonzero constant does not obey the boundary condition. Thus positive solutions can occur only for $\lambda_1 < \lambda < \mu\lambda_1$.

To prove that solutions for $\lambda \leq \lambda_1$ can not be positive anywhere, suppose the contrary to be true. Then there exists a proper subdomain $D' \subset D$ with $u > 0$ in D', $\frac{\partial u}{\partial \nu} + hu = 0$ on $\partial D' \cap \partial D$ and $u = 0$ on the rest of D'. Let λ_1' and ϕ_1' be the fundamental eigenvalue and positive eigenfunction of $-\Delta$ for D' with the boundary conditions just described. For $\lambda \leq \lambda_1$, the hypothesis gives $f(z)-\lambda_1 z \leq 0$ for $z > 0$ and hence $f(z)-\lambda_1' z < 0$ since $\lambda_1' > \lambda_1$. Subtracting $\lambda_1' u$ from both sides of (5a) and applying the Fredholm alternative, we obtain $\int_{D'} [f(u)-\lambda_1' u]\phi_1' dx = 0$, which is a contradiction.

B. Existence. We shall use monotone iteration schemes to construct maximal and minimal positive solutions. Recall that an upper solution $\bar{u}(x)$ satisfies the inequalities

$$-\Delta\bar{u}-f(\bar{u}) \geq 0 \quad , \quad x \in D \quad ; \quad \frac{\partial \bar{u}}{\partial \nu} + h\bar{u} \geq 0 \quad , \quad x \in \partial D \quad ,$$

whereas for a lower solution $\underline{u}(x)$ both inequalities are reversed.

For $A > 0$, the function $v = A\phi_1$ satisfies

$$-\Delta v - f(v) = \lambda A\phi_1\left[\frac{\lambda_1}{\lambda} - \frac{g(A\phi_1)}{A\phi_1}\right] \quad ; \quad \frac{\partial v}{\partial \nu} + hv = 0 \text{ on } \partial D \quad .$$

Since $g'(0) = 1$, we can, for each $\lambda > \lambda_1$, choose A sufficiently small so that v is a lower solution. We also know that $\phi_1 \geq \delta > 0$ on D so that we can, for each $\lambda < \mu\lambda_1$, choose A so large that v is an upper solution. (The argument must be modified in the Dirichlet case because ϕ_1 now vanishes on ∂D; an upper solution can then be found in the form $A\tilde{\phi}_1$ where $\tilde{\phi}_1$ is the positive eigenfunction of (2) corresponding to a large positive value of h).

Starting from the lower solution just described we use a standard iteration procedure to construct a monotonically increasing sequence $\underline{u}_n$ which converges to the minimal positive solution u_* of (5). Similarly, from the upper solution, we construct a decreasing sequence $\bar{u}_n$ converging to the maximal positive solution u^* of (5).

C. Uniqueness. Let u_* and u^* be the minimal and maximal positive solutions constructed in part B. Since u_* exceeds $A\phi_1$ for some positive A, we have $u_* > 0$ in D. We know that $u^* \geq u_*$; suppose the strict inequality occurs on D' whereas equality holds on the remainder of D. Applying Green's theorem to D', we find

$$\int_{D'} u_* u^* \left[\frac{f(u^*)}{u^*} - \frac{f(u_*)}{u_*}\right] dx = \int_{\sigma_1+\sigma_2} \left[u^* \frac{\partial u_*}{\partial \nu} - u_* \frac{\partial u^*}{\partial \nu}\right] ds \quad ,$$

Where $\sigma_1 = \partial D' \cap \partial D$ and σ_2 is the remainder of $\partial D'$.

The integral over σ_1 vanishes by the boundary condition (5b). On σ_2, we have $u^* = u_*$ and $\frac{\partial u_*}{\partial \nu} - \frac{\partial u^*}{\partial \nu} \geq 0$. By (3b), $f(z)/z$ is strictly decreasing which implies that the integral over D' is negative. This contradiction then shows that D' has zero measure; by continuity it follows that $u_* = u^*$ in D, completing the proof of Theorem 1.

We conclude this section with a simple result.

Lemma 1. Let f satisfy conditions (3) and let u(x) be a positive solution of $-\Delta u = f(u)$ on Ω (no boundary conditions specified). Then $f(u(x)) \geq 0$ on Ω.

Proof. At any point in Ω where the maximum is attained, $\Delta u \leq 0$ so that $f(u_m) \geq 0$. But $f(z_0) \geq 0$ implies $f(z) > 0$ for $0 < z < z_0$.

3. GRADIENT BOUNDS.

From here on we shall assume that f satisfies conditions (3) and therefore Lemma 1 is applicable. Although we wish to obtain gradient bounds for the unique positive solution of (5) when $\lambda_1 < \lambda < \mu\lambda_1$, we shall have to proceed by steps.

Lemma 2. Let u(x) be a positive solution of $-\Delta u = f(u)$ on a domain Ω and let

(6) $$J = |\text{grad } u|^2 + 2F(u) \quad ,$$

where

(7) $$F(u) = \int_0^u f(z)dz \quad ;$$

then J satisfies the elliptic inequality

(8) $$0 \leq \Delta J + \left[\sum_{k=1}^{n} a_k \frac{\partial J}{\partial x_k}\right] / |\text{grad } u|^2;$$

where the coefficients a_k are continuous and bounded on Ω.

Proof. Straightforward calculation and use of the Schwarz inequality (see [8]). The fact that $f(u(x)) \geq 0$ on Ω plays an essential role.

Theorem 2. Let u(x) be a positive solution of $-\Delta u = f(u)$ in a convex domain Ω with boundary $\partial\Omega$ (of class $C^{2+\varepsilon}$ for the time being) on which u is constant. Then

(9) $$|\text{grad } u|^2 \leq 2\left[F(u_m) - F(u)\right] .$$

Proof. Introduce a normal-tangential coordinate system in a neighbourhood of $\partial\Omega$. We have

(10) $$J = \left(\frac{\partial u}{\partial \nu}\right)^2 + |\text{grad}_t\ u|^2 + 2F(u) ,$$

where $\text{grad}_t\ u$ is the tangential component of grad u. Then

(11) $$\frac{\partial J}{\partial \nu} = 2\left(\frac{\partial u}{\partial \nu}\right)\left(\frac{\partial^2 u}{\partial \nu^2}\right) + 2\text{grad}_t\ u.\ \frac{\partial}{\partial \nu}\ \text{grad}_t\ u + 2f(u)\ \frac{\partial u}{\partial \nu}$$

and, since $\text{grad}_t\ u = 0$ on $\partial\Omega$,

(12) $$\frac{\partial J}{\partial \nu}_{\partial\Omega} = 2\frac{\partial u}{\partial \nu}\left[\frac{\partial^2 u}{\partial \nu^2} + f(u)\right] .$$

By the smoothness assumption on $\partial\Omega$ we may apply the differential equation at the boundary where it takes the form

(13) $$\frac{\partial^2 u}{\partial \nu^2} + (n-1)K\ \frac{\partial u}{\partial \nu} + \Delta' u + f(u) = 0 ,$$

where Δ' is the surface Laplacian and K the mean curvature. Since u is constant on $\partial\Omega$, $\Delta' u = 0$, and, substituting for $\frac{\partial^2 u}{\partial \nu^2}$ in (12), we find

(14) $$\left(\frac{\partial J}{\partial \nu}\right)_{\partial\Omega} = -2(n-1)K\ \left(\frac{\partial u}{\partial \nu}\right)^2 \leq 0$$

because $K \geq 0$ for a convex domain.

Since J satisfies (8) its maximum occurs either where grad u = 0 (this includes the case $J \equiv$ constant) or at a point on $\partial\Omega$ where $\frac{\partial J}{\partial \nu} > 0$ (by Hopf's second maximum principle). This latter possibility is ruled out by (14) so that J has its maximum where grad u vanishes and hence (9) follows.

Remarks. 1. The bound (9) is exact for one-dimensional problems.

2. By approximating with smooth boundaries, we can extend Theorem 2 to a convex domain Ω with a Lipschitz boundary.

3. Consider (5) for a convex domain D and $\lambda_1 < \lambda < \mu\lambda_1$. If $h = \infty$, u vanishes on ∂D and Theorem 2 is immediately applicable with $\Omega = D$. If $h < \infty$, u is not necessarily constant on ∂D, but its maximum value τ on ∂D is certainly less than u_m since $\frac{\partial u}{\partial \nu}$ is negative on ∂D. Thus (9) holds for the domain $\Omega \subset D$ where $u > \tau$ under the reasonable (but unproved) assumption that Ω is convex if D is convex. For the purpose of deriving isoperimetric inequalities in the sequel it is sufficient to know that (9) is valid for $u > \tau$. Alternatively, for $n = 2$, we have been able to modify the proof of Theorem 2 to take into account the fact that for $h < \infty$ neither $\mathrm{grad}_t\, u$ in (11) nor $\Delta' u$ in (13) vanishes on ∂D.

Theorem 3. Consider problem (5) with D convex, $n = 2$, $\lambda_1 < \lambda < \mu\lambda_1$. Then (9) holds in D.

Proof. Let s be the tangential coordinate which coincides with the arc length on ∂D. The corresponding metric coefficient $k(x)$ is then identically equal to 1 on ∂D. We then have from (10) that

$$J = \left(\frac{\partial u}{\partial \nu}\right)^2 + \left(\frac{1}{k}\frac{\partial u}{\partial s}\right)^2 + 2F(u) \quad , \tag{15}$$

and (12) becomes

$$\left(\frac{\partial J}{\partial \nu}\right)_{\partial D} = 2\,\frac{\partial u}{\partial \nu}\left(\frac{\partial^2 u}{\partial \nu^2} + f\right) - 2\left(h + \frac{\partial k}{\partial \nu}\right)\left(\frac{\partial u}{\partial s}\right)^2 \tag{16}$$

whereas (13) takes the form

$$\frac{\partial^2 u}{\partial \nu^2} + K\frac{\partial u}{\partial \nu} + \frac{\partial^2 u}{\partial s^2} + f(u) = 0 \quad . \tag{17}$$

Substituting for $\frac{\partial^2 u}{\partial \nu^2}$ in (16), we find, on observing that $\frac{\partial k}{\partial \nu} = K \geq 0$,

$$\left(\frac{\partial J}{\partial \nu}\right)_{\partial D} \leq 2\,\frac{\partial u}{\partial \nu}\left[-K\,\frac{\partial u}{\partial \nu} - \frac{\partial^2 u}{\partial s^2}\right] \quad , \tag{18}$$

which differs from (14) by the second derivative term. We also have

$$\frac{\partial J}{\partial s} = 2\,\frac{\partial u}{\partial \nu}\,\frac{\partial^2 u}{\partial s \partial \nu} + 2\left(\frac{1}{k}\,\frac{\partial u}{\partial s}\right)\frac{\partial}{\partial s}\left(\frac{1}{k}\,\frac{\partial u}{\partial s}\right) + 2f(u)\,\frac{\partial u}{\partial s}$$

which on ∂D reduces to

$$\left(\frac{\partial J}{\partial s}\right)_{\partial D} = \frac{\partial u}{\partial s}\left(2h^2u + f + 2\,\frac{\partial^2 u}{\partial s^2}\right) \quad . \tag{19}$$

A further calculation yields

$$\frac{1}{2}\left(\frac{\partial^2 J}{\partial s^2}\right)_{\partial D} = \left(\frac{\partial u}{\partial s}\right)\left(h^2\frac{\partial u}{\partial s} + \frac{\partial^3 u}{\partial s^3} + f'\frac{\partial u}{\partial s}\right) + \left(\frac{\partial^2 u}{\partial s^2}\right)\left(h^2 u + f + \frac{\partial^2 u}{\partial s^2}\right) \quad . \tag{20}$$

Let P be the point on the boundary at which J is supposed to have a maximum. Then $\frac{\partial J}{\partial s} = 0$ and $\frac{\partial^2 J}{\partial s^2} \leq 0$ at P. If $\frac{\partial^2 u}{\partial s^2} > 0$ at P, (19) shows that $\frac{\partial u}{\partial s} = 0$ and (20) would give $\frac{\partial^2 J}{\partial s^2} > 0$, a contradiction. Therefore $\frac{\partial^2 u}{\partial s^2} \leq 0$ at P, and (18) gives $\left(\frac{\partial J}{\partial \nu}\right)_P \leq 0$ in conflict with Hopf's second maximum principle. Thus the maximum of J must occur where grad u = 0 which establishes (9) once more.

<u>Corollary</u>. Let τ be the maximum of u on ∂D and τ^* the unique positive root of

$$h^2\tau^2 + 2F(\tau) = 2F(u_m) \quad ; \tag{21}$$

then

$$\tau \leq \tau^* \quad .$$

<u>Proof</u>. At the point on the boundary where $u = \tau$, we have

$$h^2\tau^2 = \left(\frac{\partial u}{\partial \nu}\right)^2 \leq |\text{grad } u|^2 \quad ,$$

but this last term has the upper bound in (9), by Remark 3 following Theorem 2.

4. NORM ESTIMATES.

Using the preceding Corollary and the bound (9) we can obtain some rough global estimates by integrating (5a) over D. It is, however, more fruitful to introduce the level surfaces for u(x). Let D(t) be the domain where u exceeds t; its boundary, which may include part of ∂D, is denoted by $\partial D(t)$, and its volume by v(t). Clearly v(t) is a decreasing function of t with maximum value V and minimum value 0 at $t = u_m$.

We first note some elementary relations between v(t) and u. For any continuous function a(z), define

$$\alpha(t) = \int_{D(t)} a(u)dx \quad .$$

Then

$$\alpha'(t) = a(t)v'(t) \quad , \quad \alpha(t) = -\int_t^{u_m} a(z)v'(z)dz \quad , \tag{22}$$

and

$$v'(t) = -\int_{\partial D(t)} |\text{grad } u|^{-1} ds \quad , \quad t \geq \tau \quad . \tag{23}$$

<u>Lemma</u> 3. Let

$$\phi(t) = \int_{D(t)} f(u)dx \quad .$$

Then, for $t \geq \tau$,

$$\phi(t) \leq \phi(\tau)\left[F(u_m)-F(t)\right]^{\frac{1}{2}}\left[F(u_m)-F(\tau)\right]^{-\frac{1}{2}} \quad . \tag{24}$$

<u>Proof</u>. For $t \geq \tau$, $\partial D(t)$ is the level surface u = t so that $-\frac{\partial u}{\partial \nu} = |\text{grad } u|$ on $\partial D(t)$. Integration of (5a) over D(t) then gives

$$\phi(t) = \int_{\partial D(t)} |\text{grad } u| \, ds \quad ,$$

while (22) and (23) yield

$$\phi'(t) = f(t)v'(t) = -f(t)\int_{\partial D(t)} |\text{grad } u|^{-1} ds \quad .$$

Combining these equations for ϕ and ϕ', and using (9), we find

$$-\phi'/\phi \geq \frac{f(t)}{\max\limits_{x\in\partial D(t)} |\operatorname{grad} u|^2} = \frac{f(t)}{2\left[F(u_m)-F(t)\right]} .$$

Integrating this inequality from τ to t then gives (24).

Lemma 3 is the basis for the following theorem which yields a variety of norm estimates.

Theorem 4. Let $a(z)$ be an arbitrary continuous function increasing for $z \geq 0$. Then

$$(25)\qquad \frac{\int_D f(u)a(u)dx}{\int_D f(u)dx} \leq \left[2F(u_m)-2F(\tau)\right]^{-\frac{1}{2}} I(\tau) \leq \frac{1}{h\tau^*}\, I(\tau^*) ,$$

where

$$(26)\qquad I(z) = \int_z^{u_m} a(t)f(t)\left[2F(u_m)-2F(t)\right]^{-\frac{1}{2}}dt .$$

Proof. Since $a(t)$ is increasing, we can write

$$(27)\qquad \int_D f(u)a(u)dx = a(\tau)\int_{D-D(\tau)} f(u)dx + \int_{D(\tau)} f(u)a(u)dx .$$

To estimate the last term, let us multiply (24) by $a'(t)$ and integrate from τ to u_m, the integration on the left side being done by parts. Using (22), we find

$$(28)\qquad \int_{D(\tau)} f(u)a(u)dx \leq \phi(\tau)a(\tau) + \phi(\tau)\left[2F(u_m)-2F(\tau)\right]^{-\frac{1}{2}} \int_\tau^{u_m} a'(t)\left[2F(u_m)-2F(t)\right]^{\frac{1}{2}}dt .$$

When substituting in (27) the first terms on the right of (28) and (27) combine to give $a(\tau)\int_D f(u)dx$; in the remaining term we replace $\phi(\tau)$ by its upper bound $\int_D f(u)dx$ and integrate by parts to obtain the first inequality in (25). The second inequality follows from (21) and the observation that $\left[2F(u_m)-2F(z)\right]^{-\frac{1}{2}}I(z)$ is an increasing function of z for $z \geq 0$.

Some consequences of (25) are worth noting. If $a(u) = u^p/f(u)$, with p a positive integer, we obtain

$$\int_D u^p dx \leq \frac{\int_D f(u)dx}{h\tau^*} \int_{\tau^*}^{u_m} t^p \left[2F(u_m)-2F(t)\right]^{-\frac{1}{2}} dt \quad . \tag{29}$$

As $h \to \infty$, τ^* tends to 0 and $h\tau^* \to \left[2F(u_m)\right]^{\frac{1}{2}}$, so that, we find for the Dirichlet problem

$$\int_D u^p dx \leq \int_D f(u)dx \int_0^{u_m} t^p \left[2F(u_m)\right]^{-\frac{1}{2}} \left[2F(u_m)-2F(t)\right]^{-\frac{1}{2}} dt \quad .$$

This last inequality can be applied in the limiting linear case $f(u) = \lambda_1 u$; we then recover, for $p = 2$, a result of [8]:

$$\int_D u^2 dx \leq \frac{\pi}{4} u_m \int_D u dx \quad .$$

We conclude this section by deriving a Payne-Rayner type of inequality complementary to (25), see [3]. We confine ourselves to the 2-dimensional problem. Multiplying the expressions for ϕ and ϕ' in the proof of Lemma 3, we find, by using the Schwarz inequality,

$$-\frac{d}{dt}\phi^2(t) \geq 2f(t)S^2(t) \geq 8\pi f(t)v(t) \quad , \quad t \geq \tau \quad ,$$

where $S(t)$ is the length of the boundary $\partial D(t)$, and the classical isoperimetric inequality was used in the last step. An integratio from $t = \tau$ to $t = u_m$ then gives

$$\phi^2(\tau) \geq 8\pi \int_{D(\tau)} F(u)dx - 8\pi F(\tau)v(\tau) \quad ,$$

which is used in the chain of inequalities

$$\int_D F(u)dx \leq \int_{D(\tau)} F(u)dx + F(\tau)\left[V-v(\tau)\right]$$

$$\leq \frac{1}{8\pi} \phi^2(\tau) + F(\tau)V$$

$$\leq \frac{1}{8\pi} \left[\int_D f(u)dx\right]^2 + F(\tau^*)V \quad .$$

In the Dirichlet case the last inequality becomes

(30) $$\int_D F(u)dx \leq \frac{1}{8\pi}\left[\int_D f(u)dx\right]^2 \quad .$$

It is perhaps worth noting that inequalities such as (21), (25) (29) all become equalities in the one-dimensional case.

5. ISOPERIMETRIC INEQUALITY FOR THE TOTAL FLUX

To simplify the calculations we confine ourselves in this Section to the Dirichlet problem. As in Section 4 we let v(t) be the volume enclosed by the level surface u = t. Since v(t) is a decreasing function of t, we may use v as a new independent variable.

With $\phi(t)$ as in Lemma 3 of the preceding Section, we define

$$\Phi(v) = \phi(t(v)) \quad ,$$

from which it follows that

(31) $$\Phi'(v) = f(t(v))$$

(32) $$\Phi''(v) = f't' = \frac{f'}{v'} \quad ,$$

where v' can be expressed in terms of |grad u| from (23). Multiplying (32) and the equation obtained from integrating (5a) over D(t(v)), we find

$$-\Phi\Phi'' \leq 2f'(t(v))\left[F(u_m)-F(t(v))\right] .$$

We establish the inequality

(33) $$2f'(z)\left[F(u_m)-F(z)\right] - 2f'(0)F(u_m) \leq -f^2(z)$$

by noting that both sides vanish at z = 0 and that the derivative on the left side is smaller than on the right. Hence

(34) $$(\Phi')^2 - \Phi\Phi'' \leq 2f'(0)F(u_m) = \alpha^2 \quad .$$

If we multiply this inequality by the positive quantity Φ'/Φ^3, we find

$$-\left[(\Phi'/\Phi)^2\right]' \leq -\alpha^2 \left[1/\Phi^2\right]' \quad ,$$

which we now integrate from v to V to obtain

$$\Phi'(v) \leq \alpha\left[1 - \frac{\Phi^2(v)}{\Phi^2(V)}\right]^{\frac{1}{2}} \quad .$$

Integrating once more, this time from 0 to V, we find

$$\Phi(V) \leq \frac{2}{\pi}\alpha V \quad ,$$

or

(35) $$\frac{\int_D f(u)dx}{V} \leq \frac{2}{\pi}\left[2f'(0)F(u_m)\right]^{\frac{1}{2}} \quad .$$

In the one-dimensional problem the two sides can be computed explicitly, the ratio of the right side to the left being $\sqrt{\lambda/\lambda_1}$ which will be small if we are close to criticality (the usual situation in applications). For the linear case, $f(u) = \lambda_1 u$, $F(u) = \lambda_1 u^2/2$, and we recover the isoperimetric inequality of [4],

(36) $$\frac{\int_D u dx}{V u_m} \leq \frac{2}{\pi} \quad .$$

ACKNOWLEDGEMENT

This research was supported in part by A.E.C. grant AT(11-1) -2280 MOD. 1. The author also wishes to express his thanks to the members of the Mathematical Institute at Oxford for their hospitality.

BIBLIOGRAPHY

1. E. HOPF, Proc. Amer. Math. Soc. 3 (1952), 291-293.

2. H.B. KELLER, Bull. Amer. Math. Soc. 74 (1968), 887-891

3. L.E. PAYNE & M.E. RAYNER, Z. Angew. Math. Phys. 23 (1972), 13-15.

4. L.E. PAYNE & I. STAKGOLD, Applicable Analysis, to appear

5. L.E. PAYNE & I. STAKGOLD, to appear

6. D.H. SATTINGER, Topics in Stability and Bifurcation Theory Lecture Notes in Mathematics, no. 309, Springer, 1973.

7. R.B. SIMPSON & D.S. COHEN, J. Math. Mech. 19 (1970), 895-910.

8. I. STAKGOLD & L.E. PAYNE, Nonlinear problems in nuclear reactor analysis, in Nonlinear Problems in the Physical Sciences and Biology, Lecture Notes in Mathematics, no. 322, Springer, 1973.

PERTURBATION THEORY FOR SOBOLEW SPACES

F. Stummel

This lecture surveys a new perturbation theory for the Sobolew spaces $W^{m,p}(G)$ (cf. [3]). The functional analysis for the treatment of general perturbations of linear operators has been established in [1]. One finds a corresponding perturbation theory for elliptic sesquilinear forms on subspaces of a Hilbert space and an application to the Dirichlet problem in [2]. Within this framework, a perturbation theory for Sobolew spaces has been developed in [3] which permits the treatment of boundary value problems in partial differential equations under perturbations of coefficients and inhomogeneous terms as well as of boundary conditions and domains of definition. The theory studies the basic concepts and methods, in particular, the convergence of sequences of open sets G_ι, $\iota \in \mathbb{N}$, to G in $\mathbb{R}^n$, the strong and weak convergence of the sequence of Sobolew spaces $W^{m,p}(G_\iota)$, $\iota \in \mathbb{N}$, to $W^{m,p}(G)$ and of $W_o^{m,p}(G_\iota)$ to $W_o^{m,p}(G)$, the discrete compactness of the sequence of natural embeddings of Sobolew spaces, and the continuous convergence of continuous linear functions, boundary integrals and trace operators for the sequence of Sobolew spaces $W^{m,p}(G_\iota)$, $\iota \in \mathbb{N}$.

References

1. Stummel, F. : Diskrete Konvergenz Linearer Operatoren. I. Math. Ann. 190, 45-92 (1970). II. Math. Z. 120, 231-264 (1971). III Proc. Conference on Linear Operators and Approximation, Oberwolfach 1971. Int. Series of Numerical Mathematics 20, 196-216, Basel: Birkhäuser 1973.

2. Stummel, F. : Singular perturbations of elliptic sesquilinear forms. Proc. Conference on Differential Equations, Dundee 1972. Lecture Notes in Mathematics 280, 155-180. Berlin-Heidelberg-New York: Springer 1972.

3. Stummel, F. : Perturbation theory for Sobolew spaces. To appear in the Proceedings of the Royal Society of Edinburgh.

PERTURBATIONS OF SELF-ADJOINT OPERATORS IN $L_2(G)$ WITH APPLICATIONS TO DIFFERENTIAL OPERATORS

Joachim Weidmann

1. Introduction

For operators T and V in $L_2(\mathbb{R}^m)$ the notion of T-smallness near infinity of V has been introduced by Jörgens and the author in [8]. It has been shown that the T-smallness near infinity of the perturbation is essentially sufficient for the invariance of the essential spectrum of a Schrödinger operator T. Meanwhile this notion has also proved to be useful in scattering theory ([9],[11],[12]).

If one considers operators in $L_2(G)$ for an open subset G of $\mathbb{R}^m$ it would be quite natural to define T-smallness near the boundary (of G). Actually it is not difficult to develop such a theory. But one easily finds examples which show that for the invariance of the essential spectrum it is not necessary that the perturbation is small near all points of the boundary.

Consider for example in $L_2(0,1)$ the operator T with

$$D(T) = \{u \in L_2(0,1): u \text{ and } u' \text{ are locally absolutely continuous in } (0,1], (x^2u')' \in L_2(0,1), u(1)=0\},$$

$$Tu(x) = -(x^2u')' \text{ for } u \in D(T).$$

It is not difficult to show that T is self-adjoint and $\sigma_e(T) = [\frac{1}{4},\infty)$ (see [13], Satz 1.1b) and Satz 3.1b)). We now consider two perturbations V_1 and V_2 of T.

V_1 is the operator of multiplication by a continuous function

$q_1: [0,1] \to \mathbb{R}$ with $q_1(x) \equiv 0$ near 0 and $q_1(x) \equiv 1$ near 1.

V_2 is the operator of multiplication by a continuous function $q_2: [0,1] \to \mathbb{R}$ with $q_2(x) \equiv 1$ near 0 and $q_2(x) \equiv 0$ near 1.

Then by the same methods as for T one shows

$$\sigma_e(T+V_1) = [\tfrac{1}{4},\infty),$$

$$\sigma_e(T+V_2) = [\tfrac{5}{4},\infty).$$

Therefore it seems that in this case the perturbations V need not be small near the boundary point 1.

This motivates the notion of T-smallness near A (T-A-smallness of Definition 2.5), where A is a subset of the boundary of G. We shall see that for many cases similar results as in [8] can be proved if the perturbation is T-A-small for suitable A. The essential condition on T and A is the T-compactness of the operators

$$u \mapsto \varphi u \text{ and } u \mapsto \varphi Tu - T\varphi u$$

for $\varphi \in C_o^\infty(\mathbb{R}^m \setminus A)$. This condition is satisfied for large classes of differential operators.

The abstract theory is developed in sections 2 and 3. In section 4 we indicate some applications to differential operators. Several known results can be recovered.

2. Relatively compact perturbations and perturbations which are small near the boundary

Let G be an open subset of $\mathbb{R}^m$, T an operator in $L_2(G)$, A a closed subset of $\overline{G}$. We consider operators V in $L_2(G)$, which are T-small near A and near infinity (T-A-small, Definition 2.5). It would also be possible to consider operators V which are small near A and only near some (or no) points at infinity; some technical changes are needed to do this. But for differential operators this seems to be not so important.

Our first theorem is a more abstract and more general version of Theorem 3.2 of [8].

<u>2.1. Theorem.</u> Let T and V be closable operators in $L_2(G)$ with $D(T) \subset D(V)$.

(i) If V is T-compact, then for every $\varepsilon > 0$ there is a compact subset K of G and a constant $C \geq 0$, such that

$$(2.2) \qquad \|Vu\| \leq \varepsilon(\|u\|+\|Tu\|)+C\|\chi_K u\|, \quad u \in D(T).$$

(ii) Let A be a closed subset of $\overline{G}$. If for every compact subset K of $\overline{G} \setminus A$ the operator $u \mapsto \chi_K u$ is T-compact and if for every $\varepsilon > 0$ there exists a compact subset K of $\overline{G} \setminus A$ and $C \geq 0$ such that (2.2) holds, then V is T-compact.

<u>2.3. Remark.</u> The statements of 2.1 may be restated in the following form, which is more similar to Theorem 3.2 of [8].

(i) If V is T-compact, then for every $\varepsilon > 0$ there is a $\varphi \in C_o^\infty(G)$ such that

$$(2.4) \qquad \|Vu\| \leq \varepsilon(\|u\|+\|Tu\|)+\|\varphi u\|, \quad u \in D(T).$$

(ii) Let A be a closed subset of $\overline{G}$. If for every $\varphi \in C_o^\infty(\mathbb{R}^m \setminus A)$ the operator $u \mapsto \varphi u$ is T-compact and if for every $\varepsilon > 0$ there is a $\varphi \in C_o^\infty(\mathbb{R}^m \setminus A)$ such that (2.4) holds, then V is T-compact.

<u>Proof</u> of Theorem 2.1. (i) Assume that the statement is not true. Then there is an $\varepsilon > 0$ such that for every compact subset K of G and every $C \geq 0$ there is a $u \in D(T)$ such that

$$\|Vu\| > \varepsilon(\|u\|+\|Tu\|) + C\|\chi_K u\|.$$

We may assume that $\|u\|+\|Tu\| = 1$, hence

$$\|Vu\| > \varepsilon + C\|\chi_K u\|.$$

Now choose $K_n := \{x \in G: |x| \leq n, d(x, G) \geq \frac{1}{n}\}$, $C_n := n$. Then there are elements $u_n \in D(T)$ such that $\|u_n\|+\|Tu_n\| = 1$ and

$$\|Vu_n\| > \varepsilon + n\|\chi_{K_n} u_n\|, \quad n \in \mathbb{N}.$$

This implies for every $m \in \mathbb{N}$ and $n > m$ (note that V is T-bounded, $\|Vu\| \leq a(\|u\| + \|Tu\|)$)

$$\|\chi_{K_m} u_n\| \leq \|\chi_{K_n} u_n\| < \frac{1}{n}\|Vu_n\| \leq \frac{a}{n} \to 0$$

for $n \to \infty$; hence $u_n \rightharpoonup 0$ for $n \to \infty$.

It remains to show that $Tu_n \rightharpoonup 0$; then the T-compactness of V implies $Vu_n \to 0$ (see [8], Lemma 3.1), which contradicts $\|Vu_n\| > \varepsilon$.

Let us first assume that D(T) is dense; then T* exists and, since T is closable, D(T*) is dense. For every $v \in D(T^*)$ we have

$$\langle v, Tu_n \rangle = \langle T^* v, u_n \rangle \to 0, \ n \to \infty.$$

Since (Tu_n) is bounded this implies $Tu_n \rightharpoonup 0$. If D(T) is not dense, then we consider T as an operator from the Hilbert space $\overline{D(T)}$ into $L_2(G)$ and reason as before. Therefore in every case $Tu_n \rightharpoonup 0$ for $n \to \infty$.

(ii) Let (u_n) be a sequence in D(T) with $u_n \rightharpoonup 0$ and $Tu_n \rightharpoonup 0$; we have to show $Vu_n \to 0$ (see [8], Lemma 3.1). For every $\varepsilon > 0$ choose a compact subset K of $\overline{G}\setminus A$ and $C \geq 0$ such that (2.2) holds. Then by assumption $\chi_K u_n \to 0$ and therefore

$$\limsup_{n \to \infty} \|Vu_n\| \leq \varepsilon \limsup_{n \to \infty} (\|u_n\| + \|Tu_n\|).$$

This holds for every $\varepsilon > 0$, i.e. $Vu_n \to 0$. Q.E.D.

The above theorem motivates the following definition.

2.5. Definition. Let T and V be operators in $L_2(G)$ with $D(T) \subset D(V)$ and let A be a closed subset of $\overline{G}$. V is called T-A-small if for every $\varepsilon > 0$ there exists a compact subset K of $\overline{G}\setminus A$ such that for every $u \in D(T)$ with supp $u \subset \overline{G}\setminus K$ we have

$$\|Vu\| \leq \varepsilon(\|u\| + \|Tu\|).$$

(T-A-smallness of V means actually that V is small relative to T in neighbourhoods of A and of infinity; supp always stands for essential support).

For later reference we note the following simple Lemma.

2.6. Lemma. If V is T-A-small, then V is also (T+V)-A-small.

Proof. Let $\varepsilon > 0$; without restriction we may assume $\varepsilon < 1$. There exists a compact subset K of $\overline{G}\setminus A$ such that for $u \in D(T)$ with $\operatorname{supp} u \subset \overline{G}\setminus K$ we have

$$\begin{aligned}\|Vu\| &\leq \tfrac{\varepsilon}{2}(\|u\|+\|Tu\|)\\ &\leq \tfrac{\varepsilon}{2}(\|u\|+\|(T+V)u\|+\|Vu\|).\end{aligned}$$

Therefore

$$\|Vu\| \leq \frac{\varepsilon}{2-\varepsilon}(\|u\|+\|(T+V)u\|) \leq \varepsilon(\|u\|+\|(T+V)u\|),$$

i.e. V is (T+V)-A-small. Q.E.D.

2.7. Theorem. Let T and V be closable operators in $L_2(G)$ with $D(T) \subset D(V)$ and let A be a closed subset of $\overline{G}$.

(i) If V is T-compact, then V is T-bounded with T-bound 0 and V is T-∂G-small (where ∂G is the boundary of G).

(ii) Assume that for every $\varphi \in C_o^\infty(\mathbb{R}^m\setminus A)$ we have $\varphi D(T) \subset D(T)$ and that the operators

$$u \mapsto \varphi u \text{ and } u \mapsto \varphi Tu - T\varphi u$$

are T-compact. If V is T-bounded with T-bound 0 and T-A-small, then V is T-compact.

Proof. The proof is an abstract and generalized version of the proof of Theorem 3.12 of [8].

(i) It is well known that a T-compact operator V has T-bound 0. It remains to show that V is T-∂G-small. Let $\varepsilon > 0$ and choose K and C by means of Theorem 2.1 (i); K is a compact subset of $G = \overline{G}\setminus\partial G$. Then for every $u \in D(T)$ with $\operatorname{supp} u \subset \overline{G}\setminus K$

$$\|Vu\| \leq \varepsilon(\|u\|+\|Tu\|) + C\|\chi_K u\| = \varepsilon(\|u\|+\|Tu\|).$$

(ii) For every sequence (u_n) in D(T) with $u_n \rightharpoonup 0$ and $Tu_n \rightharpoonup 0$

we have to show that $Vu_n \to 0$. Let (u_n) be such a sequence.

For every $\varepsilon > 0$ choose the compact subset K of $\overline{G}\setminus A$ by means of the assumption (V is T-A-small). Choose also $a \geq 0$ such that

$$\|Vu\| \leq a\|u\| + \varepsilon\|Tu\| \quad \text{for} \quad u \in D(T).$$

Let now $\varphi \in C_o^\infty(\mathbb{R}^m\setminus A)$ be such that $\varphi(x) = 1$ for $x \in K$. Then we have for every $u \in D(T)$

$$\begin{aligned}\|Vu\| &\leq \|V\varphi u\|+\|V(1-\varphi)u\| \\ &\leq a\|\varphi u\|+\varepsilon\|T\varphi u\|+\varepsilon(\|(1-\varphi)u\|+\|T(1-\varphi)u\|) \\ &\leq \varepsilon\{\|\varphi Tu\|+\|(1-\varphi)u\|+\|(1-\varphi)Tu\|\} + a\|\varphi u\| + 2\varepsilon\|T\varphi u-\varphi Tu\| \\ &\leq 2\varepsilon\{\|u\|+\|Tu\|\} + a\|\varphi u\| + 2\varepsilon\|T\varphi u-\varphi Tu\|.\end{aligned}$$

From $u_n \rightharpoonup 0$, $Tu_n \rightharpoonup 0$ and the T-compactness of $u \mapsto \varphi u$ and $u \mapsto T\varphi u-\varphi Tu$ it follows that

$$\varphi u_n \to 0 \quad \text{and} \quad T\varphi u_n - \varphi Tu_n \to 0$$

and therefore

$$\limsup_{n \to \infty} \|Vu_n\| \leq 2\varepsilon \limsup_{n \to \infty} (\|u_n\|+\|Tu_n\|).$$

Since this holds for every $\varepsilon > 0$ we have $Vu_n \to 0$. Q.E.D.

<u>2.8. Remark.</u> It is clear that in Theorem 2.4 (ii) $C_o^\infty(\mathbb{R}^m\setminus A)$ could be replaced by any linear space M of functions with compact support in $\mathbb{R}^m\setminus A$ with the property: for every compact subset K of $\overline{G}\setminus A$ there is a $\varphi \in M$ with $\varphi(x) = 1$ for $x \in K$. We shall always use $M = C_o^\infty(\mathbb{R}^m\setminus A)$, since this choice is most useful for differential operators.

The next theorem gives a close connection between T-A-smallness and T^2-compactness for self-adjoint T.

<u>2.9. Theorem.</u> Let T be a self-adjoint operator in $L_2(G)$ and A a closed subset of $\overline{G}$, such that for every $\varphi \in C_o^\infty(\mathbb{R}^m\setminus A)$ the operators

$$u \mapsto \varphi u \quad \text{and} \quad u \mapsto \varphi Tu - T\varphi u$$

are T-compact. If V is an operator in $L_2(G)$ which is T-bounded and T-A-small, then V is T^2-compact.

<u>Proof.</u> Let (u_n) be a sequence in $D(T)$ with $u_n \rightharpoonup 0$ and $T^2u_n \rightharpoonup 0$; we have to show $Vu_n \to 0$. From $u_n \rightharpoonup 0$ and $T^2u_n \rightharpoonup 0$ it follows that $Tu_n \rightharpoonup 0$, since T is a continuous linear operator from $D(T^2)$ (with its T^2-topology) into $L_2(G)$. Now, by assumption, $u_n \rightharpoonup 0$ and $Tu_n \rightharpoonup 0$ imply $\varphi u_n \to 0$ for every $\varphi \in C_o^\infty(\mathbb{R}^m \setminus A)$; in the same way $Tu_n \rightharpoonup 0$ and $T^2u_n \rightharpoonup 0$ imply $\varphi Tu_n \to 0$.

For $\varepsilon > 0$ let K be a compact subset of $\overline{G} \setminus A$ such that for $u \in D(T)$ with $\operatorname{supp} u \subset \overline{G} \setminus K$

$$\|Vu\| \leq \varepsilon(\|u\|+\|Tu\|).$$

Then for every $\varphi \in C_o^\infty(\mathbb{R}^m \setminus A)$ with $\varphi(x) = 1$ for $x \in K$ we have

$$\begin{aligned}
\|Vu_n\| &\leq \|V\varphi u_n\|+\|V(1-\varphi)u_n\| \\
&\leq a\|\varphi u_n\|+ b\|T\varphi u_n\|+ \varepsilon(\|(1-\varphi)u_n\|+\|T(1-\varphi)u_n\| \\
&\leq a\|\varphi u_n\|+ b\|\varphi Tu_n\|+ (b+\varepsilon)\|\varphi Tu_n-T\varphi u_n\|+ \varepsilon(\|u_n\|+\|Tu_n\|).
\end{aligned}$$

Since $\varphi u_n \to 0$, $\varphi Tu_n \to 0$ and $\varphi Tu_n - T\varphi u_n \to 0$ for $n \to \infty$, we conclude

$$\limsup_{n \to \infty} \|Vu_n\| \leq \varepsilon \limsup_{n \to \infty} (\|u_n\|+\|Tu_n\|).$$

This holds for every $\varepsilon > 0$ and therefore $Vu_n \to 0$. Q.E.D.

<u>2.10. Remark.</u> As a by-product the proof of Theorem 2.9 shows that the T-compactness of $u \mapsto \varphi u$ implies the T^2-compactness of $u \mapsto \varphi T$ and that the T-compactness of $u \mapsto \varphi u$ and $u \mapsto \varphi Tu - T\varphi u$ imply the T^2-compactness of $T\varphi$.

3. Essential spectra and singular sequences for self-adjoint operators in $L_2(G)$.

In this section we study the invariance of the essential spectrum of self-adjoint operators T under perturbations of the type considered in section 2. First we prove the invariance of the essential spectrum via T^2-compactness of the perturbation. Then we show that, under somewhat more restrictive conditions, even the singular sequences remain invariant, which gives a little more information about spectral properties then merely the invariance of the essential spectrum. Finally we show that in many cases the essential spectrum of self-adjoint operators in $L_2(G)$ is determined by the "behaviour" of T near certain subsets of the boundary of G.

3.1. Theorem. Let T and T+V be self-adjoint operators in $L_2(G)$ with D(T) = D(T+V), A a closed subset of $\overline{G}$ such that for every $\varphi \in C_o^\infty(\mathbb{R}^m \setminus A)$ the operators

$$u \mapsto \varphi u \quad \text{and} \quad u \mapsto \varphi Tu - T\varphi u$$

are T-compact. If V is T-A-small, then $\sigma_e(T+V) = \sigma_e(T)$.

Proof. From D(T) = D(T+V) it follows that V is T-bounded. Therefore Theorem 2.9 gives the T^2-compactness of V. By a theorem which is due to Schechter ([10], Theorem 2.5) or its modification of Gustafson and Weidmann ([4], Theorem 3.2) this implies $\sigma_e(A) = \sigma_e(A+B)$. Q.E.D.

It is well known that a real number λ belongs to the essential spectrum of a self-adjoint operator T if and only if there exists a sequence (u_n) in D(T) with $u_n \rightharpoonup 0$, $u_n \not\to 0$ and $(T-\lambda)u_n \to 0$. Such a sequence is called a singular sequence for T and λ. We note that $u_n \rightharpoonup 0$ and $Tu_n \rightharpoonup 0$ hold if and only if u_n converges weakly to 0 in D(T) with its T-topology.

3.2. Theorem. Let T and V satisfy the conditions of Theorem 3.1.

(i) Every singular sequence for T and λ is also a singular sequence for T+V and λ.

(ii) If in addition the operator $u \mapsto \varphi Vu - V\varphi u$ is T-compact for every $\varphi \in C_o^\infty(\mathbb{R}^m \setminus A)$, then the singular sequences for T and T+V coincide (i.e. (u_n) is a singular sequence for T and λ if and only if it is a singular sequence for T+V and λ).

Proof. (i) Let (u_n) be a singular sequence for T and λ, i.e. $u_n \rightharpoonup 0$ $(u_n \not\to 0)$ and $(T-\lambda)u_n \to 0$; this implies $Tu_n \rightharpoonup 0$. From the T-compactness of $u \mapsto \varphi u$ we get $\varphi u_n \to 0$ for every $\varphi \in C_o^\infty(\mathbb{R}^m \setminus A)$. Together with $(T-\lambda)u_n \to 0$ this implies $\varphi T u_n \to 0$ for every $\varphi \in C_o^\infty(\mathbb{R}^m \setminus A)$. Exactly as in the proof of Theorem 2.9 we conclude $Vu_n \to 0$ and therefore $(T+V-\lambda)u_n \to 0$. This says that (u_n) is a singular sequence for T+V and λ.

(ii) In this case the assumptions are symmetric in T and T+V such that the roles of T and T+V can be interchanged and the proof follows as in (i) (remember Lemma 2.6 and notice that $u \mapsto \varphi(T+V)u - (T+V)\varphi u$ is (T+V)-compact).

3.3. Remark. Theorem 3.2(ii) gives another proof of Theorem 3.1 for the case that $u \mapsto \varphi Vu - V\varphi u$ is T-compact. Of course the conclusion of Theorem 3.2(ii) contains more information about the spectral properties of T and T+V than Theorem 3.1. We were not able to show that the condition "$u \mapsto \varphi Vu - V\varphi u$ is T-compact" cannot be dropped.

3.4. Theorem. Let G_i (i=1,2) be open subsets in $\mathbb{R}^m$, T_i self-adjoint operators in $L_2(G_i)$, A a closed subset of $\mathbb{R}^m$ such that for every $\varphi \in C_o^\infty(\mathbb{R}^m \setminus A)$ the operators

$$u \mapsto \varphi u \quad \text{and} \quad u \mapsto \varphi T_j u - T_j \varphi u$$

are T_j-compact (j=1,2). Assume that there is a compact subset K of $\mathbb{R}^m \setminus A$ such that

$$\overline{G}_1 \setminus \{K \cap \overline{G}_1\} = \overline{G}_2 \setminus \{K \cap \overline{G}_2\} =: M$$

$$D(T_1) \cap L_2(M) = D(T_2) \cap L_2(M) =: D$$

and that for every $\varepsilon > 0$ there is a compact subset K_ε of $\mathbb{R}^m \setminus A$ with $K \subset K_\varepsilon$ such that

$$\|(T_1 - T_2)u\| \le \varepsilon(\|u\| + \|T_1 u\|) \tag{3.5}$$

for every $u \in D$ with $K_\varepsilon \cap \operatorname{supp} u = \emptyset$. Then $\sigma_e(T_1) = \sigma_e(T_2)$. (Notice that in the right hand side of (3.5) T_1 can be replaced by T_2).

Proof. Since the conditions are completely symmetric in T_1 and T_2, it suffices to prove $\sigma_e(T_1) \subset \sigma_e(T_2)$.

Let $\lambda \in \sigma_e(T_1)$, (u_n) a singular sequence for T_1 and λ; we may assume that $\|u_n\| = 1$ for all n. Then for every $\varphi \in C_o^\infty(\mathbb{R}^m \setminus A)$ $\varphi u_n \to 0$ and $\varphi T u_n \to 0$ (see the proof of Theorem 3.2).

For every $\varepsilon > 0$ let $\varphi \in C_o^\infty(\mathbb{R}^m \setminus A)$ be such that $\varphi(x) = 1$ for $x \in K_\varepsilon$. Then $\varphi u_n \to 0$ and therefore $\|(1-\varphi_\varepsilon)u_n\| \to 1$, but $\varphi u_n \rightharpoonup 0$. We estimate

$$\begin{aligned}\|(T_2-\lambda)(1-\varphi)u_n\| &\leq \|(T_1-\lambda)(1-\varphi)u_n\|+\|(T_1-T_2)(1-\varphi)u_n\| \\ &\leq \|(1-\varphi)(T_1-\lambda)u_n\|+\|(T_1\varphi-\varphi T_1)u_n\| \\ &\quad + \varepsilon(\|(1-\varphi)u_n\|+\|T_1\varphi u_n\|) \\ &\leq \|(T_1-\lambda)u_n\|+(1+\varepsilon)\|(T_1\varphi-\varphi T_1)u_n\| \\ &\quad + \varepsilon(\|u_n\|+\|T_1 u_n\|).\end{aligned}$$

This implies (remember: $(T_1-\lambda)u_n \to 0$ and $(T_1\varphi-\varphi T_1)u_n \to 0$)

(3.6) $$\limsup_{n\to\infty} \|(T_2-\lambda)(1-\varphi)u_n\| \leq \varepsilon \limsup_{n\to\infty} (\|u_n\|+\|Tu_n\|).$$

Assume now that λ is not in $\sigma_e(T_2)$. Then there is a $\delta > 0$ such that (for the spectral resolution E of T_2)

$$\dim(E(\lambda+\delta) - E(\lambda-\delta)) < \infty.$$

Then for every sequence (v_n) in $D(T_2)$ with $\|v_n\| \to 1$ and $v_n \rightharpoonup 0$ we have

$$\liminf_{n\to\infty} \|(T_2-\lambda)v_n\| \geq \delta.$$

For $v_n = (1-\varphi)u_n$ and $\varepsilon > 0$ with

$$\varepsilon < \delta\,[\limsup_{n\to\infty} (\|u_n\|+\|Tu_n\|)]^{-1}$$

this contradicts inequality (3.6). This implies $\lambda \in \sigma_e(T_2)$. Q.E.D.

4. Applications to differential operators

In this section we will not state any explicit theorem but only indicate in which way our results are applicable to differential operators, and how some known results may be recovered.

Let G be an open subset of $\mathbb{R}^m$. We consider operators of Schrödinger type

$$(4.1) \qquad Tu(x) = \sum_{j,k=1}^{m} (i\partial_j + b_j(x)) a_{jk}(x)(i\partial_k + b_k(x))u(x) + q(x)u(x)$$

in $L_2(G)$, where $\partial_j := \frac{\partial}{\partial x_j}$ and

a_{jk}, b_j and q are real valued,

(4.2) $a_{jk} \in C^2(G)$, $b_j \in C^1(G)$, $q \in L_{2,loc}(G)$,

the matrix $(a_{jk}(x))$ is positive definit for $x \in G$.

(see [5], or for the case $G = \mathbb{R}^m$ and $a_{jk} = \delta_{jk}$ [8]). Jörgens [5] gave further conditions which guarantee the essential self-adjointness of T in $C_o^\infty(G)$.

The special case $G = \mathbb{R}^m$, $a_{jk} = \delta_{jk}$ has been studied in detail by Jörgens and the author in [8]. It has been shown that, if T and T+V are self-adjoint on the same domain, the essential spectra of T and T+V coincide if V is T-small at infinity (Definition 3.3 of [8]). Actually an additional condition was needed (either T+V had to be a Schrödinger operator [8] too, or T had to be bounded below and $\sigma_e(T)$ a half line). By means of the above theorems and some compactness results [8] (section 3) it is now easy to show that such an additional condition is not needed. This has been proved directly by Böcker [2].

The above results are also applicable if $a_{jk} \neq \delta_{jk}$. For this purpose one has to show that for $\varphi \in C_o^\infty(\mathbb{R}^m)$ the operators

$$(4.3) \qquad u \mapsto \varphi u \text{ and } u \mapsto \varphi Tu - T\varphi u$$

are T-compact. The proof of this fact is quite involved, but not difficult.

Let us now consider T in $L_2(G)$, where $G = \mathbb{R}^m \setminus G_o$ and G_o is compact. If we assume that the conditions of Jörgens [5] hold and in addition

(4.4) a_{jk} and b_j are continuous on $\overline{G}$,

the matrix $(a_{jk}(x))$ is positive definite for $x \in \overline{G}$,

Then one can show that for $\varphi \in C_o^\infty(\mathbb{R}^m)$ the operators (4.3) are T-compact (a proof appears between the lines in [1]). Therefore a perturbation V of T needs not to be small near ∂G (but only near infinity) in order to leave the essential spectrum unchanged. Of course the T-compactness of the operators (4.3) depends essentially on these quite restrictive conditions. For more general cases we have to expect that the T-smallness near some part of ∂G is needed.

The T-compactness of the operators (4.3) implies also (Theorem 3.4) that the essential spectrum of T depends only on the behaviour of T near infinity (as long as Jörgens' conditions and (4.4) hold). This has been proved by Beck [1] (see [7], Theorem 9). Again, in more general cases we have to expect, that the behaviour of T near some part of ∂G would be essential for the determination of the essential spectrum of T.

The main advantage of our abstract results lies in the fact that the application is not limited to operators of Schrödinger type. For example all results concerned with perturbations which are T-small near infinity (T-$\emptyset$-small) are immediately applicable to the Dirac operator. This is due to the fact that D(T) is a subspace of the local Sobolev space $W^1_{2,loc}(\mathbb{R}^m)$ (see [6], section 2) and that $\varphi T - T\varphi$ is the operator of multiplication by a finite (bounded) function; this important observation lead to the work of Jörgens [6] on the Dirac operator.

5. Additional remarks

It should be noted that for semi-bounded T our theorems 3.1 and 3.2 do not imply the semi-boundedness of T+V. It is not difficult to give examples which show that this actually is not true.

It is remarkable that in all our theorems the T-compactness of $u \mapsto \varphi u$, $u \mapsto \varphi Tu - T\varphi u$ and $u \mapsto \varphi Vu - V\varphi u$ is only needed for those $\varphi \in C_o^\infty(\mathbb{R}^m \setminus A)$ which are equal to 1 on an arbitrary (fixed) compact subset K_o of $\overline{G} \setminus A$.

References

[1] Beck,V.: Über die Unabhängigkeit des wesentlichen Spektrums eines elliptischen Differentialoperators vom Randverhalten der Koeffizienten. Diplomarbeit, Heidelberg 1968.

[2] Böcker,U.: Invarianz des wesentlichen Spektrums bei Schrödingeroperatoren. (to appear).

[3] Evans,W.D.: On the essential spectrum of second order degenerate elliptic operators. Lecture delivered at the Conference on Spectral Theory and Asymptotics of Differential Equations, The Hague 1973.

[4] Gustafson,K. and J.Weidmann: On the essential spectrum. J.Math. Anal.Appl. 25, 121-127 (1969).

[5] Jörgens,K.: Wesentliche Selbstadjungiertheit singulärer elliptischer Differentialoperatoren zweiter Ordnung in $C_o^\infty(G)$. Math.Scand. 15, 5-17 (1964).

[6] ___: Perturbations of the Dirac Operator. Conf. Ordinary Partial Diff. Equ., Dundee/Schottland 1972, Lecture Notes Math. 280, 87-102 (1972).

[7] ___: Spectral Theory of Schrödinger Operators. University of Colorado, Dept. of Mathematics 1970.

[8] Jörgens,K. and J.Weidmann: Spectral Properties of Hamiltonian Operators. Lecture Notes in Mathematics 313, 1973.

[9] ___: Zur Existenz der Wellenoperatoren. Math.Z. 131, 141-151 (1973).

[10] Schechter,M.: On the Essential Spectrum of an Arbitrary Operater I. J.Math.Anal.Appl. 13, 205-215 (1966).

[11] Veselić,K. and J.Weidmann: Existenz der Wellenoperatoren für eine allgemeine Klasse von Operatoren. Math.Z. 134, 255-274 (1973).

[12] ___: Asymptotic estimates of wave functions and the existence of wave operators. J. Functional Anal. (to appear).

[13] Weidmann,J.: Zur Spektraltheorie von Sturm-Liouville-Operatoren. Math.Z. 98, 268-302 (1967).

SQUARE INTEGRABLE SOLUTIONS OF L^p PERTURBATIONS OF SECOND ORDER LINEAR DIFFERENTIAL EQUATIONS[1]

James S. W. Wong

1. We are here concerned with the square integrability of solutions of second order linear differential equations on the half line $[0,\infty)$. Consider the second order linear equation

$$x'' + a(t)\, x = 0, \qquad t \geqslant 0, \tag{1}$$

and its perturbed equation

$$y'' + (a(t)+b(t))\, y = 0, \qquad t \geqslant 0, \tag{2}$$

where $a(t)$, $b(t)$ are real valued locally integrable functions on $[0,\infty)$. Following Weyl [20], we say that equation (1), and similarly equation (2), is in the limit circle case if all solutions of (1) belong to $L^2[0,\infty)$. On the other hand, we say that equation (1) is in the limit point case if it is not in the limit circle case, i.e. there is at least one solution of (1) which does not belong to $L^2[0,\infty)$. The classification of second order equations as being either in the limit circle or in the limit point cases are of fundamental importance in the study of singular boundary value problems, and more generally in the study of the spectrum of second order linear differential operators, see e.g. Naimark [15], Hellwig [10], Hille [11], and

[1] Research supported in part by Army Research Office, Durham through Contract No. DA-ARO-D-31-124-72-G95.

Everitt and Chaudhuri [5].

The basic result of Weyl states that if $b(t) \in L^{\infty}[0,\infty)$, then equation (2) retains the limit circle or limit point classification of equation (1). This immediately raises the question of whether the result is still ture if one assumes instead that $b(t) \in L^p[0,\infty)$ for some $p \geq 1$, see [17]. The purpose of this talk is to present some partial answers and likely conjectures to this question, thus lending credence to the validity of the general conjecture.

2. In this section, we first describe the situation with respect to the limit circle case. It seems natural to examine first of all certain known criteria on the coefficient function $a(t)$ so that equation (1) is in the limit circle case and ask whether equation (2) enjoys the same property provided that $b(t) \in L^p[0,\infty)$, for some $p \geq 1$. The well known limit circle criteria of Titchmarsh [19], Hille [11] and others as given in [4] (see also [21] and the references therein), all guarantee that solutions of (1) are also bounded. Therefore, one may ask the same question concerning equation (2) under the weaker assumption that all solutions of equation (1) are bounded. This has been answered in the affirmative in the following

THEOREM A. ([17]) *Suppose that equation* (1) *is in the limit circle case and all solutions are bounded. If* $b(t) \in L^p[0,\infty)$, *for some* $p \geq 1$, *then equation* (2) *is also in the limit circle case and all of its solutions are also bounded.*

Theorem A raises the following interesting conjecture, which we formulate as

CONJECTURE I *If equation (1) is in the limit circle case, then all solutions must necessarily be bounded.*

We are unable to settle this conjecture. Although it seems unlikely that this statement is true in general, it may be valid if one imposes some additional condition on $a(t)$, such as $a(t) \geqslant 0$. If conjecture I is answered in the affirmative then this settles the general conjecture in as far as limit circle case is concerned, whilst a negative answer does not have a direct bearing on the validity of the general conjecture.

Theorem A admits further generalization to equations of higher order. We mention here an extension to the self-adjoint equation of order 2n:

$$(3) \qquad (-1)^n \frac{d^n}{dt^n}\left(p_0 \frac{d^n}{dt^n} y\right) + (-1)^{n-1} + - + \ldots\ldots + p_n y = 0,$$

where $p_0, p_1, \ldots, p_n$ are locally integrable on $[0,\infty)$ and sufficiently smooth so that the differential expression in (3) makes sense. Denote by $\mathcal{L}$ the differential expression in (3) and consider the following perturbed nonlinear equation

$$(4) \qquad \mathcal{L}y = f(t,y), \qquad t \geqslant 0,$$

where $f(t,y)$ satisfies

(5) $$|f(t,y)| \leqslant \lambda(t)\,|y|^{\sigma}, \qquad 0 \leqslant \sigma \leqslant 1,$$

for all $t \geqslant 0$, and $|y| < \infty$ and $\lambda(t) \in L^p[0,\infty)$ for some $p \geqslant 1$. We now have the following

THEOREM B. *Suppose that all solutions of* (3) *are in* $L^2 \cap L^\infty[0,\infty)$ *and* $f(t,y)$ *satisfies* (5) *with* $\lambda(t) \in L^p[0,\infty)$, *for any* p, $1 \leqslant p \leqslant 2/1-\sigma$. *Then, all solutions of* (4) *belong to* $L^2 \cap L^\infty[0,\infty)$.

This generalizes some earlier results of Bellman [1], [2;p.116], Bradley [3], and Zettl [25].

3. The situation with regard to the limit point case is somewhat more complicated. First of all, in contrast to Theorem A, we have the following result:

THEOREM C ([17]). *Suppose that all solutions of* (1) *are bounded. If equation* (1) *is in the limit point case then equation* (2) *is also in the limit point case, provided that* $b(t) \in L^p[0,\infty)$, $p \geqslant 1$.

Theorem C was proved in [17] only for the case when $1 \leqslant p \leqslant 2$. However, for $p > 2$, we consider the decomposition $b(t) = b_1(t) + b_2(t)$, where $b_1(t) \in L^1[0,\infty)$ and $b_2(t) \in L^\infty[0,\infty)$. This can be achieved by defining

$$b_2(t) = \begin{cases} b(t), & \text{whenever } |b(t)| \leqslant 1, \\ 0, & \text{otherwise.} \end{cases}$$

By Weyl's result, we know that equation (2) is in the same classification as the following equation

$$u'' + (a(t) + b_1(t))\, u = 0, \qquad t \geqslant 0. \tag{6}$$

Since $b_1(t) \in L^1[0, \infty)$, equation (6) is in the limit point case by our earlier result [17].

Let us now examine two well known limit point criteria of Levinson [13] and Titchmarsh [18], and ask whether equation (2) remains in the limit point case provided that $b(t) \in L^p[0, \infty)$ and $a(t)$ satisfies these limit point criteria. This was answered in the affirmative by the following results:

THEOREM D ([17]). <u>Suppose that</u> $a(t)$ <u>satisfies</u> : $a(t) \leqslant M(t)$, <u>where</u> $M(t)$ <u>is positive and satisfies</u>

$$\int_0^\infty \frac{dt}{\sqrt{M(t)}} = \infty .$$

<u>Either</u> (a) $M(t)$ <u>is differentiable and satisfies</u>

$$\left| \frac{M'(t)}{M^{3/2}(t)} \right| \leqslant K,$$

<u>or</u> (b) $M(t)$ <u>is nondecreasing; then equation</u> (2) <u>is in the limit point case provided that</u> $b(t) \in L^p[0, \infty)$, $p \geqslant 1$.

On the other hand, if a(t) satisfies the limit point criterion of Hartman and Wintner [9], i.e.

$$\int_0^t a_+(s)\, ds = O(t^3), \qquad t \to \infty ,$$

and $b(t) \in L^p[0,\infty)$, $p \geq 1$, then equation (2) is also in the limit point case.

It is well known that if equation (1) is non-oscillatory then it is in the limit point case, Hartman [8]. (See also [16], [17], [23] for shorter proofs.) One might then ask whether equation (2) is in the limit point case if equation (1) is non-oscillatory and $b(t) \in L^p[0,\infty)$, for some $p \geq 1$. In this regard, we have the following

THEOREM E. *If equation (1) is non-oscillatory with* $a(t) \geq 0$, *and* $b(t) \in L^p[0,\infty)$, $p \geq 1$; *then equation (2) is in the limit point case.*

It is also known that if equation (1) possesses a positive nondecreasing solution and $b(t) \in L^2[0,\infty)$, then equation (2) is in the limit point case, see [22], [24]. Thus, it seems natural to formulate

CONJECTURE II. *Theorem E remains valid without the additional assumption that* $a(t) \geq 0$.

For earlier results relating nonoscillation to limit point classification of (1), we mention also the paper by Kurss [12].

The analogue of limit point case for higher order self-adjoint equation is the so called limit (n,n) case for equation (3), i.e. there exist at least n linearly independent solutions of (3) which are not square integrable. In this regard, Theorem D(a) admits further extension to equation (3), see Everitt, Hinton and Wong [7]. Generalization of Theorem D(b) to the higher order case seems not yet known. Before one can prove an extension of Theorem E, we would like to settle the following

CONJECTURE III. Suppose that equation (3) is non-oscillatory, i.e. no solution of (3) can have more than n zeros in the neighborhood of infinity counting multiplicities. Then, equation (3) is in the limit (n,n) case.

Everitt, Giertz, and Weidman [6] also introduced the concept of strong limit point case for equation (1). Consider the linear maniforld $\Delta(a) \in L^2[0,\infty)$, which depends on the choice of the coefficient function a(t), defined by $f(t) \in \Delta(a)$ if

(i) f(t) and f'(t) are locally absolutely continuous on $[0,\infty)$,

(ii) $f'' + a(t)f$ belong to $L^2[0,\infty)$.

Equation (1) is said to be in the strong limit point at infinity if

$$\lim_{T\to\infty} f(T)\,g'(T) = 0, \quad \text{for all } f,g \in \Delta(a).$$

It is easy to see that if equation (1) is in the strong limit point case then it is also in the limit point case. The converse is not true and was given a counterexample in [6]. A major step towards answering the general conjecture raised in the introduction is to settle the following

CONJECTURE IV. _Suppose that equation_ (1) _is in the strong limit point case and_ $b(t) \in L^p[0,\infty)$, $p \geq 1$; _then equation_ (2) _is also in the strong limit point case._

We refer the reader to [6], [7], for further details. Proofs of Theorems B and E and their extensions will appear elsewhere.

We hope that these results presented above will be of sufficient interest to warrant a closer study towards resolving the proposed conjecture.

REFERENCES

[1] R. Bellman, " A stability principle of solutions of linear differential equations ", Duke Math. J., 11(1944), 513-516.

[2] R. Bellman, Stability Theory of Differential Equations, McGraw-Hill, New York, 1953.

[3] J. S. Bradley, " Comparison Theorems for the square integrability of solutions of $(r(t)y')' + q(t)y = f(t,y)$", Glasgow Math. J., 13(1972), 75-79.

[4] N. Dunford and J.T. Schwartz, Linear Operators, Part II: Spectral Theory, Interscience, New York, 1963.

[5] W. N. Everitt and J. Chaudhuri, " On the spectrum of ordinary second-order differential operators ", Proc. Royal Soc. Edinburgh, 68(1969), 95-119.

[6] W. N. Everitt, M. Giertz, and J. Weidmann, "Some remarks on a separation and limit point criterion of second order ordinary differential expressions ", Math. Ann., 200(1973), 335 - 346.

[7] W. N. Everitt, D.B. Hinton and J.S.W. Wong, " On the strong limit-n classification of linear ordinary differential expressions of order 2n ", Proc. London Math. Soc., (1974), (to appear).

[8] P. Hartman, " The number of L^2-solutions of $x'' + p(t)x = 0$ ", Amer. J. Math., 73(1951), 635 - 645.

[9] P. Hartman and A. Wintner, " A criterion for the nondegeneracy of the wave equation ", Amer. J. Math., 71(1949), 206 - 213.

[10] G. Hellwig, <u>Differential Operators of Mathematical Physics,</u> Addison-Wesley, Reading, Massachusetts, 1964.

[11] E. Hille, <u>Lectures on Ordinary Differential Equations,</u> Addison-Wesley, Reading, Massachusetts, 1969.

[12] H. Kurss, " A limit point criterion for nonoscillatory Sturm-Liouville differential operators ", Proc. Amer. Math. Soc., 18(1967), 445 - 449.

[13] N. Levinson, " Criteria for the limit point case for second order linear differential operators ", Casopis Pest. Mat. Fys., 74(1949), 17 - 20.

[14] A. Yu Levin, " Nonoscillation of solutions of the equation $x^{(n)} + p_1(t)\, x^{(n-1)} + \ldots\ldots + p_n(t)x = 0$ ", Russian Math. Survey, 24(1969), 43 - 99.

[15] M. A. Naimark, <u>Linear Differential Operators</u>, part II, Ungar, New York, 1968.

[16] W. T. Patula and P. Waltman, " Limit point classification of second order linear differential equations ", J. London Math. Soc., (1973), to appear.

[17] W. T. Patula and J.S. W. Wong, " An L^p- analogue of the Weyl's alternative ", Math. Ann., 197(1972), 9 - 27.

[18] E.C. Titchmarsh, " On the uniqueness of the Green's function associated with a second order differential equation ", Canadian J. Math., 1(1949), 191 - 198.

[19] E.C. Titchmarsh, <u>Eigenfunction expansions associated with second order differential equations</u>, (Second Edition), Oxford Press, Oxford ,1962.

[20] H. Weyl, " Uber gewohnliche Differential-gleichungen mit Singularitaten und die zugehorige Entwicklung will kurlicker Functionen ", Math. Ann., 68(1910), 220 - 269.

[21] J.S.W. Wong, " Remarks on the limit circle classification of second order differential operators ", Quarterly J. Math., 24(1973), 423 - 425.

[22] J.S.W. Wong, " On L^2-solutions of linear ordinary differential equations ", Duke Math. J., 38(1971), 93 - 97.

[23] J.S.W. Wong and A. Zettl, " On the limit point classification of second order differential equations ", Math. Zeitschrift, 132(1973), 297 - 304.

[24] A. Zettl, " A note on square integrable solutions of linear differential equations ", Proc. Amer. Math. Soc., 21(1969), 671 - 672.

[25] A. Zettl, " Square integrable solutions of Ly = f(t,y) ", Proc. Amer. Math. Soc., 26(1970), 635 - 639.

Deficiency Indices of Polynomials in Symmetric Differential Expressions

Anton Zettl

Abstract. Given a symmetric (formally self-adjoint) ordinary linear differential expression L which is regular on the interval $[0,\infty)$ and has complex C^{∞} coefficients, we investigate the relationship between the deficiency indices of L and those of $p(L)$ where $p(x)$ is any real polynomial of degree $k > 1$. Our main results are the inequalities: (a) For k even, say $k = 2m$, $N_+(p(L)), N_-(p(L)) \geq m[N_+(L) + N_-(L)]$ and (b) for k odd, say $k = 2m + 1$, $N_+(p(L)) \geq (m+1)N_+(L) + mN_-(L)$ and $N_-(p(L)) \geq mN_+(L) + (m+1)N_-(L)$. Here $N_+(M)$, $N_-(M)$ denote the deficiency indices of the symmetric expression M associated with the upper and lower half-planes, respectively.

1. In this paper we report results which have been obtained in recent months on the relationship between the deficiency indices of L and p(L) for L any symmetric differential expression and p(x) any real polynomial.

Let S denote the set of all symmetric (formally self-adjoint) regular ordinary linear differential expressions with C^{∞} coefficients on the interval $[0,\infty)$. Let L be an element of S of order n, then L has the representation (see [3]):

$$(1)\quad Ly = \sum_{j=0}^{[n/2]} (p_j y^{(j)})^{(j)} + i \sum_{j=0}^{[(n-1)/2]} (-1)^j [(q_j y^{(j)})^{(j+1)} + (q_j y^{(j+1)})^{(j)}]$$

where p_j, q_j are real C^{∞} functions and the leading coefficient $p_{[n/2]}$ or $q_{[(n-1)/2]}$ is positive on $t \geq 0$. Clearly S is closed under the operation of addition and also under multiplication by real scalars. Moreover, for any L in S, L^2, L^3, ... are in S [3] so that for any polynomial p(x) with real coefficients p(L) is in S.

Let L_0 be the minimal operator associated with a given L in S in the Hilbert space $H = L^2[0,\infty)$. This operator L_0 is defined as the minimal closed extension of the restriction of L to C_0^{∞} --the set of C^{∞} functions which have compact support in $(0,\infty)$. The deficiency indices of L are a pair of integers $N_+(L)$ and $N_-(L)$ defined as the dimension of the orthogonal complements of the ranges of the operators $L_0 - \lambda I$ and $L_0 + \lambda I$, respectively, for λ a complex number with $\mathcal{I}(\lambda) > 0$. . It is known that $N_+(L)$ and $N_-(L)$ are independent of the particular number λ--as long as $\mathcal{I}(\lambda) > 0$. We mention the general classification results: For n even, $n = 2r$, $r \leq N_+(L), N_-(L) \leq 2r$ and for n odd, say $n = 2r +$ with $r > 0$, $r \leq N_+(L) \leq 2r + 1 = n$ and $r + 1 \leq N_-(L) \leq 2r + 1 = n$. In addition $N_+(L) = n$ if and only if $N_-(L) = n$. In the real case i.e. when

L has only real coefficients n must be even, say $n = 2r$, the classification reduces to $r \leq N_+(L) = N_-(L) \leq 2r$. We say that L is in the limit point case when $N_+(L)$ and $N_-(L)$ assume their minimum values and in the limit circle case when $N_+(L)$ and $N_-(L)$ take on the maximum values. For a discussion of the basic theory of linear differential operators we refer the reader to [1], [3] and [10]. For the general classification results, see [5].

Unless explicitly stated otherwise all our differential expressions are considered on the interval $[0,\infty)$ and our underlying space is the Hilbert space $H = L^2_{[0,\infty)}$. Let $p(x) = a_k x^k + a_{k-1} x^{k-1} + \dots + a_1 x + a_0$ with a_i real for $i = 0, 1, \dots, k$ and $a_k > 0$.

Our main result is:

Theorem 1. (a) Suppose k is even, say $k = 2m$. Then $N_+(p(L)) \geq m[N_+(L) + N_-(L)]$ and $N_-(p(L)) \geq m[N_+(L) + N_-(L)]$.

(b) Suppose k is odd, say $k = 2m + 1$. Then $N_+(p(L)) \geq (m+1)N_+(L) + mN_-(L)$ and $N_-(p(L)) \geq mN_+(L) + (m+1)N_-(L)$.

In [2] Chaudhuri and Everitt have an example of a second order expression L in S satisfying $N_+(L) = N_-(L) = 1$ and $N_+(L^2) = N_-(L^2) = 3$. This shows that strict inequality can occur in theorem 1.

As an immediate consequence of theorem 1 and the general classification results for deficiency indices we obtain two corollaries.

Corollary 1. If, for some polynomial $p(x)$ of degree $k > 1$, either $N_+(p(L))$ or $N_-(p(L))$ takes on its minimum value, then both $N_+(L)$ and $N_-(L)$ take on their minimum value i.e. L is in the limit point case.

Proof. We consider four cases.

(a) Suppose $k = 2m$ and $n = 2r + 1$. If $N_+(p(L)) = mn$, then $mr + m(r+1) \leq mN_+(L) + mN_-(L) \leq mn$. Hence $N_+(L) = r$ and $N_-(L) = r + 1$. The argument in case $N_-(p(L)) = mn$ is entirely similar.
The other cases follow similarly.

In particular if $p(L)$ is in the limit point case for some polynomial $p(x)$, then L is in the limit point case. This result was obtained by R. M. Kauffman in [8] by different methods.

Corollary 2. If L is in the limit circle case, then $p(L)$ is in the limit circle case for any real polynomial $p(x)$.

The converse of Corollary 2 is also valid. Moreover Corollary 2 and its converse hold even for non-symmetric expressions. Let $My = y^{(n)} + r_{n-1}y^{(n-1)} + \ldots + r_1y' + r_0y$ with $r_i \in C^\infty$ for $i = 0, \ldots, n-1$. Recall that

$$M^*y = (-1)^n y^{(n)} + (-1)^{n-1}(\bar{r}_{n-1}y)^{(n-1)} + \ldots + (-1)(\bar{r}_1y)' + \bar{r}_0y.$$

Definition. We say that M is in the limit circle case if all solutions of $My = 0$ and all solutions of $M^*y = 0$ are in $L^2[0,\infty)$.

Notice that this definition of the limit circle case for non-symmetric expressions reduces to the usual one if M is symmetric, i.e. if $M = M^*$.

Theorem 2. For any polynomial $p(x)$ with real coefficients, $p(M)$ is limit circle if and only if M is limit circle.

Proof. For M a real symmetric expression this is theorem 3 in [11]. The same proof used there can be used to prove theorem 2 once we have established

Lemma 1. If all solutions of $My = 0$ and of $M^*y = 0$ are in $L^2[0,\infty)$, then all solutions of $My = \lambda y$ are in $L^2[0,)$ for any complex number λ.

Lemma 1 is a special case of theorem 1 in [12] for real r_i, however the same proof given there works for complex r_i.

2. Before proceeding to the proof of theorem 1, we state some related results. These are either theorems from the papers [8] and [11] of the references or are extensions of such theorems. The results in [11] were first obtained by Everitt and Giertz in [4] for the case when L is a real second order operator and $p(x) = x^k$.

The writer is indebted to R. M. Kauffman for the private communication of his results from [8].

Definition. Let L be a symmetric differential expression and $p(x)$ a real polynomial of degree k. We say that $p(L)$ is partially separated if the conditions: $f \in L^2[0,\infty)$, $f^{(2nk-1)}$ absolutely continuous on compact subintervals of $[0,\infty)$ and $p(L)f \in L^2_{[0,\infty)}$ together imply that $L^r f \in L^2_{[0,\infty)}$ for all $r = 1, 2, \ldots, k-1$.

This definition was used in [11] and is an extension of the concept of partial separation as used by Everitt and Giertz in [4].

Theorem 3. Suppose L in S is real and $p(x)$ is a real polynomial.

(a) If L is in the limit point case and $p(L)$ is partially separated, then $p(L)$ is in the limit point case.

(b) If $p(L)$ is in the limit point case, then $p(L)$ is partially separated.

Theorem 3 (a) is established in [11]. Also 3 (b) is proven in [11] under the additional hypothesis that L is in the limit point case. But this assumption is superfluous since $P(L)$ limit point implies L limit point by Corollary 1.

Theorem 4. (R. M. Kauffman [8]). If $p(x)$ and $q(x)$ are real polynomials of the same degree, then deficiency index $p(L)$ = deficiency index $q(L)$ for any L in S. (Actually Kauffman's result is somewhat more general than this--see the remark below.)

The next two results give sufficient conditions on the coefficients of an expression L for $p(L)$ to be in the limit point case for any real polynomial $p(x)$. Theorem 5 is from [11] and is due to Everitt-Giertz for L of second order and $p(x) = x^k$. Theorem 6 is a result of R. M. Kauffman from [8].

Theorem 5. Let L be given by (1) with $q_j = 0$, i.e. L is a real symmetric expression of order $n = 2m$. If there exist real constants $b_o, b_1, \ldots, b_m$ with $b_m \neq 0$ such that $p_i - b_i \in L^1[0,\infty)$ for each $i = 0, 1, \ldots, m-1$ and $1/p_m - 1/b_m \in L^1[0,\infty)$, then $p(L)$ is in the limit point case (and hence partially separated) for every polynomial $p(x)$ with real coefficients.

Theorem 6. (R. M. Kauffman [8]). Suppose V is a real C^∞ function on $[0,\infty)$ which is bounded below and satisfies: For each integer $i = 0, 1, \ldots, 2k$ there is a $\lambda > 0$ such that $\exp[-\lambda x]V^{(i)}(x)$ is bounded on $[0,\infty)$. Let L be defined by $Ly = -y'' + Vy$. Then $p(L)$ is in the limit point condition for every polynomial $p(x)$ of degree k or less.

In [8] Kauffman establishes a stronger version of Corollary 1 which we state as:

Theorem 7. If $p(L)$ is in the limit point case for L in S and $p(x)$ a polynomial of degree k, then $q(L)$ is in the limit point case for any polynomial $q(x)$ of degree k or less.

This result combined with theorem 3 yields:

Theorem 8. For any real L in S and any real polynomial $p(x)$ of degree k if $p(L)$ is in the limit point case, then $q(L)$ is partially separated for any polynomial $q(x)$ of degree k or less

3. Proof of theorem 1. In view of theorem 4 we need only establish the special case $p(x) = x^k$. This is done with the help of theorem A from I. M. Glazman's fundamental paper [6]. Recall that a complex number λ is called a regular point of an operator A if there exists a positive number $r(\lambda)$ such that $||(A-\lambda I)f|| \geq r(\lambda)||f||$ for all f in the domain of A. By def. A we mean the dimension of the orthogonal complement of the range of A in H.

Theorem A. Let A and B be closed linear operators with dense domains in H. Suppose A and B have finite deficiencies and $\lambda = 0$ is a regular point of both. Then AB is a closed operator with dense domain and $\lambda = 0$ as a regular point and def. (AB) = def. A + def. B.

Let $w_1, \ldots, w_k$ be the kth roots of $i = \sqrt{-1}$. Note that for k even, say $k = 2m$, m of these roots lie in the upper half-plane and m lie in the lower half-plane and for k odd, say $k = 2m + 1$, $m + 1$ roots are in the upper half-plane and m are in the lower half-plane.

Since $x^k - i = \Pi_{j=1}^k (x - w_j)$ we have $(L_0)^k - iI = \Pi_{j=1}^k (L_0 - w_j I)$. By theorem A, def. $[(L_0)^k - iI] = \sum_{j=1}^k$ def. $(L_0 - w_j I)$.

Claim. $(L_0)^k - iI \supset (L^k)_0 - iI$.

From theorem A we know that $(L_0)^k - iI$ is a closed operator. The inclusion then follows from $[(L_0)^k - iI](y) = [(L^k)_0 - iI]y = (L^k - i)y$ for all C_0^∞ functions y and from the fact that $(L^k)_0 - iI$ is the minimal closed extension of $(L^k - i)$ restricted to C_0^∞.

From this inclusion it follows that

$$\text{def. } [(L_0)^k - iI] \leq \text{def. } [(L^k)_0 - iI].$$

Hence $\sum_{j=1}^{k}$ def. $(L_0 - w_j I) \leq N_+(L^k)$.

For $k = 2m$ the left hand side of the above inequality reduces to $mN_+(L) + mN_-(L)$ and for $k = 2m + 1$ it becomes $(m+1)N_+(L) + mN_-(L)$. This concludes the proof of the first inequality in parts (a) and (b) of theorem 1. The second inequalities are proven similarly: replace i by $(-i)$ and proceed as above.

In [6] R. M. Kauffman has extended the notion of the limit point case to expressions which are not necessarily symmetric. Thus he can consider the limit point case for $p(L)$ even when $p(x)$ has non-real complex coefficients. He shows that--with this extended notion of the limit point concept--theorem 7 is valid with $p(x)$ a polynomial with complex coefficients and L a symmetric expression with complex coefficients.

The assumption that the coefficients of L are all C^∞ is made mainly for convenience. It is used only to insure that we can form $p(L)$. Furthermore by using quasi-differential expressions differentiability assumptions can be avoided entirely.

References

1. Akhiezer, N. I. and I. M. Glazman. "Theory of Linear Operators in Hilbert Space," Vol. II, Ungar, New York (1963).

2. Chaudhuri, J. and W. N. Everitt. "On the Square of a Formally Self-Adjoint Differential Expression," J. London Math. Soc. (2), 1 (1969), 661-673.

3. Dunford, N. and Schwartz, J. T., "Linear Operators," Part II, Interscience, New York (1963).

4. Everitt, W. N. and M. Giertz. "On Some Properties of the Powers of a Formally Self-Adjoint Differential Expression," Proc. London Math. Soc. (3), 24 (1972), 149-170.

5. Everitt, W. N., "Integrable-Square Solutions of Ordinary Differential Equations," Quart. J. Math. (2) (1959), 145-55.

6. Glazman, I. M., "On the theory of singular differential operators," Uspehi Mat. Nauk. (N.S.) 5, No. 6 (40), 102-135 (1950). (Russian). Amer. Math. Soc. Translation no. 96 (1953).

7. Goldberg, S. "Unbounded Linear Operators," McGraw-Hill, New York (1966).

8. Kauffman, R. M. "Polynomials and the Limit Point Condition," Trans. Amer. Math. Soc. (to appear).

9. Miller, K. S. "Linear Differential Equations in the Real Domain," Norton & Co., New York (1963).

10. Naimark, M. A. "Linear Differential Operators," part II, Ungar, New York (1968).

11. Zettl, Anton. "The Limit Point and Limit Circle Cases for Polynomials in a Differential Operator," Proc. Royal Soc. Edinburgh, series A (to appear).

12. ________. "Square Integrable Solutions of $Ly = f(t,y)$," Proc. Amer. Math. Soc. Vol. 26, No. 4 (1970) 635-639.

TRANSFORM THEOREMS FOR TWO-PARAMETER EIGENVALUE PROBLEMS IN HILBERT SPACE

F.M.Arscott

1. Background and Introduction

In [1] , and more fully in [2] , it has been shown that certain eigenvalue problems in differential equations can be transformed into integral equations, the kernel of which is not the usual Green's function but a solution of a certain partial differential equation.

Discretization of this problem yields, naturally, a theorem relating to eigenvalues of matrices. The work of this paper arose from an attempt to formulate a result in Hilbert space which would incorporate both of those mentioned. In the process of so doing, it became clear that a wider theorem could be stated with very little extra difficulty.

For convenience of reference, we mention briefly here the one-parameter eigenvalue theorem; the two-parameter case will be given later.

Let $w(z)$ be a solution of the eigenvalue problem

$$L_z w \equiv \left\{ q(z) - \frac{d^2}{dz^2} \right\} w = \lambda w , \quad w(a) = w(b) = 0 \tag{1a,b}$$

and let $K(z, \zeta)$ satisfy the partial differential equation

$$(L_z - L_\zeta) K = 0, \tag{2}$$

with $K(z,a) = K(z,b) = 0$, i.e. K, regarded as a function of ζ, satisfies the boundary conditions (1b). We define

$$W(z) = \int_a^b K(z, \zeta)\, w(\zeta)\, d\zeta . \tag{3}$$

Then, formally, $W(z)$ also satisfies (1a).

Moreover, if K is symmetric, then $W(z)$ also satisfies (1b) and so is either identically zero or an eigenfunction corresponding to λ. Finally, if λ is a simple eigenvalue then $W(z)$ is necessarily a multiple of $w(z)$, say $W(z) = \Lambda w(z)$, so that $w(z)$ satisfies the integral equation

$$\int_a^b K(z, \zeta)\, w(\zeta)\, d\zeta = \Lambda w(z) \tag{4}$$

where possibly $\Lambda = 0$. Proof of this involves only straightforward manipulation .

The matrix theorem arising from discretization is the following: Let x be an eigenvector of the matrix A corresponding to an eigenvalue λ, let K be any matrix which commutes with A, and let $\xi = Kx$. Then ξ is either zero or also an eigenvector of A corresponding to λ. Moreover, if λ is simple then x is an eigenvector of K, possibly corresponding to a zero eigenvalue. Verification of these statements is simple.

2. The one-parameter transform theorem

(a) Let $\mathcal{H}_i$, $i = 1,2$ be real Hilbert spaces, the inner product being denoted by $\langle\ ,\ \rangle_i$, and let bases be $\{u_{ir}\}$, $r = 1,2,\ldots$ Let $\mathcal{A}_i$ be linear operators in $\mathcal{H}_i$ and let $\mathcal{U}_i \subset \mathcal{H}_i$ be subspaces in which the $\mathcal{A}_i$ are self-adjoint.

(b) Now we form the tensor product $\mathcal{H}_1 \otimes \mathcal{H}_2$ which we denote by $\mathcal{H}_{12}$. There are two operations which we need to carry out in $\mathcal{H}_{12}$ and which we now define.

(i) Let $G \in \mathcal{H}_{12}$, and let G be expressed in terms of the natural basis in $\mathcal{H}_{12}$ as

$$G = \sum_r \sum_s a_{rs}\, u_{1r} \otimes u_{2s}\ .$$

The operator $\mathcal{A}_1$, which we have defined as acting in $\mathcal{H}_1$ has an induced operator acting in $\mathcal{H}_{12}$, which we continue to denote by $\mathcal{A}_1$. Its action on G is defined by

$$\mathcal{A}_1 G = \sum_r \sum_s a_{rs}\, (\mathcal{A}_1 u_{1r}) \otimes u_{2s} \qquad (5)$$

and similarly for $\mathcal{A}_2 G$. Naturally $\mathcal{A}_1 G,\ \mathcal{A}_2 G \in \mathcal{H}_{12}$.

(ii) Let $y \in \mathcal{H}_2$, then we need to consider the inner product $\langle G,y \rangle_2$, by which we mean

$$\langle G,y \rangle_2 = \sum_r \sum_s a_{rs}\, u_{1r} \langle u_{2s},y \rangle_2$$

$$= \sum_r \Big(\sum_s a_{rs} \langle u_{2s},y \rangle_2 \Big)\, u_{1r}; \qquad (6)$$

Clearly $\langle G,y \rangle_2 \in \mathcal{H}_1$.

(c) We must postulate a property of commutativity between the operator $\mathcal{A}_1$ acting in $\mathcal{H}_{12}$ and the inner product $\langle\ ,\ \rangle_2$, which we shall call "property C", namely

$$\mathcal{A}_1 \langle G,y \rangle_2 = \langle \mathcal{A}_1 G,y \rangle_2 \qquad (7)$$

for all $G \in \mathcal{H}_{12}$, $y \in \mathcal{H}_2$.

If the spaces $\mathcal{H}_i$ are both finite-dimensional, this condition is automatically satisfied, but in general it is a non-trivial postulate; in the differential-equation problem, for instance, it is equivalent to the validity of applying a differential operator under an integral sign.

We can now state the required general theorem.

<u>Theorem 1</u> Let $\phi \in \mathcal{H}_2$ be an eigenvector of $\mathcal{A}_2$ corresponding to the eigenvalue λ.

Let $K \in \mathcal{H}_{12}$ be such that $\mathcal{A}_1 K = \mathcal{A}_2 K$. (8)

Define $\psi \in \mathcal{H}_1$ by $\psi = \langle K, \phi \rangle_2$. (9)

Then $\psi = 0$ or ψ is an eigenvalue of $\mathcal{A}_1$ corresponding to λ.

<u>Proof:</u> $\mathcal{A}_1 \psi \equiv \mathcal{A}_1 \langle K, \phi \rangle_2 = \langle \mathcal{A}_1 K, \phi \rangle_2$ (property C)

$= \langle \mathcal{A}_2 K, \phi \rangle_2$ (by (8))

$= \langle K, \mathcal{A}_2 \phi \rangle_2$ (self-adjointness of $\mathcal{A}_2$)

$= \langle K, \lambda\phi \rangle_2 = \lambda \psi$.

<u>Observations</u>

(i) For simplicity, the Hilbert spaces have been taken as real, but there seems no serious difficulty over extension to complex spaces.

(ii) An important special case is when $\mathcal{H}_1, \mathcal{H}_2$ are isometric. We need now to take care over notation because while $\mathcal{H}_1, \mathcal{H}_2$ are 'the same' in the sense of isometry they are nevertheless 'distinct' and we need the tensor product of the two spaces still. We therefore use the same letter, namely $\mathcal{H}$, in place of $\mathcal{H}_1, \mathcal{H}_2$ but insert a superscript $^{(1)}, ^{(2)}$ as necessary to distinguish between them. We also use the same letter to denote corresponding elements, operators and subspaces in $\mathcal{H}^{(1)}, \mathcal{H}^{(2)}$, employing the same superscripts where necessary. The above theorem then becomes:

Let $\phi \in \mathcal{U}$ be an eigenvector of $\mathcal{A}$ corresponding to the eigenvalue λ. Let $K \in \mathcal{H}^{(1)} \otimes \mathcal{H}^{(2)}$ be such that

$$\mathcal{A}^{(1)} K = \mathcal{A}^{(2)} K . \tag{10}$$

Then ψ, defined as $\langle K, \phi^{(2)} \rangle_2$ is either zero or also an eigenvector of $\mathcal{A}$ corresponding to λ, and if λ is a simple eigenvalue of $\mathcal{A}$, then

$$\langle K, \phi^{(2)} \rangle_2 = \Lambda \phi^{(1)} \tag{11}$$

for some constant Λ, possibly zero. The formulae (10),(11) are, of course, analogues respectively of the partial differential equation (2) and the integral equation (4).

3. Two-parameter problems

In [1], [2] a study is also made of the two-parameter eigenvalue problem consisting of the differential equation

$$w''(z) + (\lambda + \mu f(z) + g(z))\, w = 0, \tag{12a}$$

with conditions $\quad w(a) = w(b) = w(c) = 0. \qquad (12b)$

The related integral formula states that if $w(z)$ is an eigenfunction of (12a,b) corresponding to the eigenvalue pair λ, μ, $H(\alpha,\beta,\gamma)$ is a solution of the partial differential equation

$$\sum_{\alpha,\beta,\gamma} \{f(\beta) - f(\gamma)\}\{H_{\alpha\alpha} + g(\alpha)\, H\} = 0 \tag{13}$$

such that $H(\alpha,a,\gamma) = H(\alpha,b,\gamma) = H(\alpha,\beta,b) = H(\alpha,\beta,c) = 0$, then formally

$$W(\alpha) \equiv \int_a^b \int_b^c H(\alpha,\beta,\gamma)\, w(\beta) w(\gamma)\, \{f(\beta)-f(\gamma)\}\, d\beta\, d\gamma \tag{13'}$$

is also a solution of (12a). Under further assumptions on the symmetry of H and simplicity of the eigenvalue pair we can deduce an integral equation analogous to (4) for $w(z)$ which is interesting in that it is non-linear; the unknown function w appears twice under the integral sign but once outside it.

To obtain the Hilbert space analogue of this result, we take three spaces $\mathcal{H}_i$, $i = 1,2,3$, form tensor products $\mathcal{H}_{ij}$ as above, and also the triple tensor product $\mathcal{H}_{123}$.

Next, let $\mathcal{J}_i, \mathcal{P}_i$, $i = 1,2,3$ be operators in $\mathcal{H}_i$, self-adjoint in $\mathcal{U}_i \subset \mathcal{H}_i$. Again, we use the same letters to denote induced operators acting in the various tensor product spaces.

The inner product $\langle\ ,\ \rangle_j$ can be taken in $\mathcal{H}_{ij}$ as before, while in $\mathcal{H}_{123}$ we can take inner products twice, as follows: let $H \in \mathcal{H}_{123}$, $y \in \mathcal{H}_2$, $z \in \mathcal{H}_3$, then $\langle H,y\rangle_2 \in \mathcal{H}_{13}$ is defined and we can further form $\langle\langle H,y\rangle_2, z\rangle_3$ which yields, of course, an element of $\mathcal{H}_1$. For brevity we write such a double inner product as

$$\langle\langle H,y,z\rangle\rangle_{23}\ . \tag{14}$$

We further postulate that the operators $\mathcal{J}_i, \mathcal{P}_i$ have the commutative property C, and recall that operators with different suffices commute when acting in a tensor product space (although, of course, $\mathcal{J}_i$ and $\mathcal{P}_i$ will not normally commute.)

With this apparatus, we may formulate

<u>Theorem 2</u> Let $\phi \in \mathcal{U}_2$, $\psi \in \mathcal{U}_3$ be such that

$$A_2 \phi = (\lambda + \mu P_2)\phi \ , \tag{15a}$$

$$A_3 \psi = (\lambda + \mu P_3)\psi \ . \tag{15b}$$

Let $K \in \mathcal{H}_{123}$ be such that
$$\begin{vmatrix} A_1 & A_2 & A_3 \\ P_1 & P_2 & P_3 \\ I_1 & I_2 & I_3 \end{vmatrix} K = 0 \tag{16}$$

where I_i denotes the identity operator in $\mathcal{H}_i$. (The determinantal formulation of (16) is unambiguous because of the commutativity of operators with different suffices).

Let θ be defined by

$$\theta = \langle\langle (P_2 - P_3)K, \phi, \psi \rangle\rangle_{23} \ . \tag{17}$$

Then θ satisfies $\quad A_1\theta = (\lambda + \mu P_1)\theta.$ (18).

<u>Proof</u>

From (16) we have

$$\begin{vmatrix} A_1 - \lambda I_1 - \mu P_1 & A_2 - \lambda I_2 - \mu P_2 & A_3 - \lambda I_3 - \mu P_3 \\ P_1 & P_2 & P_3 \\ I_1 & I_2 & I_3 \end{vmatrix} K = 0$$

whence

$$\begin{aligned}
-(A_1 - \lambda I_1 - \mu P_1)\theta &= -(A_1 - \lambda I_1 - \mu P_1)\langle\langle (P_2 - P_3)K, \phi, \psi\rangle\rangle_{23} \\
&= \langle\langle (A_2 - \lambda I_2 - \mu P_2)(P_3 - P_1)K, \phi, \psi\rangle\rangle_{23} + \\
&\quad + \langle\langle (A_3 - \lambda I_3 - \mu P_3)(P_1 - P_2)K, \phi, \psi\rangle\rangle_{23} \\
&= \langle\langle (P_3 - P_1)K, (A_2 - \lambda I_2 - \mu P_2)\phi, \psi\rangle\rangle_{23} + \\
&\quad + \langle\langle (P_1 - P_2)K, \phi, (A_3 - \lambda I_3 - \mu P_3)\psi\rangle\rangle_{23} \\
&= \langle\langle (P_3 - P_1)K, 0, \psi\rangle\rangle_{23} + \langle\langle (P_1 - P_2)K, \phi, 0\rangle\rangle_{23} \\
&= 0.
\end{aligned}$$

This establishes (18) and the proof is complete.

To deduce the analogue of the integral equation (13') is a little more subtle than might be expected. We cannot merely let $\mathcal{H}_1, \mathcal{H}_2, \mathcal{H}_3$ coincide since the presence of the factor $P_2 - P_3$ in (17) makes the result trivial; $\mathcal{H}_2$ must consist of functions defined on (a,b) and $\mathcal{H}_3$ of functions defined on (b,c) so they are essentially distinct and $P_2 \neq P_3$ in $\mathcal{H}_{123}$.

This reflects the observation that if, in the integral equation (13') we took the integration with respect to γ over (a,b) instead of (b,c) the result would be true but trivial since the integral would be identically zero. We shall not, therefore, trouble to write down the Hilbert space analogue of (13') explicitly.

4. The discrete analogue of the two-parameter problem.

It is easy to see that the discrete analogue of the two-parameter differential equation problem (12a,b) consists of the two matrix equations

$$Ax = (\lambda I + \mu P_1)x, \qquad Ay = (\lambda I + \mu P_2)y, \tag{19}$$

where λ, μ are to be so chosen that non-trivial solutions x,y both exist. For this problem, no simple formulation of the analogous treatment has so far been given, i.e. involving analogues of the partial differential equation and of the integral formula. The above analysis explains why this should be so; the analogue of the function H would be, not a matrix but a third-order tensor (which can be thought of as a 'cubic matrix') to which we have to apply three multiplication operators in place of the familiar two operations of pre- and post-multiplication. It remains to be seen whether use of such a tensor would yield useful results in practice.

5. Extension

We conclude with the remark that extension of the above theory, set out in paragraph 3, to the case of n (> 2) parameters appears straightforward since the essential step of expanding the determinant according to its first row can still be carried out.

References

[1] Integral-equation formulation of two-parameter eigenvalue problems, Proc. Scheveningen conference on Spectral Theory and Asymptotics of Differential Equations, 1973 (Ed.E.M. de Jager, North-Holland, 1974.

[2] Two-parameter eigenvalue problems in Differential Equations, F.M.Arscott, Proc.Lond.Math.Soc., 14, 1964,459-70.

* Bivariational bounds on $\langle\phi,g\rangle$, when $A\phi = f$

M.F. Barnsley and P.D. Robinson

Abstract

Complementary (upper and lower) bivariational bounds are presented on the inner product $\langle\phi,g\rangle$ associated with the linear equation $A\phi = f$ in a Hilbert space, where the operator A is self-adjoint. The vector g is arbitrary. Possible applications are mentioned, including the derivation of point-wise bounds on ϕ. Variational bounds on $\langle\phi,f\rangle$ are taken as a starting-point.

1. Variational bounds on $\langle\phi,f\rangle$.

$A : \mathcal{H} \to \mathcal{H}$ is a positive self-adjoint operator in a real Hilbert space $\mathcal{H}$ with symmetric inner product $\langle\, ,\, \rangle$. Suppose that A has a positive lower bound, taken as unity, so that

$$\langle\Phi,A\Phi\rangle \geq \langle\Phi,\Phi\rangle \quad \text{for all } \Phi \in \mathcal{H} . \tag{1}$$

Then associated with the equation

$$A\phi = f \quad (\text{solution } \phi = A^{-1} f \in \mathcal{H}) \tag{2}$$

are the dual variational functionals

$$J(\Phi) = -\langle\Phi,A\Phi\rangle + 2\langle\Phi,f\rangle \tag{3}$$

and

$$G(\Phi') = \langle\Phi',A(A-1)\Phi'\rangle - 2\langle\Phi',(A-1)f\rangle + \langle f,f\rangle , \tag{4}$$

which are each stationary around

$$J(\phi) = G(\phi) = \langle\phi,f\rangle = \langle A^{-1} f,f\rangle \tag{5}$$

and furnish complementary bounds

$$J(\Phi) \leq \langle\phi,f\rangle \leq G(\Phi') . \tag{6}$$

In fact

$$G(\Phi) = J(\Phi) + S(\Phi) \tag{7}$$

where

$$S(\Phi) = \langle A\Phi-f,\ A\Phi-f\rangle . \tag{8}$$

Adding a suffix f to the J and S functionals, the variational bounds are

$$\boxed{J_f(\Phi) \leqslant \langle\phi,f\rangle \; = \; \langle A^{-1}f,f\rangle \leqslant J_f(\Phi') + S_f(\Phi')} \tag{9}$$

2. Bounds on $\langle\phi,g\rangle$.

It is possible to infer complementary bounds on

$$\langle\phi,g\rangle \; = \; \langle A^{-1}f,g\rangle \tag{10}$$

for arbitrary $g \in \mathcal{H}$, if complementary bounds on $\langle A^{-1}f,f\rangle$ [such as (9)] are available for arbitrary f. The identity

$$\pm 2\langle A^{-1}f,g\rangle \; = \; \langle A^{-1}(bf \pm b^{-1}g), (bf \pm b^{-1}g)\rangle - b^2\langle A^{-1}f,f\rangle - b^{-2}\langle A^{-1}g,g\rangle \tag{11}$$

is employed, together with three pairs of complementary bounds on $\langle A^{-1}f,f\rangle$, $\langle A^{-1}g,g\rangle$ and $\langle A^{-1}(bf \pm b^{-1}g), (bf \pm b^{-1}g)\rangle$ respectively. Optimization with respect to the parameter b tightens the bounds on $\langle A^{-1}f,g\rangle$, and either the positive or the negative signs can be chosen throughout (11).

3. Bivariational bounds on $\langle\phi,g\rangle$.

The three pairs of complementary variational bounds

$$J_f(\Phi) \leqslant \langle A^{-1}f,f\rangle \leqslant J_f(\Phi') + S_f(\Phi'), \tag{9}$$

$$J_g(\Psi) \leqslant \langle A^{-1}g,g\rangle \leqslant J_g(\Psi') + S_g(\Psi'), \tag{12}$$

$$J_{bf\pm b^{-1}g}(\Theta) \leqslant \langle A^{-1}(bf \pm b^{-1}g), (bf \pm b^{-1}g)\rangle \leqslant J_{bf\pm b^{-1}g}(\Theta') + S_{bf\pm b^{-1}g}(\Theta'), \tag{13}$$

together with the help of the natural simplifying choices of trial vector

$$\Theta \; = \; b\Phi' \pm b^{-1}\Psi', \quad \Theta' \; = \; b\Phi \pm b^{-1}\Psi \tag{14}$$

lead on b-optimization to the complementary bivariational bounds

$$\boxed{\begin{aligned} \mathcal{J} - (S_f S_g)^{\frac{1}{2}} + \tfrac{1}{2}(|\mathcal{S}| + \mathcal{S}) \leqslant \langle\phi,g\rangle \; = \; \langle A^{-1}f,g\rangle \leqslant \\ \mathcal{J} + (S_f S_g)^{\frac{1}{2}} - \tfrac{1}{2}(|\mathcal{S}| - \mathcal{S}) \end{aligned}} \tag{15}$$

Here

$$\mathcal{J}(\Phi,\Psi) \; = \; -\langle\Phi,A\Psi\rangle + \langle\Phi,g\rangle + \langle\Psi,f\rangle \tag{16}$$

and

$$\mathcal{S}(\Phi,\Psi) \; = \; \langle A\Phi - f, A\Psi - g\rangle \tag{17}$$

are the bivariational forms of $J(\Phi)$ and $S(\Phi)$.

4. Generalizations.

(a) If $\alpha^{-1} \geqslant A \geqslant \beta^{-1} > 0$ instead of simply $A \geqslant 1$, then (9) becomes

$$J_f + \alpha S_f \leqslant \langle\phi,f\rangle \leqslant J_f + \beta S_f \tag{9'}$$

and (15) becomes

$$\mathcal{J} - (\beta-\alpha)\big(S_f S_g\big)^{\frac{1}{2}} + \tfrac{1}{2}(\beta+\alpha)|\mathcal{S}| + \tfrac{1}{2}(\beta-\alpha)\mathcal{S} \leqslant \langle\phi,g\rangle \leqslant$$
$$\mathcal{J} + (\beta+\alpha)\big(S_f S_g\big)^{\frac{1}{2}} - \tfrac{1}{2}(\beta+\alpha)|\mathcal{S}| + \tfrac{1}{2}(\beta-\alpha)\mathcal{S}. \tag{15'}$$

(b) Boundary terms can be included if necessary if A is only formally self-adjoint.

(c) Results are being investigated when A is <u>not</u> self-adjoint.

5. Applications.

(a) Pointwise bounds on solutions of linear differential equations.

{Transform to Fredholm integral equation with Green's function $K(x,y)$; take $g(x) = K(x,y)$ to give bounds on $\phi(y)$}.

(b) Bounds on Fourier coefficients of ϕ, in an expansion solution of $A\phi = f$.

(c) Generalization of Padé approximant result $[N,N-1] \leqslant \langle\phi,f\rangle \leqslant [N,N]$.

(d) Problems in potential theory, diffusion theory and perturbation theory.

These bounds are of more practical use than those of T. Kato (Math. Ann. **126**, 253-262, 1953), which depend on a decomposition $A = T^*T$ and involve trial vectors U, V constrained to satisfy $T^*U = f$, $T^*V = g$. Such constraints can impose severe handicaps.

* A paper on this topic is to appear in Proc. Roy. Soc. (London) <u>A</u> in mid 1974.

On the Limit point and Strong Limit point classification of $2n^{th}$ order differential expressions with wildly oscillating coefficients

B. Malcolm Brown and W. Desmond Evans

Let M(f) be the formally self adjoint differential expression $M(f) = \sum_{r=o}^{n} (-1)^r (p_{n-r} f^{(r)})^{(r)}$, where the coefficients p_{n-r} satisfy the following conditions on an interval $I = [a\ \infty)$

(i) p_{n-r} is real $0 \leq r \leq n$

(ii) $p_o > 0$

(iii) $p_{n-r}^{(r-1)} \in AC_{Loc}(I)$ for $1 \leq r \leq n$ and $p_o \in L_{Loc}^{(I)}$.

It is well known [10] that, for complex λ the number m of $L^2(I)$ solutions of $M(f) = \lambda f$ satisfies $n \leq m \leq 2n$ and is also independent of λ. One of our aims in this paper is to obtain criteria for m = n. In anology with the Weyl classification, for n = 1, we shall call this the Limit Point case LP.

This problem has attracted much attention over the years, and we mention below a few of the well known results which are relevant to what follows.

M is LP in each of the following cases

I when $n = 1$, $p_o = 1$, $p_1 > -K x^2$ [9]

II when $n = 2$ $p_o = 1$ $p_2 > -K$ and $0 \leq p_1 \leq K x^2(1+|p_2|^{\frac{1}{2}})$ [4]

III when $n = 2$ $p_o = 1$ $p_2 \geq K x^{4/3}$ and $|p_1| < K x^{2/3}$ [5]

Results for general n have been mainly of an asymptotic nature, thus requiring stringent smoothness conditions on the coefficients. A notable exception is Hinton's result [8] which, inter alia, generalises I, II above to the general equation.

In this paper we obtain a generalisation of both of Everitt's results II and III to the general equation, and in addition we require the condition on the coefficients to hold only on a sequence of intervals, thus allowing for oscillatory behaviour.

A result of this type was first obtained by Hartman in [7] for $n = 1$; he proved that M is LP if $p_o = 1$ and p_1 is bounded below on a sequence of intervals $[a_m\ b_m]$ with $b_m - a_m \geq \delta > 0$. More general results for $n = 1$ are proved in [1] and for $n = 2$ analogous results may be found in [3]. Generalisations of the following results, together with detailed proofs may be found in [2].

§2 The Main Lemma

In this section we give some integral inequalities from which all the subsequent results follow.

We shall need the following notation. Let $v(x)$ be a real non-negative function with support in an interval I $(a\ \infty)$ and having a piecewise continuous n^{th} derivative. Also assume that

$$\text{(i)}\quad v^{(1)} = O(1),\ (v^{2n-1})^{(j)} = O(v^{2n-1-j}),$$
$$(v^{2n})^{(j)} = O(v^{2n-j}) \text{ for } 2 \leq j \leq n.$$

We shall also be assuming that the coefficients of M satisfy a condition of the following type: -

(ii) there exists a constant $K > 0$ such that for $1 \leq r \leq n-1$

$$q_{n-r} \equiv p_{n-r}\, v^{2(n-r)} + K\, p_o > 0$$
$$|v\, q_{n-r}^{(1)}|^2 < K\, q_{n-r}$$

Let $F_\ell(q) = \int_I q|v^\ell f^{(\ell)}|^2$ for $0 \leq \ell \leq n$ and similarly for $G_\ell(q)$, with f replaced by g.

The Lemma which we give below gives estimates for $F_\ell(q)$, $G_\ell(q)$ in terms of $M(f)$, and the Dirichlet form $\{f\ g\} = \sum_{r=1}^{n} f^{(r-1)}\, \bar{g}^{[2n-r]}$. Here we have introduced the quasi derivatives $f^{[i]}$ defined by

$$f^{[i]} = f^{(i)} \qquad 0 \leq i \leq n-1$$
$$f^{[n]} = p_o\, f^{(n)}$$
$$f^{[2n-i]} = p_{n-i}\, f^{(i)} - (f^{[2n-i-1]})^{(1)} \qquad 0 \leq i \leq n-1$$

Note that $\{f\ g\}$ is related to the bilinear form $[f\ g]$ that appears in the Lagrange identity by $[fg] = \{fg\} - \overline{\{gf\}}$.

Lemma

Under the assumptions given above there exists a positive constant K such that

$$\int_I (f\,M(f) - p_n(f)^2)\, v^{2n}\, dx \geq \frac{1}{2}\sum_{r=1}^{n} F_r(q_{n-r}) - K \sum_{r=1}^{n} F_o)q_{n-r}) \qquad (2.1)$$

$$\left|\int_I \{fg\}\, v^{2n-1}\, dx\right| \leq K \sum_{r=1}^{n} \{F_r(q_{n-r}) + G_r(q_{n-r}) + F_o(q_{n-r}) + G_o(q_{n-r})\} \qquad (2.2)$$

The proof of the Lemma involves integration by parts, and the use of some integral inequalities.

The main results of the paper follow from this Lemma by suitably choosing the function v. The method of approach, in common with [1] and [3] is by use of the result: m = n iff

$$\lim_{x\to\infty} [fg]\,(x) = 0 \quad \text{for all } f,\ g \in D =$$

$$\{f \mid f,\ M(f) \in L^2|a\ \infty),\quad f^{(2n-1)} \in A\ C_{Loc}\ |a\ \infty)$$

§3 Limit Point Criteria

The extension of Everitt's result II is achieved by choosing v to be of the form

$$V(X,x) = \begin{cases} \phi(x) & 0 \leq x \leq X \\ 2X-x & X \leq x \leq 2X \\ 0 & x \geq 2X \end{cases}$$

where ϕ is chosen in such a way that the conditions of the Lemma are satisifed on $[02X]$. For each interval I_m below we define v_m by translation from $v(X,x)$, so that in each of the intervals I_m Lemma 1 can be applied. The result is

Theorem I Let $I_m = [a_m, b_m]$ $1 \leq m \leq \infty$ be a sequence of intervals in $(a\ \infty)$ with $\lim_{m\to\infty} a_m = \infty$ and $b_m < K(b_m - a_m)$ for some constant $K > 0$.

Suppose that on I_m there exists a constant $K > 0$ such that

(i) $q_{n-r} \equiv x^{2(n-r)}\ p_{n-r} + K\,p_o > 0$, $|x\, q^{(1)}_{n-r}| < K\,p_{n-r}$ $1 \leq r \leq n$.

(ii) $p_n > -K$

(iii) $p_{n-r} \leq K \begin{cases} x^{2r}\ (|p_n|^{\frac{1}{2}} + 1) & 1 \leq r \leq \left[\frac{n}{2}\right] \\ x^{2r} & \left[\frac{n}{2}\right] < r \leq n \end{cases}$

An example of a wildly oscillating p_n which is covered by Theorem I is $p_n = x^{\alpha} \sin(\pi \operatorname{Log} x)$ for any real α.

The following extension of Everitt's result III comes about by choosing $v_m = u_m x^{-\frac{1}{2n-1}}$, where u_m is a suitable smoothing factor for each interval I_m.

<u>Theorem II</u>

Let $I_m = [a_m\ b_m]$ $1 \leq m\ \infty$ be a sequence of mutually disjoint intervals in $[a\ \infty)$ satisfying $\lim_{m\to\infty} = \infty$, $\liminf_{m\to\infty} (b_m - a_m) > 0$. Also let $J_m = [c_m\ d_m] \in I_m$ be such that for some $\delta > 0$ $c_m - a_m \geq \delta$ and $b_m - d_m \geq \delta$ for all m.

Suppose there exists a positive constant K such that the following conditions are satisfied on $\bigcup_{m=1}^{\infty} I_m$

(i) $\quad q_{n-r} \equiv x^{-\frac{2(n-r)}{2n-1}} p_{n-r} + Kp_o > 0 \qquad q_o \equiv p_o$

$\quad \left| x^{-\frac{1}{2n-1}} q^{(1)}_{n-r} \right| < K\, q_{n-r} \qquad 1 \leq r \leq n$

(ii) $\quad p_n > -K\, x^{\frac{2n}{2n-1}}$

(iii) $\quad p_{n-r} < K\, x^{\frac{2(n-r)}{2n-1}} \qquad 1 \leq r \leq n$

(iv) $\quad \sum_{m=1}^{\infty} \int_{J_m} \frac{dx}{x} = \infty$

then M is LP at infinity.

If the conditions (i)-(iv) are required to hold on $[a\ \infty)$, i.e. not merely on the intervals I_m then the above result is weaker than that of Hinton [8] as we require p_o bounded.

IV Strong Limit Point

We shall define M(f) to be SLP at infinity if and only if $\lim_{x\to\infty} \{fg\} = 0$ for all f, g ε D. For n = 1 this notion was discussed in [6].

<u>Theorem III</u>

Let the conditions of Theorem I be satisfied on the whole of $[a\ \infty)$ and in addition let $p_{n-r} \geq 0$ for $1 \leq r \leq n$. Then

(i) $|p_n|^{\frac{1}{2}} f, \ p_{n-r}^{\frac{1}{2}} f^{(r)} \in L^2(a\ \infty)$

(ii) M is SLP at infinity.

The result (i) implies the existence of the Direchlit integral associated with M on D and this is equivalent to the existence of $\lim_{x\to\infty} \{fg\}$, for all $f, g \in D$.

The proof that the value of the limit is zero follows in the same way as the proof in Theorem I.

The techniques that we have used appear to be applicable to the most general formally self adjoint differential equation.

References

1. Atkinson, F.V., Evans, W.D.: On solutions of a differential equation which are not of integrable square. Math Z. 127, 323-332 (1972).

2. Brown, B.M., Evans, W.D.: On the limit-point and strong limit-point classification of 2nth order differential expressions with wildly oscillating coefficients. Math Z. 134, 351-368 (1973).

3. Evans, W.D.: On non-integrable square solutions of a fourth order differential equation and the limit-2 classification. J. London Math. Soc. (2) 7, 343-354 (1973).

4. Everitt, W.N.: Some positive definite differential operators. J. London Math. Soc. 43, 465-473 (1968).

5. Everitt, W.N.: On the limit point classification of fourth order differential equations. J. London Math Soc. 44, 273-281 (1969).

6. Everitt, W.N.: Giertz, M., Weidmann, J.: Some remarks on a separation and limit-point criterion of second-order, Ordinary Differential Expressions. Math. Ann. 200, 335-346 (1973).

7. Hartman, P.: The number of L^2 solutions of $x'' + q(t)\ x = 0$. Amer. J. Math 73, 635-645 (1951).

8. Hinton, D.: Limit point criteria for differential equations. Can. J. Math. 24, 293-305 (1972).

9. Levinson, N.: Criteria for the limit point case for second order linear differential operators. Casopis pro pestovani matematiky a fysiky 74, 17-20 (1949).

10. Naimaik, M.A.: Linear Differential Operators, Volume II. Harrap (London) 1968.

Determination of Weyl's m-coefficient for a continuous spectrum

by

Erkki Brändas and Michael Hehenberger

Abstract.

The spectral properties of a self-adjoint second order operator of the limit-point type are discussed for the case of a continuous spectrum. A method that employs real numerical integration of the initial solutions in combination with the knowledge of a fundamental system of exponential solutions is presented. The latter are conveniently calculated from Riccati's differential equation.

The method is applied to the Stark effect in the hydrogen atom and agreement with previous results based on Airy functions in the asymptotic region is found.

1. Introduction

Through its intimate connection with the spectral density, the Weyl[1] - Titchmarsh[2] m-function is the key quantity in the solution of boundary-value problems for differential operators. Its analytic properties can be used to classify the spectrum[3]. In this contribution we will confine ourselves to self-adjoint second order operators of the limit point type for the case of a continuous spectrum. An important field of application is the quantum theory of scattering. Whereas the calculation of the limit point for the discrete part of spectrum yields the position of the eigenvalues, its knowledge in the continuum can be used to get information about socalled resonance levels and their associated lifetimes.

In order to discuss the techniques used to obtain the m-coefficient we will consider the differential equation

$$L[y] = -(py')' + qy = \lambda y \tag{1}$$

where p and q are real-valued and continuous on $[0,\infty)$, $p(x)>0 \ \forall \ x\in[0,\infty)$ and $\lambda = E+i\epsilon$. A general solution χ can be written

$$\chi(x,\lambda) = \varphi(x,\lambda) + m(\lambda)\,\psi(x,\lambda) \tag{2}$$

where φ and ψ are given by

$$\underline{Z}(0) = \begin{pmatrix} 1 & 0 \\ 0 & 1 \end{pmatrix}, \tag{3}$$

using the matrix notation

$$\underline{Z}(x) = \begin{pmatrix} \varphi(x,\lambda) & \psi(x,\lambda) \\ p(x)\,\varphi'(x,\lambda) & p(x)\,\psi'(x,\lambda) \end{pmatrix} \tag{4}$$

A numerical method for the determination of $m(\lambda)$, $\epsilon\neq 0$, based on Weyl's theory in combination with a generalization of a formula due to Titchmarsh[4] has been developed and applied to the hydrogen atom perturbed by an electric field[5]. The formula, which assumes the knowledge of the logarithmic derivative of the asymptotic solution $f(x,\lambda)$ of (1), takes the form[5][6]

$$m = \frac{\varphi(x_0)f'(x_0) - \varphi'(x_0)f(x_0)}{f(x_0)\psi'(x_0) - f'(x_0)\psi(x_0)} = \frac{W(\varphi,f)}{W(f,\psi)} \tag{5}$$

where the Wronskian has to be evaluated at a point where f and χ are proportional. In general it is necessary to work with complex λ and hence complex solution matrices (4), but for the continuous spectrum in case that m exists, it is simple to take the limit $\epsilon \to 0$ and then to work with real-valued functions. When doing so, we have to remember that $m(\lambda)$ takes on different limiting values

$$\lim_{\epsilon \to 0^{+}_{-}0} m(\lambda) = m_1(E) \pm i m_2(E) \tag{6}$$

on the real axis.

In the following section we briefly comment on the relation between the m-function and the spectral density. Section 3 is devoted to a description of a new approach for the actual numerical computation of m based on Riccati's differential equation. Finally this method is applied to the Stark effect in the hydrogen atom and comparisons are made with results previously obtained by the use of Airy integrals.[5)]

2. The spectral density

Consider the resolvent

$$(L - \lambda I)^{-1} = \psi(x_{<}, \lambda)\, \chi(x_{>}, \lambda) \tag{7}$$

where $\lambda = E + i\epsilon$ belongs to the resolvent set of L and $x_{<}$ ($x_{>}$) means the smaller (larger) of the two independent variables occuring in ψ and χ. Defining

$$\Phi(x, \lambda) = (L - \lambda I)^{-1} f(x) \tag{8}$$

with $f \in \mathcal{L}^2[0, \infty)$, we see that for E such that $\lim_{\epsilon \to 0^{+}_{-}0} m(\lambda)$ exists

$$\lim_{\epsilon \to 0^{+}_{-}0} \operatorname{Im} \Phi(x, \lambda) = \pm m_2\, \psi(x, E) \int_0^{\infty} \psi(\xi, E) f(\xi)\, d\xi \tag{9}$$

Defining a spectral function $\rho(w)$, m can be represented, disregarding the case of point-continuous spectrum, as

$$m(\lambda) = \int_{-\infty}^{+\infty} \frac{d\rho(w)}{w - \lambda} \tag{10}$$

from which it follows[6)]

$$m_2(E) = \pi \left(\frac{d\rho}{dw}\right)_{w=E} \tag{11}$$

whenever the derivative $\frac{d\rho}{dw}$ exists. Note that the formula (10) also includes the discrete part of the spectrum. In the latter case, however, $m(\lambda)$ has a simple real pole, the residue being the reciprocal value of the normalization integral. For the continuous spectrum the poles of $m(\lambda)$ may be encountered by analytical continuation across the real axis.

3. Calculation of logarithmic derivatives.

In the following we restrict ourselves to the one-dimensional Schrödinger equation

$$y'' + Q(x)y = 0 \tag{12}$$

where in an obvious notation

$$Q(x) = \frac{2m}{\hbar^2}(E - V(x)) = \frac{p^2(x)}{\hbar^2} \tag{13}$$

$V(x)$ is assumed to be analytic, but even a piecewise analytic potential can be handled in a similar way.
Using (13), equation (12) can be transformed into a Riccati equation

$$\frac{\hbar}{i} z'(x) = p^2(x) - z^2(x) \tag{14}$$

for the logarithmic derivative

$$\frac{y'(x)}{y(x)} = \frac{i}{\hbar} z(x) \tag{15}$$

Expanding

$$z(x) = \sum_{k=0}^{\infty} \left(\frac{\hbar}{i}\right)^k z_k(x) \tag{16}$$

we obtain the recursion relation

$$z'_{k-1} + \sum_{l=0}^{k} z_l z_{k-l} = 0 \tag{17}$$

from which the coefficients z_k can be determined as functions of $p(x)$ and its derivatives. The convergence of the series (16) depends on the form of the potential $V(x)$ at the argument considered. Usually numerical convergence occurs in the asymptotic region. Assuming this to be the case we obtain a fundamental system of exponential solutions

$$\begin{aligned} y^{\pm} = f^{\pm}(x,\lambda) &= \exp\left(\int^{x}(z_1 - (\hbar)^2 z_3 + \dots)dx'\right) \times \\ &\times \exp\left(\frac{\pm i}{\hbar}\int^{x}(z_0 - (\hbar)^2 z_2 + \dots)dx'\right) \\ &= (Re(z))^{-1/2}\exp\left(\pm\frac{i}{\hbar}\int^{x} Re(z(x'))dx'\right) \end{aligned} \tag{18}$$

For the special case when $x^2V(x)\xrightarrow{x\to\infty}0$, $f^{\pm}(x,\lambda)$ reduce to the well-known Jost solutions[7], which are of importance in scattering theory.

Given the solution matrix (3) and z(x) at some x_0, m is easily determined from (5)

$$m = \frac{\varphi(x_0)\frac{i}{\kappa}\, z(x_0) - \varphi'(x_0)}{\psi'(x_0) - \frac{i}{\kappa}\, z(x_0)\, \psi(x_0)} \qquad (19)$$

Finally, from the initial conditions (3), we deduce

$$m^{\pm} = \frac{i}{\kappa}\, z^{\pm}(0) \qquad (20)$$

4. Application and results

The formulas of the preceding section have been tested numerically for the equation

$$y'' + \left(\frac{1-m^2}{4x^2} + \frac{Z_2}{x} + \frac{F}{4}\,x + \frac{E}{4}\right)y = 0 \qquad (21)$$

which occurs for the Stark effect in the hydrogen atom after separation into parabolic coordinates. For details we refer to[5], in which Airy functions were used for the calculation of the asymptotic solutions. These results are compared with the Riccati approach in tables I and II for two different electric fields. The energies chosen are the respective resonance levels and the m-function is evaluated at points where the regular solution ψ has nodes. Even the convergence behaviour of the z_k's defined in equ. (16) is displayed. We conclude that

1) both methods yield reliable results, with somewhat quicker convergence for the formula (19);

2) the higher terms in the expansion (16) decrease for increasing distances x and increasing fields.

Table I : Field strength = 0.05 a.u., Energy = -1.0122107 Ry, m = 0
Interval $[0.1, \infty)$

x (a.u.) (nodes of ψ)	32.6 (2)	72.3 (10)	115.0 (23)
Re(m;Airy) $\times 10^1$	-7.73050	-1.17727	-1.01480
Re(m;Ric.) $\times 10^1$	-0.37962	-0.98350	-0.98347
Im(m;Airy) $\times 10^4$	4.50783	4.50628_3	4.50627_4
Im(m;Ric.) $\times 10^4$	4.50785	4.50627_7	4.50627_6
$z_0(x)$	0.411893	0.811011	1.090424
$z_1(x)$	-0.017742	-0.004716	-0.002621
$z_2(x)$	-0.001859	-0.000068	-0.000016
$z_3(x)$	-0.000463	-0.000002	-0.000000_2

Table II : F = 0.25 a.u., E = -1.104 Ry, m = 0
Interval $[0.1, \infty)$

x (a.u.) (nodes of ψ)	12.0 (2)	35.1 (10)	60.0 (23)
Re(m;Airy)	6.42316	6.45497	6.45543
Re(m;Ric.)	6.45656	6.45550	6.45550
Im(m;Airy) $\times 10^1$	1.74322	1.74231	1.74231
Im(m;Ric.) $\times 10^1$	1.74275	1.74231	1.74231
$z_0(x)$	0.712323	1.389694	1.866213
$z_1(x)$	-0.029310	-0.008048	-0.004478
$z_2(x)$	-0.002833	-0.000116	-0.000027
$z_3(x)$	-0.000658	-0.000004	-0.000000_4

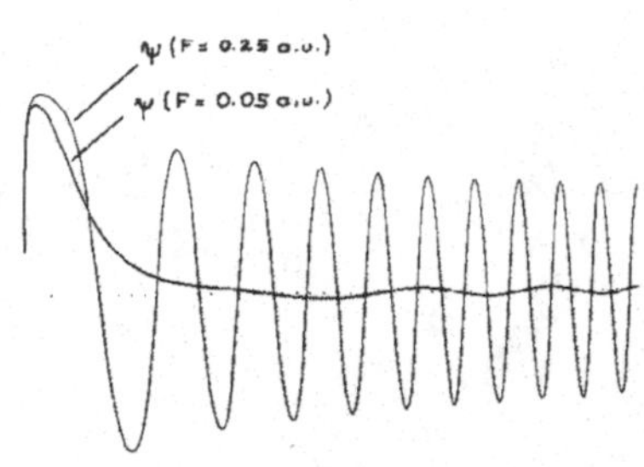

5. References.

1) H. Weyl, Math. Ann. 68, 220 (1910).

2) E. C. Titchmarsh, Eigenfunction expansions associated with second order differential equations, Oxford University Press. Part I; ibid, Part II (1946,1958).

3) J. Chaudhuri and W. N. Everitt, Proc. Royal Soc. Edinburgh (A) 68, 95 (1968).

4) E. C. Titchmarsh, Proc. Roy. Soc. A 207, 321 (1951).

5) M. Hehenberger, H. V. McIntosh and E. Brändas, Preliminary Research Report No 380, 1973. Quantum Chemistry Group, Uppsala University. (to be published in Phys. Rev.)

6) E. Brändas, M. Hehenberger and H. V. McIntosh, Preliminary Research Report No 379, 1973. Quantum Chemistry Group, Uppsala University. (to be published in Int. J. Quant. Chem.)

7) R. Jost, Physica 12, 509 (1946).

Acknowledgement.

We would like to thank Professor Per-Olov Löwdin and Professor Harold V. McIntosh for numerous discussions on the subject.

Support from the Swedish Natural Sciences Research Council and from Uppsala University is gratefully acknowledged.

The Matrix Functional-Differential Equation $\underline{y}'(x) = A\underline{y}(\lambda x) + B\underline{y}(x)$

J. CARR AND J. DYSON

1. Introduction

The matrix functional differential equation

$$\underline{y}'(x) = A\underline{y}(\lambda x) + B\underline{y}(x) \qquad (0 \leq x < \infty), \qquad (1.1)$$

where A and B are constant n x n matrices, $\underline{y}$ an n-dimensional vector and $0 < \lambda < 1$ arises from an industrial problem involving the dynamics of an overhead current collection system for an electric locomotive [1], [2]. The asymptotic behaviour of the scalar form of (1.1) has been discussed by Kato and McLeod [3]. [4], and the object here is to generalise their results to the matrix case. In the industrial problem A and B are real matrices but we allow them to have complex entries.

By a solution of (1.1) we mean a complex-valued continuous vector function $\underline{y}(x)$ defined in some subinterval of $0 \leq x < \infty$ and satisfying (1.1). If the original interval of definition is of the form $[\lambda x_o, x_o]$, $x_o > 0$, then the solution can be extended uniquely for all $x > x_o$ and becomes increasingly smooth as x increases. If the initial condition is a constant vector at $x = 0$ then the solution exists for all $x \geq 0$ and is an analytic function of x. Thus it makes sense to discuss the asymptotics of all solutions as $x \to \infty$.

2. Algebraic asymptotic behaviour

As might be expected the asymptotic behaviour of the solutions of (1.1) depends on the signs of the eigenvalues of B. If the real parts of the eigenvalues are all negative then by analogy with the scalar equation we expect algebraic behaviour. If $\underline{y}(x) \sim x^k \underline{c}$ for a constant vector $\underline{c}$, then if $\underline{y}'$ is negligible compared with $\underline{y}$, we determine k and $\underline{c}$ from the equation $(A\,\lambda^k + B)\,\underline{c} = 0$. A typical result is the following in which all asymptotic expressions refer to the limit $x \to \infty$.

Theorem 1 Suppose that the real parts of the eigenvalues of B are negative and that $B^{-1}A$ is diagonalisable. Let $-\lambda^{-k_1},\ldots,-\lambda^{-k_n}$ be the eigenvalues of $B^{-1}A$ with corresponding eigenvectors $\underline{v}_1,\ldots,\underline{v}_n$ such that $h_1 \geq h_2 \ldots \geq h_n$ where $h_i = \mathrm{Re}\,(k_i)$. [Note that the k_i are determined only up to an integral multiple of $2\pi i/\mathrm{Log}\,\lambda$ but that the h_i are uniquely determined].

(i) Every solution of (1.1) is $O(x^{h_1})$.

(ii) If $\underline{y}(x) = o(x^{h_i})$ then $\underline{y}(x) = O(x^{h_{i+1}})$, $(i=1,\ldots,n-1)$, and if $\underline{y}(x) = o(x^{h_n})$, then $\underline{y}(x) \equiv 0$.

(iii) Every solution of (1.1) has the asymptotic form

$$\underline{y}(x) = \sum_{i=1}^{n} \underline{y}_i(x) \qquad (2.1)$$

with $\underline{y}_i(x) = (x^{h_i}\, g_i\,(\ln x) + o(x^{k_i}))\underline{v}_i$,

where $g_i\,(s)$ is a C^∞ function, periodic of period $|\mathrm{Log}\,\lambda|$ and satisfying

$$|g_i^{(m)}(s)| \leq MK^m \lambda^{-m^2/2} , \ (m = 0,1,\ldots)$$

for some positive constants M and K and for all s.

(iv) Given n functions $g_1(s),\ldots,g_n(s)$ with the above properties there is a unique solution of (1.1) with asymptotic behaviour (2.1).

We give a sketch of the proof of (i). Set $s = \text{Log } x$, $d = \text{Log } \lambda$ (<0), $\underline{w}_r(s) = x^{-(p-r)}\underline{y}^{(r)}(x)$ $(r = 0,1,\ldots)$, with p to be specified later. $\underline{w}_r$ satisfies the equation

$$A\lambda^p \, \underline{w}_r \, (s+d) + B \, \underline{w}_r \, (s) \ = e^{-s}\underline{w}_{r+1}(s) \qquad (2.2)$$

We first show that for some p $\underline{w}_r(s)$ is bounded, where $p \geqslant h_1$. Equation (2.2) is then considered as a difference equation

$$B^{-1} \, A \, \lambda^p \underline{w}_r \, (s+d) + \underline{w}_r \, (s) \ = \ \phi(s) \ , \qquad (2.3)$$

where $\phi\,(s) = O(e^{-s})$. The asymptotic behaviour of $\underline{w}_r(s)$ will be that of the corresponding homogeneous problem plus a term which in general will be $O(e^{-s})$. The behaviour of the solution of the homogeneous problem is determined by the eigenvalues of $B^{-1} A \lambda^p$. If the $O(e^{-s})$ term dominates then $\underline{w}_r(s) = O(e^{-s})$ and we substitute this into the right of (2.2). This procedure is repeated until the term involving the homogeneous problem dominates.

If $B^{-1}A$ is not diagonalisable then the solution of the

difference equation (2.3) will contain powers of s. Consequently powers of Log x enter into the asymptotic form for $\underline{y}(x)$.

3. Exponential asymptotic behaviour

We now allow some of the eigenvalues of B to have positive real parts. Here the same behaviour as before is found, modulo certain large solutions. More precisely:

Theorem 2 Suppose the same conditions as Theorem 1 hold except that now B is a diagonalisable matrix whose eigenvalues $b_1,\ldots,b_n$ satisfy $\mathrm{Re}(b_1) > \ldots > \mathrm{Re}(b_r) > 0 > \mathrm{Re}(b_{r+1}) > \ldots > \mathrm{Re}(b_n)$. Let T be such that $T^{-1}BT$ is diagonal and denote the ith component of $T^{-1}\underline{y}$ by $(T^{-1}\underline{y})_i$.

(i) $\underline{y}(x) = O(e^{b_1 x})$ and there is a constant L_1 such that

$$(T^{-1}\underline{y})_1 \, e^{-b_1 x} \to L_1$$

and
$$(T^{-1}\underline{y})_j \, e^{-b_j x} \to 0 \quad (j = 2,\ldots,n), \tag{3.1}$$

(ii) If $1 < j < r-1$ and $\underline{y}(x) = o(e^{b_j x})$, then $\underline{y}(x) = O(e^{b_{j+1} x})$ and there are constants L_{j+1} such that

$$(T^{-1}\underline{y})_{j+1} \, e^{-b_{j+1}} \to L_{j+1}$$

and
$$(T^{-1}\underline{y})_k \, e^{-b_{j+1} x} \to 0, \quad (k \neq j+1). \tag{3.2}$$

(iii) If $\underline{y}(x) = o(e^{b_r x})$ then $\underline{y}(x)$ has the same behaviour as that exhibited in Theorem 1.

(iv) Given r constants $L_1,\ldots,L_r$ there are solutions of (1.1) with asymptotic behaviour (3.1), (3.2).

If B is not diagonalisable we get a similar theorem.

4. B has eigenvalues on the imaginary axis

The results here are incomplete compared with the other cases. The scalar equation $y'(x) = ay(\lambda x)$ was discussed by de Bruijn in [5]. His method of proof which depends on writing particular solutions as a Fourier Integral seems to carry over to the matrix equation

$$\underline{y}'(x) = A\underline{y}(\lambda x) \qquad (4.1)$$

although to date this has not been done. If a complete analysis of (4.1) were available, it would then be possible to obtain complete results for (1.1) when B has zero eigenvalues.

The asymptotic behaviour of the scalar equation

$$y'(x) = a\, y\,(\lambda x) + b\, y\,(x) \qquad (4.2)$$

when b is imaginary is not covered by [3] or [4]. We have some results for (4.2), but they are incomplete compared with other cases.

References

[1] L. Fox. D.F. Mayers, J.R.Ockendon and A.B.Tayler 'On a functional differential equation' J. Inst. Math. Appl. 8 (1971) 271-307.

[2] J.R.Ockendon and A.B.Tayler. 'The dynamics of a current collection system for an electric locomotive' Proc. Roy. Soc. Lond. A 322 (1971) 447-468.

[3] Tosio Kato and J.B. McLeod, 'The functional-differential equation $y'(x) = ay(\lambda x) + by(x)$' Bull. Amer. Math. Soc. 77 (1971). 891-937.

[4] Tosio Kato 'Asymptotic behaviour of solutions of the functional-differential equation' $y'(x) = ay(\lambda x) + by(x)$ O.D.E. 1971 NRL-MRC conference, (proc); ed. by L. Weiss.

[5] N.G. de Bruijn, 'The difference-differential equation $F'(x) = e^{\alpha x+\beta} F(x-1)$, I, II, Nederl. Akad. Wetensch. Proc. Set. A56 = Indag Math. 15 (1953), 449-464.

A TETRAD APPROACH FOR A SYSTEM OF PDE'S

G. Debney

and

D. Farnsworth

The field equations for general relativity comprise a system of non-linear, second-order PDE's for the ten components of the bilinear metric form (or metric) $g_{\mu\nu}$ of a four-dimensional Lorentz manifold (or "space-time"). These equations are analogous to the Poisson equation for a potential function. If Greek subscripts run from 1 to 4 and indicate tensor components with respect to any coordinate system $\{x^{\mu}\}$ the metric $g_{\mu\nu}$ has signature $(-,-,-,+)$. The dual components, $g^{\mu\nu}$, are simply the matrix inverse of $g_{\mu\nu}$. If $R_{\alpha\mu\nu\beta}$ is the Riemannian curvature tensor, $R_{\mu\nu} \equiv R_{\alpha\mu\nu\beta}g^{\alpha\beta}$ is the Ricci curvature, and $R \equiv R_{\mu\nu}g^{\mu\nu}$ is the scalar curvature, Einstein's field equations with galaxies as "dust particles" take the form

$$R_{\mu\nu} - \frac{1}{2}g_{\mu\nu}R + \lambda g_{\mu\nu} = -\rho u_{\mu}u_{\nu}.$$

(Here, λ is the cosmological constant; ρ is the density function on the dust; and $\underset{\sim}{u} = u^{\mu}\frac{\partial}{\partial x^{\mu}}$ is tangent to the necessarily geodesic trajectories of the particles -- $u_{\mu} = g_{\mu\nu}u^{\nu}$.)

To allow solutions to be obtained in this cosmological system, directions of isometry are usually introduced. In the following we impose a four-parameter group of Bianchi Type V ([1] and references therein). In a coordinate system $\{x,y,z,t\}$ the metric becomes

$$ds^2 = g_{\mu\nu} dx^\mu dx^\nu = -\exp(2U(t))\,dx^2 - \exp(2V(t)+x)(dy^2+dz^2) + dt^2.$$

The four Killing vector fields expressing the Bianchi V symmetry group are then

$$\underset{\sim}{X}_1 = \frac{\partial}{\partial y} \qquad \underset{\sim}{X}_3 = -\frac{\partial}{\partial x} + y\frac{\partial}{\partial y} + z\frac{\partial}{\partial z}$$

$$\underset{\sim}{X}_2 = \frac{\partial}{\partial z} \qquad \underset{\sim}{X}_4 = -z\frac{\partial}{\partial y} + y\frac{\partial}{\partial z}\,.$$

By normalizing $\underset{\sim}{u}$ to be a unit vector, the invariance of $\underset{\sim}{u}$ under the group implies that $\underset{\sim}{u} = \alpha(t)\frac{\partial}{\partial x} + \beta(t)\frac{\partial}{\partial t}$, that $\alpha \exp(2U) = -C$ ($C < 0$, a constant), and that $C\alpha + \beta^2 = 1$.

As a standard procedure in avoiding complicated DE's a tetrad (vierbein) of basis vector fields is aligned over the manifold ([2], [3]). Advantages of this formalism in general are that (1) the potential for "spurious nonlinearities" is reduced; (2) the bilinear metric form is constant; (3) coordinate and parameter changes often evolve naturally. Briefly, the procedure is to take tetrad components of the tensor equations and to find first integrals in this context.

The orthonormal tetrad $\{\underset{\sim}{e}_a = e_a{}^\mu \frac{\partial}{\partial x^\mu} \mid a = 1,\dots,4\}$ is chosen to be one of those for which $\underset{\sim}{e}_4 = \underset{\sim}{u}$ and $\underset{\sim}{e}_1$, $\underset{\sim}{e}_2$ are proportional to $\underset{\sim}{X}_1$, $\underset{\sim}{X}_2$, respectively:

$$\underset{\sim}{e}_1 = \exp(-V-x)\frac{\partial}{\partial y} \qquad \underset{\sim}{e}_3 = -\beta\exp(-U)\frac{\partial}{\partial x} - \alpha\exp(U)\frac{\partial}{\partial t}$$

$$\underset{\sim}{e}_2 = \exp(-V-x)\frac{\partial}{\partial z} \qquad \underset{\sim}{e}_4 = \alpha\frac{\partial}{\partial x} + \beta\frac{\partial}{\partial t}.$$

The dual one-forms $\{\epsilon^a = \epsilon^a_{\ \mu}\, dx^\mu\}$ consist of

$$\epsilon^1 = \exp(V+x)\, dy \qquad \epsilon^3 = -\beta\exp(U)\, dx + \alpha\exp(U)\, dt$$

$$\epsilon^2 = \exp(V+x)\, dz \qquad \epsilon^4 = C\, dx + \beta\, dt.$$

The tetrad components of the metric $ds^2 = g_{\mu\nu}\, dx^\mu dx^\nu = g_{ab}\epsilon^a\epsilon^b$ are easily seen to be $g_{ab} = \mathrm{diag}(-1,-1,-1,+1)$.

The Ricci rotation coefficients $\Gamma_{abc} = -e_{a\mu;\nu}e_b^{\ \mu}e_c^{\ \nu}$ are found from the first Cartán structure equations $d\epsilon^a = \Gamma^a_{\ bc}\epsilon^b \wedge \epsilon^c$ ([2] and [4]), where $\Gamma^a_{\ bc} \equiv g^{ad}\Gamma_{dbc}$ and $\Gamma_{abc} = -\Gamma_{bac}$. From the second Cartán structure equations $d(\Gamma_{abc}\epsilon^c) + \Gamma_{amc}\Gamma^m_{\ bd}\epsilon^c \wedge \epsilon^d = \frac{1}{2}R_{abcd}\epsilon^c \wedge \epsilon^d$ one finds the tetrad components of the Riemann form R_{abcd}. The Ricci components R_{ab} and R are easily computed so that the field equations may be written explicitly:

$$R_{ab} - \frac{1}{2}g_{ab}R + g_{ab}\lambda = -\rho\delta^4_a\delta^4_b.$$

At this stage a useful simplification results if the following substitution is made: let $X \equiv \beta\exp(U)$ and $Y \equiv \exp(V+x)$. Then the metric becomes

$$ds^2 = (C^2 - X^2)\, dx^2 - Y^2(dy^2 + dz^2) + dt^2,$$

so that the entire desired solution is given through X and Y. The Bianchi V group conditions give $\underset{\sim}{D}X = 0$ and $\underset{\sim}{D}Y = -Y$ where $\underset{\sim}{D} \equiv X\underset{\sim}{e}_3 - C\underset{\sim}{e}_4$, corresponding to $X = X(t)$ in the former. The integrability condition $\underset{\sim}{D}\underset{\sim}{e}_4 - \underset{\sim}{e}_4\underset{\sim}{D} = 0$ implies the existence of a function $P(x,t)$ for which $\underset{\sim}{D}P = 1$ and $\underset{\sim}{e}_4 P = 0$.

Hence, the general solution for Y is given by

$$Y = w(t)\exp(-P).$$

The first field equation ($R_{34} = 0$) is equivalent to $e_4(e_3 Y) = 0$ so that $e_3 Y = C_1 \exp(-P)$ where C_1 is a constant. It follows that

$$X = C_1^{-1}(Cw_{,4} - w),$$

where $e_4 w \equiv w_{,4}$, $e_4(w_{,4}) \equiv w_{,44}$, etc.

Two field equations, one for w, and one for ρ in terms of w complete the system. The former of these has a first integral, the <u>Friedmann</u> equation:

$$(w_{,4})^2 = (C_1)^2 + C_2 w^{-1} + (\tfrac{\lambda}{3})w^2. \quad (C_2 \text{ constant})$$

The equation for ρ comes from R_{44} and takes the explicit form

$$\rho = 6(Cw_{,4} - w)^{-1}[w_{,44} - (\tfrac{\lambda}{3})w].$$

To complete the task in terms of coordinates it is best to notice that there also exists a function T(x,t) such that $e_4 T = 1$ and $e_3 T = 0$. The functions P and T are then well-defined coordinates for which $Xe_3 = \frac{\partial}{\partial P}$ and $e_4 = \frac{\partial}{\partial T}$, and the metric assumes the form

$$ds^2 = -X^2\, dP^2 - Y^2(dy^2 + dz^2) + dT^2.$$

Solutions to the Friedmann equation then determine the metric and the density up to the constants $C(< 0)$, C_1 and C_2.

The kinematics of $u = e_4$ is easily investigated by this approach [3]. Since $d\epsilon^4 = 0$ the trajectories form orthogonal hypersurfaces and the rotation of $\underset{\sim}{u}$ is always identically zero. These surfaces are not the same as the hypersurfaces of transitivity where the group X_i must act, but they correspond to the instantaneous rest frames of observers moving with the particles. The shear and expansion of the evolution in space-time are easily calculated from the tetrad equations

$$\sigma_{ab} = -\Gamma_{4(ab)} - \frac{1}{3}\theta(g_{ab} - \delta_{4a}\delta_{4b}) \qquad \text{(shear)}$$

$$\theta = \Gamma_{4\ a}^{\ a} \qquad \text{(expansion)}.$$

The matrix σ_{ab} is diagonal: $\sigma_{11} = \sigma_{22} = (\frac{1}{3})\underset{\sim}{e}_4(\ell n\, Xw^{-1})$; $\sigma_{33} = (-\frac{2}{3})\underset{\sim}{e}_4(\ell n\, Xw^{-1})$; $\sigma_{44} = 0$. Furthermore, $\theta = \underset{\sim}{e}_4(\ell n\, Xw^2)$. Hence these cases have a "non-Friedmann" property, in that the shear in general does not vanish.

REFERENCES

[1] D. Farnsworth, J. Math. Phys. 8, 2315 (1967).

[2] G. Debney, R. P. Kerr, and A. Schild, J. Math. Phys. 10, 1842 (1969).

[3] G. F. R. Ellis, J. Math. Phys. 8, 1171 (1967).

[4] S. I. Goldberg, Curvature and Homology (Academic Press, New York, 1962).

QUASI-ANALYTIC SOLUTIONS OF DIFFERENTIAL EQUATIONS

J.W. de Roever

We consider a differential equation with constant coefficients

(1) $\qquad P(D)f=g$

where P is a polynomial, $D=(\frac{\partial}{\partial x_1},\ldots,\frac{\partial}{\partial x_n})$; g is an element in some class of generalized functions and we look for solutions f in the same class. In particular g is the distributional boundary value of a function holomorphic in a tubular radial domain. Namely, let C be an open convex cone in $\mathbb{R}^n$, C_k an increasing sequence of compact subcones with $\bigcup_{k=1}^{\infty} C_k=C$ and let g^+ and g^- be functions holomorphic in $\mathbb{R}^n+iC$ or $\mathbb{R}^n-iC$, respectively, satisfying for all k and $\varepsilon>0$

(2) $$|g^+(z)|\leq M_k(\varepsilon)\ K(x)\ (1+|y|^{-m})\ e^{(a+\varepsilon)|y|} \qquad \forall\ y\in C_k$$
$$|g^-(z)|\leq M_k(\varepsilon)\ K(x)\ (1+|y|^{-m})\ e^{(a+\varepsilon)|y|} \qquad \forall\ y\in -C_k,\ z=x+iy.$$

Here K is an arbitrary positive function in x, which depends on $g^{\pm}$; the nonnegative constants $M_k(\varepsilon)$ and m also depend on $g^{\pm}$, while the constant a is independent of $g^{\pm}$. Then the limits $\lim_{y\to 0,\, y\in\pm C_k} g^{\pm}(x+iy)$ exist in the space D' of Schwartz-distributions and we define the distribution $g\in D'$

$$g=\lim_{y\to 0,\, y\in C_k}\{g^+(x+iy)-g^-(x-iy)\}.$$

We call the class of such distributions $L_a^{\sim}(C)$. A particular case is the class $L_a(C)=L_a^{\sim}(C)\cap S'$, where S' is the space of tempered distributions. Elements of $L_a(C)$ are the difference of distributional boundary values of holomorphic functions g^+ and g^- satisfying

$$|g^{\pm}(z)|\leq M_k(\varepsilon)(1+|x|)^m(1+|y|^{-m})e^{(a+\varepsilon)|y|} \qquad \forall\ y\in\pm C_k.$$

In this lecture we establish the same results for the class $L_a^{\sim}(C)$ as obtained in [1] for the class $L_a(C)$.

We use the theory of holomorphic functions in several complex variables in order to derive properties of distributions g in $L_a^{\sim}(C)$. Namely these distributions are quasi-analytic, that is if g vanishes in some open set G in $\mathbb{R}^n$, it also vanishes in the larger set

$$B(G)=\{x\,|\,x\in\{x_1+C\}\cap\{x_2-C\},\ \forall\ x_1,x_2\in G\}.$$

The fact that g is zero in G means that for $x\in G$ $g^+(x+i0)=g^-(x-i0)$ in distributional sense. According to the "Edge of the Wedge Theorem" this equality also holds in ordinary sense. Hence g^+ and g^- are analytic continuations of each other. Thus there is a complex neighborhood N(G) of G and a function holomorphic in $\Omega=\{\mathbb{R}^n+iC\}\cup\{\mathbb{R}^n-iC\}\cup N(G)$ that equals $g^{\pm}$ in $\mathbb{R}^n\pm iC$. It may happen that any function f, which is holomorphic in some open set V in $\mathbb{C}^n$ $(n>1)$, is also holomorphic in a larger set H(V), the envelope of holomorphy; this envelope only depends on the shape of V and not on the function f. Here we have that $H(\Omega)$ contains a complex neighborhood of B(G).

We want to find conditions such that f belongs to $L_a^{\sim}(C)$ if g does. For that purpose we use Fourier transformation. A distribution in D' has a Fourier transform which is an element of Z'. Elements of Z' can be represented as measures in C^n which act on entire functions φ satisfying $(1+|\zeta|)^m|\varphi(\zeta)|\leq M_m e^{b|\operatorname{Im}\zeta|}$ for all $m\geq 0$ and for some nonnegative b depending on φ. We say that an element of Z' is carried by the set $\Omega\subset\mathbb{C}^n$, when it can be represented as a measure with support in Ω. We need the following theorem:

<u>Theorem 1</u> A distribution g in $L_a^{\sim}(C)$ has a Fourier transform $\hat{g}$ carried by $\Omega(a+\varepsilon;C)=\{\zeta=\xi+i\eta \mid -y\cdot\xi\leq a+\varepsilon, \forall y\in C \text{ with } |y|=1\}\cup\{\zeta=\xi+i\eta \mid y\cdot\xi\leq a+\varepsilon, \forall y\in C \text{ with } |y|=1\}$. Conversely, an element of Z' carried by $\Omega(a+\varepsilon;C)$ is the Fourier transform of the distributional boundary value in D' of a function of exponential type a in y, holomorphic in $\{\mathbb{R}^n+iC\}\cup\{\mathbb{R}^n-iC\}$.

The proof of theorem 1 is in some way similar to the proof of the theorem for the class $L_a(C)$ in [1] and it is given in full detail in [2].

In order that the solution f of the differential equation (1) is quasi-analytic (that is it belongs to $L_a^{\sim}(C)$), the polynomial P should satisfy

(3) $\qquad P(i\zeta)\neq 0 \quad \text{for } \zeta\notin\Omega(a+\varepsilon;C).$

This is clear from the following consideration:

Fourier transform of (1) gives $P(i\zeta)\hat{f}(\zeta)=\hat{g}(\zeta)$ where $\hat{g}$ is carried by $\Omega(a+\varepsilon;C)$. It follows that also $\hat{f}$ is carried by $\Omega(a+\varepsilon;C)$, so that f belongs to $L_a^{\sim}(C)$ according to theorem 1.

In case $g\in L_a(C)$ and $f\in S'$ we require that

(4) $\qquad P(i\xi)\neq 0 \quad$ for $\xi\notin O(a;C)=\{\xi \mid -y.\xi\leq a, \forall y\in C$ with $|y|=1\}$,

see [1] and [3].

For example the wave operator $(\frac{\partial}{\partial t})^2-\Delta_n$ satisfies (3) only when n=1 and (4) for all n. Thus any solution in S' of the wave equation with right member $g\in L_a(C)$ is quasi-analytic (see [1] and [3]). The same result holds for all solutions in D' when n=1 and $g\in \tilde{L}_a(C)$. For the general wave operator (n>1) condition (3) is no longer satisfied. Then for any solution f of (1) with $g\in \tilde{L}_a(C)$ there exists a solution f_o of the homogeneous equation $P(D)f_o=0$, such that the solution $f_1=f-f_o$ of (1) is quasi-analytic.

We now consider the particular distributions in $\tilde{L}_a(C)$, where the function $K(x)e^{a|y|}$ in (2) should be of exponentional type in x and y. The type of the functions $g^{\pm}$ depends on the direction of $z=x+iy$ regarded in $\mathbb{R}^{2n}=\mathbb{C}^n$; therefore we write $z=(x,y)$ as a vector in $\mathbb{R}^{2n}$. Let $a(x,y)$ be a convex function, homogeneous of degree 1 in $(x,y)\in(\mathbb{R}^n,C)\subset\mathbb{R}^{2n}$. Then the following modification of theorem 1 holds :

<u>Theorem 2</u> Let $g^{\pm}$ be holomorphic functions in $\mathbb{R}^n\pm iC$ satisfying for all k and $\varepsilon>0$ $\qquad |g^{\pm}(z)|\leq M_k(\varepsilon)(1+|y|^{-m})e^{a(x,\pm y)+\varepsilon|(x,y)|}$, $\forall y\in\pm C_k$.

Then the distributional boundary values $g(x\pm i0)$ are the Fourier transforms of elements $\hat{g}^{\pm}$ of Z' carried by

$\Omega_1(a(x,y)+\varepsilon;\pm C)=\{\zeta=\xi+i\eta \mid -y.\xi-x.\eta\leq a(x,y)+\varepsilon|(x,y)|, \forall x\in\mathbb{R}^n, \forall y\in\pm C\}$.

Conversely, the Fourier transform of an element in Z' carried by $\Omega_1(a(x,y)+\varepsilon;\pm C)$ is the distributional boundary value in D' of a function of exponential type $a(x,y)$ in x and y, holomorphic in $\mathbb{R}^n\pm iC$.

Theorem 2 is much harder to prove than theorem 1. It can be derived from theorem 1 and from Ehrenpreis' fundamental principle [4] (see [2]). Briefly, this principle yields that any solution f of a homogeneous differential equation with constant coefficients $Q(D)f=0$ is the Fourier transform of certain derivatives of measures concentrated on the zero-set of the polynomial $Q(i\zeta)$.

Let now K(x) in (2) be $K(x)=e^{(b+\varepsilon)|x|}$. Then we call the distribution $g\in\tilde{L}_a(C)$ of exponential type b. In this case the function a(x,y) in theorem 2 becomes $b|x|+a|y|$ and the set $\Omega_1(a(x,y)+\varepsilon;C)\cup\Omega_1(a(x,y)+\varepsilon;-C)$ becomes $\Omega(a+\varepsilon;C)\cap\{\zeta=\xi+i\eta \mid |\eta|\leq b+\varepsilon\}$. Without loss of generality we can take $a\geq b$. Then condition (3) is satisfied for the general wave operator P as long as $|\eta|\leq b+\varepsilon$. Hence according to theorem 2, all solutions in D' of the wave equation with right member g, that are of the same type as g, are quasi-analytic.

[1] V.S. Vladimirov Methods of the Theory of Functions of many complex variables (translated from Russian) The M.I.T. Press, 1966

[2] J.W. de Roever Fourier transforms of holomorphic functions and application to Newton interpolation series II to be published, Mathematisch Centrum, Amsterdam 1974

[3] E.M. de Jager On functions holomorfic in tube-domains $\mathbb{R}^n+iC$ Spectral theory and asymptotics of differential equations, p 137 North Holland Mathematics Studies 13, North Holland/ American Elsevier, 1974

[4] L. Ehrenpreis Fourier analysis in several complex variables Tracts in mathematics number 17, Wiley-interscience publishers, 1970

INTEGRAL INEQUALITIES AND THE LIOUVILLE TRANSFORMATION

W. N. Everitt

1. This paper is concerned with an integral inequality derived from the ordinary, formally symmetric differential expression $l[\cdot]$ where

$$l[f] = -(pf')' + qf \text{ on } [a, b) \qquad (' \equiv d/dx), \tag{1.1}$$

and the consequences of applying to $l[\cdot]$ the Liouville transformation.

In (1.1) $[a, b)$ is a half-closed, half-open interval of the real line with $-\infty < a < b \leqslant \infty$; the coefficients p and q are real-valued on $[a, b)$ and satisfy the basic conditions

$$\begin{aligned} &\text{(i)} \quad p \in AC_{loc}[a, b) \ \underline{\text{and}} \ p(x) > 0 \quad (x \in [a, b)) \\ &\text{(ii)} \quad q \in L_{loc}[a, b). \end{aligned} \tag{1.2}$$

(AC - absolute continuity; L - Lebesgue integration; 'loc' indicates a condition to be satisfied on all compact sub-intervals of $[a, b)$.)

In the sense of Naimark, see [15, Section 15.1], it is assumed throughout that b is a singular point of $l[\cdot]$, i.e. either $b = \infty$, or if $b < \infty$ then $p^{-1} \notin L(a, b)$ or $q \notin L(a, b)$. All points of $[a, b)$ are regular points of $l[\cdot]$.

In view of the symmetry of $l[\cdot]$ it follows from the general theory in [15; see Section 17.5] that $l[\cdot]$ may be classified as either limit-point (LP) or limit-circle (LC) at the singular point b.

2. Let δ denote the linear manifold of the Hilbert function space $L^2(a, b)$, defined by

$$f \in \delta \text{ if} \quad \begin{aligned} &\text{(i)} \quad f \text{ is real-valued on } [a, b) \\ &\text{(ii)} \quad f \in L^2(a, b) \\ &\text{(iii)} \quad f' \in AC_{loc}[a, b) \\ &\text{(iv)} \quad l[f] \in L^2(a, b). \end{aligned} \tag{2.1}$$

(We assume throughout that f is real-valued on [a, b); there is no subsequent loss of generality as all the results given here extend without complication to complex-valued functions on [a, b).)

It is known that $l[\cdot]$ is LP at the singular point b if and only if

$$\lim_{b-} p(fg' - f'g) = 0 \qquad (f, g \in \delta); \tag{2.2}$$

see [15, Section 18.3]. From Green's formula the limit on the left of (1.4) exists and is finite; the LP condition for $l[\cdot]$ requires this limit to be zero for all $f, g \in \delta$.

Following the definition in [9, Section 1], see also [8, Section 1], the differential expression $l[\cdot]$ is said to be strong limit-point (SLP) at b if

$$\lim_{b-} pf g' = 0 \quad (f, g \in \delta). \tag{2.3}$$

In comparison with (2.2) it should be noted that the limit on the left of (2.3) may or may not exist for $f, g \in \delta$; the SLP condition requires both the existence of the limit and that it be zero, for all $f, g \in \delta$.

Clearly if $l[\cdot]$ is SLP at b then it is LP at b; the converse of this result is false, see [9. Section 5] and [8, Section 2].

The differential expression $l[\cdot]$ is said to satisfy the Dirichlet (D) condition at the singular point b if

$$p^{\frac{1}{2}}f' \;\underline{\text{and}}\; |q|^{\frac{1}{2}}f \in L^2(a, b) \qquad (f \in \delta). \tag{2.4}$$

Recently, in a written communication to the author, Kalf [14] has shown that if $l[\cdot]$ is D at b then $l[\cdot]$ is SLP at b (and so is LP at b). The converse of this statement is false; see [8, Sections 3 and 4].

Thus if $l[\cdot]$ is D at the singular point b then integration by parts yields the Dirichlet formula, see [9, Section 1]

$$\int_a^b \{pf'g' + qf g\} = -p(a)\, f(a)g'(a) + \int_a^b f\, l[g] \qquad (f, g \in \delta). \tag{2.5}$$

3. It follows from (2.5) that if $g = f$ and $f(a) = 0$ or $f'(a) = 0$ then, from the Cauchy-Schwarz inequality,

$$\left(\int_a^b \{pf'^2 + qf^2\}\right)^2 \leqslant \int_a^b f^2 \int_a^b l[f]^2.$$

It is not to be expected that this inequality will hold for all $f \in \delta$ so that in general the following inequality is considered

$$\left(\int_a^b \{pf'^2 + qf^2\}\right)^2 \leqslant k\int_a^b f^2 \int_a^b l[f]^2 \qquad (f \in \delta) \tag{3.1}$$

where k is independent of the elements of δ and satisfies $0 < k \leqslant \infty$. When $k = \infty$ this is understood to imply there is no valid inequality of the form (1.8); when $0 < k < \infty$ this is understood to imply that k is the best possible, i.e. the smallest, positive number for which (3.1) holds.

The inequality (3.1) is the inequality of the title of this paper. It is the generalisation to symmetric differential expressions of an important inequality of Hardy and Littlewood, see [11, Section 7] and [12, Section 7.8], which is mentioned below.

The general inequality (3.1) is discussed in detail by Everitt in [4]; see also the first section of the report [6].

The following example of (3.1) is considered in [4, Section 17] and [6, Section 1]. Let

$$a = 1,\ b = \infty \qquad p(x) = x^\tau \ \underline{\text{and}}\ q(x) = 0 \qquad (x \in [1, \infty))$$

where τ is a real number. In this case (3.1) takes the form (both k and δ depend upon τ)

$$\left(\int_1^\infty x^\tau f'(x)^2\,dx\right)^2 \leqslant k(\tau)\int_1^\infty f(x)^2 dx \int_1^\infty \{(x^\tau f'(x))'\}^2 dx \qquad (f \in \delta(\tau))$$

where $\delta(\tau)$ is determined by (2.1). The function $k(\cdot)$ on the real line has been characterised in terms of the elementary functions; details are given in [6, Section 1]. The following special cases are given here:

(i) $k(-1) = 4$; the only case of equality is f is null on $[1, \infty)$

(ii) $k(0) = 4$; for cases of equality see [11, Section 7] and [12, Section 7.8]

(iii) $k(\frac{1}{2}) = 6 + 2\sqrt{5}$; the only case of equality is f is null on $[1, \infty)$

(iv) $k(1) = \infty$.

The case $\tau = 0$ is essentially the Hardy-Littlewood inequality. There is no valid inequality in the case $\tau = 1$, nor indeed when $\tau \geq 1$.

4. Consider now the application of the Liouville transformation to $l[\cdot]$. This transformation is considered in [1, Chapter X, Section 9] and [3, Section 3.9]; for a more general discussion see [2, Page 1498]. The notation adopted here is taken from the account in [5, Section 9]. Note that the essence of this notation is that lower case letters apply to the original differential expression $l[\cdot]$ and capital letters to the transformed differential expression $L[\cdot]$.

Let the coefficents p and q satisfy the conditions (1.2), and additionally let p satisfy

$$p' \in AC_{loc}[a, b). \tag{4.1}$$

Define $X(\cdot) : [a, b) \to [A, B)$ by

$$X(x) = A + \int_a^x p^{-\frac{1}{2}} \quad (x \in [a, b)) \tag{4.2}$$

where A is an arbitrary, but then fixed, real number and $B = X(b)$, i.e. $-\infty < A < B \leq \infty$. Let $x(\cdot)$ denote the well-determined inverse function to $X(\cdot)$ so that $x(\cdot) : [A, B)$ onto $[a, b)$.

Define the functions P_0 and Q_0 on $[A, B)$ by

$$P_0(X) = p(x(X)) \quad Q_0(X) = q(x(X)) \qquad (X \in [A, B)). \tag{4.3}$$

Given any $f \in L^2(a, b)$ define $F = Uf$, say, on $[A, B)$ by

$$F(X) = \{p(x)\}^{\frac{1}{4}} f(x) = \{P_0(X)\}^{\frac{1}{4}} f(x(X)) \qquad (X \in [A, B)). \tag{4.4}$$

A calculation shows that, on changing the variable of integration,

$$\int_a^b f^2 = \int_A^B (Uf)^2 = \int_A^B F^2 \qquad (f \in L^2(a, b)). \tag{4.5}$$

The inverse transformation $f = U^{-1}F$ is given by

$$f(x) = \{P_0(X)\}^{-\frac{1}{4}} F(X) = \{p(x)\}^{-\frac{1}{4}} F(X(x)) \qquad (x \in [a, b)). \tag{4.6}$$

U is a unitary map on $L^2(a, b)$ onto $L^2(A, B)$.

If now $f \in \delta$, see (2.1), then a calculation shows that

$$l[f](x) = P_0(X)^{-\frac{1}{4}} L[f](X) \qquad (X \in [A, B)) \tag{4.6a}$$

where the differential expression $L[\cdot]$ is given by

$$L[F](X) = - F''(X) + Q(X)F(X) \qquad (X \in [A, B)) \tag{4.7}$$

with

$$Q(X) = q(x(X)) + \frac{1}{4} p''(x(X)) - \frac{1}{16} \frac{\{p'(x(X))\}^2}{p(x(X))}$$

$$= Q_0(X) + \frac{1}{4}\frac{P_0''(X)}{P_0(X)} - \frac{3}{16}\frac{P_0'(X)^2}{P_0(X)^2} \qquad (X \in [A, B)).$$

Note that primes on lower case (respectively capital) letters denote differentation with respect to x (respectively X). Conditions (1.2) and (4.1) imply that $Q \in L_{loc}[A, B)$. Note also that $L[\cdot]$ is formally symmetric on $[A, B)$.

If the linear manifold Δ of $L^2(A, B)$ is defined by

$F \in \Delta$ if (i) F is real-valued on $[A, B)$

(ii) $F \in L^2(A, B)$

(iii) $F' \in AC_{loc}[A, B)$

(iv) $L[F] \in L^2(A, B)$

then it follows that U maps δ onto Δ and that $l[\cdot]$ and $L[\cdot]$ are unitarily equivalent in the sense that, using (4.6a),

$$\int_a^b l[f]^2 = \int_A^B (Ul[f])^2 = \int_A^B L[F]^2 \qquad (f \in \delta) \tag{4.8}$$

A further calculation shows that if f and g $\in \delta$, $F = Uf$ and $G = Ug$, then F and G $\in \Delta$, and

$$p(x)\big(f(x)g'(x) - f'(x)g(x)\big) = F(X)G'(X) - F'(X)G(X) \qquad (X \in [A, B)).$$

It follows that if $l[\cdot]$ is LP at b then $L[\cdot]$ is LP at B and, indirectly, that B is a singular point of $L[\cdot]$. Since $Q \in L_{loc}[A, B)$ all points of $[A, B)$ are regular points of $L[\cdot]$.

In general it has to be assumed that if $l[\cdot]$ is SLP or D at b then nothing can be said about SLP or D for $L[\cdot]$ at B. However this requires further investigation; some results in this direction are given in sections 9 and 10 below.

5. Let $L[\cdot]$ be the formally symmetric differential expression defined in the previous section. Now suppose that $L[\cdot]$ satisfies the D condition at the singular point B; sufficient conditions for this to hold will be discussed later. Thus

$$F' \text{ and } |Q|^{\frac{1}{2}} F \in L^2(A, B) \qquad (F \in \Delta). \tag{5.1}$$

Following the discussion in section 3 the inequality for $L[\cdot]$, which is equivalent to (3.1) for $l[\cdot]$, takes the form

$$\left(\int_A^B \{F'^2 + QF^2\}\right)^2 \leqslant K\int_A^B F^2 \int_A^B L[F]^2 \qquad (F \in \Delta) \tag{5.2}$$

where $0 < K \leqslant \infty$ and is independent of the elements of Δ.

All the general remarks in section 3 apply equally well to the inequality (5.2); in particular the analysis and results of [4] are valid.

The problem considered in this paper concerns the relationship between the inequality (3.1) and the inequality (5.2) obtained from the Liouville transformation of $l[\cdot]$.

6. Let $f \in \delta$ and let $F = Uf$. It follows from (4.5) and (4.8) that

$$\int_a^b f^2 \int_a^b l[f]^2 = \int_A^B F^2 \int_A^B L[F]^2. \tag{6.0}$$

Thus the connection between the two inequalities depends on the manner in which the Dirichlet integrals are related.

<u>Theorem 1</u> <u>Let the notations and definitions of previous sections hold</u>; <u>let the coefficients</u> p <u>and</u> q <u>of</u> $l[\cdot]$ <u>satisfy the conditions</u> (1.2) <u>and</u> (4.1); <u>let</u> $l[\cdot]$ <u>be singular and satisfy condition</u> D <u>at</u> b; <u>let the transformed differential expression</u> $L[\cdot]$ <u>satisfy condition</u> D <u>at the singular point</u> B; <u>let</u> $f \in \delta$ <u>and</u> $F = Uf \in \Delta$; <u>then</u>

$$\int_a^b \{pf'^2 + qf^2\} - \int_A^B \{F'^2 + QF^2\} = \frac{1}{4}\, p'(a)\ f(a)^2 \tag{6.1}$$

$$= \frac{1}{4}\frac{P_0'(A)}{P_0(A)}\ F(A)^2 \tag{6.2}$$

<u>If</u> $p'(a) = 0$, <u>equivalently</u> $P_0'(A) = 0$, <u>then</u> <u>the inequalities</u> (3.1) <u>and</u> (5.2) <u>are equivalent in the sense that either</u>

(i) $k = K = \infty$, i.e. <u>neither inequality is valid</u>

or (ii) $0 < k = K < \infty$, i.e. <u>both inequalities are valid, and</u> F <u>gives equality for</u> (5.2) <u>if and only if</u> $F = Uf$ <u>where</u> f <u>gives equality for</u> (3.1).

<u>If</u> $p'(a) \neq 0$, <u>equivalently</u> $P_0'(A) \neq 0$, <u>then in general the inequalities</u> (3.1) <u>and</u> (5.2) <u>are not related in that examples exist, with the coefficient</u> p <u>not constant on</u> $[a, b)$, <u>such that all the four combinations of</u> $k < \infty$, $k = \infty$ <u>and</u> $K < \infty$, $K = \infty$ <u>can be realised</u>.

<u>Proof</u>. This is given below.

<u>Corollary</u>. <u>Let the notations and conditions of Theorem 1 hold; if</u> $F \in \Delta$ <u>then</u>

$$F', \quad |Q|^{\frac{1}{2}}F, \quad |Q_0|^{\frac{1}{2}}F, \quad |P_0''P_0^{-1}|^{\frac{1}{2}}F \ \text{and} \ P_0'P_0^{-1}F$$

<u>all belong to</u> $L^2(A, B)$.

<u>Proof</u>. This is given below.

7. To prove Theorem 1. Let $f \in \delta$ and $F = Uf$; then suitable calculations show that

$$p(x)^{\frac{1}{2}} f'(x) = P_0(X)^{-\frac{1}{4}} F'(X) - P_0'(X)\, P_0(X)^{-\frac{5}{4}} F(X)/4 \qquad (x \in [a, b))$$

so that

$$pf'^2 = P_0^{-\frac{1}{2}} F'^2 - \tfrac{1}{2} P_0'P_0^{-\frac{3}{2}} F'F + P_0'^2 P_0^{-\frac{5}{2}} F^2/16$$

$$qf^2 = Q_0 P_0^{-\frac{1}{2}} F^2 \qquad\qquad dx = P_0(X)^{\frac{1}{2}} dX.$$

If $t \in (a, b)$ let $T = X(t) \in (A, B)$. Then

$$\int_a^t \{pf'^2 + qf^2\} = \int_A^T \{F'^2 - \tfrac{1}{2}\frac{P_0'}{P_0} FF' + \frac{1}{16}\frac{P_0'^2}{P_0^2} F^2 + Q_0F^2\}. \tag{7.1}$$

Integration by parts proves that

$$\int_A^T P_0'P_0^{-1}FF' = [\tfrac{1}{2} P_0'P_0^{-1}F^2]_A^T = \tfrac{1}{2}\int_A^T \{P_0''P_0^{-1} - P_0'^2P_0^{-2}\}F^2.$$

Thus, using (7.1),

$$\int_a^t \{pf'^2 + qf^2\} = \int_A^T \{F'^2 + (\frac{1}{4}\frac{P_0''}{P_0} - \frac{3}{16}\frac{P_0'^2}{P_0^2} + Q_0)F^2\} - \frac{1}{4}[P_0'P_0^{-1}F^2]_A^T$$

$$= \int_A^T \{F'^2 + QF^2\} - \frac{1}{4}[P_0'P_0^{-1}F^2]_A^T. \tag{7.2}$$

This last result holds for all $t \in (a, b)$.

From the conditions of Theorem 1 it follows that

$$p^{\frac{1}{2}}f' \text{ and } |q|^{\frac{1}{2}}f \in L^2(a, b), \quad F' \text{ and } |Q|^{\frac{1}{2}}F \in L^2(A, B).$$

Thus (6.2) will be established from (7.2) as $t \to b-$, i.e. $T \to B-$, provided it is the case that

$$\lim_{T\to B-} P_o'(T)P_o(T)^{-1}F(T)^2 = 0. \tag{7.3}$$

To establish this last result we use the identity, obtained from a suitable calculation,

$$p(t)f(t)f'(t) = F(T)F'(T) - \frac{1}{4}P_o'(T)P_o(T)^{-1}F(T)^2. \tag{7.4}$$

Since $l[\cdot]$ is D at b and $L[\cdot]$ is D at B, $l[\cdot]$ and $L[\cdot]$ are SLP at b and B respectively (from the result of Kalf in [14]), i.e.

$$\lim_{t\to b-} p(t)f(t)f'(t) = \lim_{T\to B-} F(T)F'(T) = 0$$

and so (7.3) follows. Thus from (7.2) the required result (6.2) is obtained. The equivalent result (6.1) follows from a suitable calculation.

If $p'(a) = 0$ then $P_o'(A) = 0$ and the equivalence of the inequalities (3.1) and (5.2) then follows from (6.0) and (6.1). Clearly $k = K$ and if $k < \infty$ then cases of equality in (3.1) and (5.2) are related as shown.

To establish the stated results when $p'(a) \neq 0$, i.e. $P_o'(A) \neq 0$, consider the following examples:

(i) $a = 1$, $b = \infty$ $p(x) = x^{-1}$ <u>and</u> $q(x) = 0$ $(x \in [1, \infty))$; in this case $k = 4$, see (i) of Section 3 above, and

$$A = 0, \quad B = \infty \qquad Q(X) = 7\{4(3X+2)^2\}^{-1} \quad (X \in [0, \infty));$$

here $K = \infty$ from the result of [4, Section 15, Theorem 2].

(ii) $a = 1$, $b = \infty$ $p(x) = x$ and $q(x) = 0$ $(x \in [1, \infty))$; in this case $k = \infty$, see (iv) of Section 3 above, and

$$A = 0, \quad B = \infty \qquad Q(x) = -\{4(X + 2)^2\}^{-1} \qquad (X \in [0, \infty));$$

here it may be shown that $K < \infty$ from [4, Section 12, Theorem 1], [4, Section 17, Example 2(b)] and the transformation formula [5, Formula (10.7)] for the m-coefficients.

(iii) $a = 0$, $b = \infty$ $p(x) = e^{-2x}$ and $q_\tau(x) = x^2 - \tau$ $(x \in [0, \infty))$ where τ is a real number. Since $\lim q_\tau = \infty$ at ∞ it follows that the spectrum of the differential operator T_0 generated in $L^2(0, \infty)$ by^{+} $l_0[\cdot]$ with boundary condition $f(0) = 0$, is discrete and bounded below; see [15, Section 24.1]. Let λ_1 be the smallest eigenvalue of T_0 then the analysis in [4, Section 16] shows that $k(\lambda_1) < \infty$ and $k(\lambda_1 + \epsilon) = \infty$ if ϵ is positive but small enough. Now for p and q_τ above

$$A = 0, \quad B = \infty \qquad Q_\tau(X) = \{\log(X + 1)\}^2 - \tau + 3\{4(X + 1)^2\}^{-1}$$
$$(X \in [0, \infty)).$$

Since $\lim Q_\tau = \infty$ at ∞ the spectrum of the differential operator S_0, generated in $L^2(0, \infty)$ by $L_0[\cdot]$ with boundary condition $F(0) = 0$, is also discrete, and is in fact identical with the spectrum of T_0. It follows then, again using the results of [4, Section 15], that $K(\lambda_1) < \infty$ and $K(\lambda_1 + \epsilon) = \infty$.

These examples complete the proof of Theorem 1.

8. To prove the Corollary to Theorem 1.

Since $L[\cdot]$ is D at the singular point B, by hypothesis, it follows that both F' and $|Q|^{\frac{1}{2}}F$ are in $L^2(A, B)$ for all $F \in \Delta$. A calculation shows that, if $F = Uf$,

$$\int_a^b |q||f|^2 = \int_A^B |Q_0||F|^2$$

+ For τ real, $l_\tau[f] = -(pf')' + q_\tau f$; similarly for $L_\tau[F]$.

and the first integral is finite since $l[\cdot]$ is D at b, again by hypothesis; thus $|Q_0|^{\frac{1}{2}}F \in L^2(A, B)$.

These results show that $|Q - Q_0|^{\frac{1}{2}}F \in L^2(A, B)$, i.e. $|4P_0''P_0^{-1} - 3P_0'^2P_0^{-2}|^{\frac{1}{2}}F \in L^2(A, B)$ for all $F \in \Delta$. Thus for $T \in (A, B)$

$$\int_0^T (4P_0''P_0^{-1} - 3P_0'^2P_0^{-2})F^2 = 4\int_0^T (P_0'P_0^{-1})'F^2 + \int_0^T P_0'^2P_0^{-2}F^2$$

$$= 4[P_0'P_0^{-1}F^2]_0^T - 8\int_0^T P_0'P_0^{-1}FF' + \int_0^T P_0'^2P_0^{-2}F^2.$$

Using the result (7.3) it follows that

$$\lim_{T\to B-}\left(\int_0^T P_0'^2P_0^{-2}F^2 - 8\int_0^T P_0'P_0^{-1}FF'\right)$$

exists and is finite. Now if $P_0'P_0^{-1}F \notin L^2(A, B)$ then this last result would yield a contradiction (use the Cauchy-Schwarz inequality on the second integral and recall that $F' \in L^2(A, B)$).

Thus $P_0'P_0^{-1}F \in L^2(A, B)$ and it then follows from above that $|P_0''P_0^{-1}|^{\frac{1}{2}}F \in L^2(A, B)$; this holds for all $F \in \Delta$.

This completes the proof of the Corollary.

9. In this and the next section two theorems are given which, from conditions on the coefficients p and q, ensure that $l[\cdot]$ and $L[\cdot]$ satisfy the D condition, at b and B respectively.

<u>Theorem 2</u>. <u>Let the notations and definitions of previous sections hold; let</u> $b = \infty$ <u>and let the coefficients</u> p <u>and</u> q <u>satisfy the conditions</u> (1.2) <u>and</u> (4.1) <u>on</u> $[a, \infty)$; <u>additionally suppose that</u>

(i) $\int_a^\infty p^{-\frac{1}{2}} = \infty$, i.e. $B = \infty$

(ii) <u>for some real number</u> σ

$q(x) \geq \sigma$ (<u>almost all</u> $x \in [a, \infty)$)

(iii) <u>either</u> (a) <u>for some real number</u> τ

$4p''(x) - p'(x)^2\, p(x)^{-1} \geq \tau$ (<u>almost all</u> $x \in [a, \infty)$)

<u>or</u> (b) <u>for some real number</u> $r \in [1, \infty)$

$p^{-1/2r}(16q + 4p'' - p'^2 p^{-1}) \in L^r(a, \infty)$;

<u>then both</u> $l[\cdot]$ <u>and</u> $L[\cdot]$ <u>satisfy the</u> D <u>condition at</u> ∞, <u>and so both are</u> SLP <u>and</u> LP at ∞.

<u>Proof</u>. Firstly condition (ii) implies that $l[\cdot]$ is D at ∞; see [4, Section 3] or [13, Example 1].

Secondly (ii) and (iii) (a) together imply that the coefficient Q (see the definition of Q in the line below (4.7)) is essentially bounded below on $[A, \infty)$. As in the previous paragraph this is sufficient to place $L[\cdot]$ in the D condition at ∞.

Thirdly (iii) (b), the definition of Q and the type of calculation which gave (4.8) show that

$$\int_A^\infty |Q|^r = \int_a^\infty p^{-\frac{1}{2}}\left|q + \frac{1}{4}p'' - \frac{1}{16}p'^2 p^{-1}\right|^r < \infty;$$

thus $Q \in L^r(A, \infty)$. This is sufficient to place $L[\cdot]$ in the D condition at ∞; see [8, Theorem 2] or [13, Example 2].

It has been noted previously that D implies SLP implies LP.

This completes the proof of the Theorem.

10. <u>Theorem 3</u> <u>Let the basic conditions for</u> $l[\cdot]$ <u>on</u> $[a, \infty)$ <u>of Theorem</u> 2 <u>hold;</u> additionally suppose that

(i) $\int_a^\infty p^{-\frac{1}{2}} < \infty$, i.e. $B < \infty$

(ii) <u>for some real number</u> σ

$$q(x) \geq \sigma \quad (\text{almost all } x \in [a, \infty))$$

(iii) <u>for some real number</u> τ

$$q(x) + \frac{1}{4} p''(x) - \frac{1}{16} p'(x)^2 p(x)^{-1} \geq \frac{3}{4} \left\{ \int_x^\infty p^{-\frac{1}{2}} \right\}^{-2} + \tau$$

$$(\text{almost all } x \in [a, \infty))$$

<u>then</u> $l[\cdot]$ <u>and</u> $L[\cdot]$ <u>satisfy the</u> D <u>condition at</u> ∞ <u>and</u> B <u>respectively</u>.

<u>Proof</u>. Firstly condition (ii) implies that $l[\cdot]$ is D at ∞, as in Theorem 2.

Secondly, with $B < \infty$ from (i), condition (iii) may be written as, see the definition of Q,

$$Q(X) \geq \frac{3}{4} (B - X)^{-2} + \tau$$

for almost all $X \in [A, B)$. This is sufficient to place $L[\cdot]$ in the D condition at B; see [7, Section 2].

This completes the proof of the Theorem.

11. In this final section it is noted that the above results may be extended to the case when the differential expression (1.1) takes the form

$$l[f] = k^{-1} \left(-(pf')' + qf\right) \text{ on } [a, b)$$

where k is an essentially positive function on $[a, b)$. In this case the space $L^2(a, b)$ has to be replaced by the k-weighted integrable-square function space $L^2_k(a, b)$.

The results in the inequality paper [4] extend to this more general differential expression. Results on the strong limit-point and Dirichlet conditions in this case are to be found in the papers by Kalf, [13], and Everitt, Hinton, Wong [10].

There is a Liouville transformation in this case; essential details are given [1, Chapter X, Section 9].

<u>Acknowledgement</u> The author gratefully acknowledges help from discussion with Professor Konrad Jörgens, University of Munich, on the application of the Liouville transformation to the integral inequality.

References

1. Birkhoff, G. and Rota, G-C. : Ordinary differential equations. Ginn and Company, New York, 1962.

2. Dunford, N. and Schwartz, J. T. : Linear operators: Part II. Interscience Publishers, New York and London, 1963.

3. Eastham, M. S. P. : Theory of ordinary differential equations. Van Nostrand Reinhold Company, London, 1970.

4. Everitt, W. N. : On an extension to an integro-differential inequality of Hardy, Littlewood and Polya. Proc. Royal Soc. Edinburgh (A) 69 (1971/72) 294-333.

5. Everitt, W. N. : On a property of the m-coefficient of a second-order linear differential equation. Journ. London Math. Soc. (2) 4 (1972) 443-457.

6. Everitt, W. N. : Report on certain integral inequalities. University of Dundee, 1974.

7. Everitt, W. N. and Giertz, M. : A Dirichlet type result for ordinary differential operators. Math. Ann. 203 (1973) 119-128.

8. Everitt, W. N., Giertz, M. and McLeod, J. B. : On the strong and weak limit-point classification of second-order differential expressions. Proc. London Math. Soc. To appear in 1974.

9. Everitt, W. N., Giertz, M. and Weidmann, J. : Some remarks on a separation and limit-point criterion of second-order, ordinary differential expressions. Math. Ann. 200 (1973) 335-346.

10. Everitt, W. N., Hinton, D. B. and Wong, J. S. W. : On the strong limit-n classification of linear ordinary differential expressions of order 2n. Proc. London Math. Soc. To appear in 1974.

11. Hardy, G. H. and Littlewood, J. E. : Some integral inequalities connected with the calculus of variations. Quart. Journal Math. Oxford Ser. (2) 3 (1932) 241-252.

12. Hardy, G. H., Littlewood, J. E. and Polya, G. : Inequalities. Cambridge, 1934.

13. Kalf, H. : Remarks on some Dirichlet type results for semibounded Sturm-Liouville operators, 1974. To be published.

14. Kalf, H. : Personal communication, February, 1974.

15. Naimark, M. A. : Linear differential operators: Part II. Ungar, New York, 1968.

Hopf Bifurcation with an Infinite Limiting Period

H.I. Freedman*

1. Introduction

In this note, we consider the ordinary differential equation

$$(1) \qquad x' = F(x,\varepsilon), \qquad (' = \frac{d}{dt})$$

where x, F are n-dimensional real vectors, and ε is a real scalar parameter. We suppose that F is as smooth as required and that for sufficiently small ε, there exists a real vector $a(\varepsilon)$ such that

$$(2) \qquad F(a(\varepsilon),\varepsilon) = 0 \ .$$

In 1943, E. Hopf [4] gave a set of sufficiency conditions in terms of the eigenvalues of the matrix

$$(3) \qquad A(\varepsilon) \equiv F_x(a(\varepsilon),\varepsilon)$$

and their derivatives at $\varepsilon = 0$ in order to guarantee a bifurcation of periodic solutions from $\varepsilon > 0$ or $\varepsilon < 0$, or all at $\varepsilon = 0$. In order to guarantee that the three posibilities are mutually exclusive he assumed that F is analytic in x and ε.

---

*Research for this paper was partially supported by the National Research Council of Canada, Grant No. NRC A4823.

Friedrichs [3] later showed that the bifurcation described above will occur if the following conditions hold in two dimensions:

$$\text{(4)} \qquad \operatorname{tr}A(0) = 0,\ \det|A(0)| > 0,\ \operatorname{tr}\frac{dA(0)}{d\varepsilon} \neq 0 \ .$$

In [1], it was shown that for $n = 2$, the Hopf conditions and conditions (4) are equivalent. Further, a sufficiency condition was given guaranteeing that the bifurcation does not all occur at $\varepsilon = 0$ A condition for bifurcation to occur in the case $\operatorname{tr}\frac{dA(0)}{d\varepsilon} = 0$ was also given. In [2], conditions for non-bifurcation of periodic solutions were given under the assumption that $\det|A(0)| = 0$.

The Hopf bifurcation theorem has been extended in other ways as well (see the references in [1]). None of the work up to the present, however, includes the case where

$$\text{(5)} \qquad F(x,\varepsilon) = \varepsilon G(x,\varepsilon) \ .$$

This is because it is usually assumed that the period $\tau(\varepsilon)$ of the sought for periodic solution has a finite positive limit as $\varepsilon \to 0$. We shall show that this is not always the case if (5) holds.

2. An example.

The ideas in this paper are motivated by the following simple example. Consider the system

$$\text{(6)} \qquad x' = \varepsilon y, y' = -\varepsilon x \ ,$$

the general solution of which for $\varepsilon \neq 0$ is

$$\text{(7)} \qquad x = \alpha \cos \varepsilon t + \beta \sin \varepsilon t \ , \ y = \beta \cos \varepsilon t - \alpha \sin \varepsilon t \ .$$

We see that for fixed $\varepsilon \neq 0$ all solutions are periodic of period $\frac{2\pi}{\varepsilon}$ and as $\varepsilon \to 0$, the period becomes unbounded.

3. Main results.

We consider now the system

$$(8) \qquad \frac{dx(t)}{dt} = \varepsilon G(x,\varepsilon) \quad .$$

We transform variables by

$$(9) \qquad s = \varepsilon t \quad ,$$

and for $\varepsilon \neq 0$, system (9) is equivalent to

$$(10) \qquad \frac{dy(s)}{ds} = G(y,\varepsilon) \quad ,$$

where $y(s) = x(\frac{s}{\varepsilon})$. We suppose that there exists for sufficiently small ε , $b(\varepsilon)$ such that

$$(11) \qquad G(b(\varepsilon),\varepsilon) = 0$$

and we define

$$(12) \qquad B(\varepsilon) = G_y(b(\varepsilon),\varepsilon) \quad .$$

Then on the basis of [1] and [4] the following theorem is valid.

Theorem: Let $G(x,\varepsilon)$ be such that (11) holds for sufficiently small ε and $G(x,\varepsilon)$ be analytic in a neighborhood of $(b(0),0)$. Let $B(\varepsilon)$ be such that it has two eigenvalues of the form $u(\varepsilon) \pm iv(\varepsilon)$ such that $u(0) = 0$, $\frac{du(0)}{d\varepsilon} \neq 0$, and that $Re(\lambda(0)) \neq 0$ if $\lambda(\varepsilon)$ is any other eigenvalue of $B(\varepsilon)$. Further, let $G(y,\varepsilon)$ be such that the bifurcation of periodic solutions of system (10) (guaranteed by the conditions on the eigenvalues of $B(\varepsilon)$ above) is such that it occurs either for $\varepsilon > 0$ or $\varepsilon < 0$ (see [1] for when this happens in two dimensions). Then there is a bifurcation of periodic solutions of system (8) from $\varepsilon = 0$ valid for either positive or negative ε, whose least periods are of the form $\tau(\varepsilon) = \frac{2\pi}{|v(0)|\varepsilon} + o(\frac{1}{\varepsilon})$ as $\varepsilon \to 0$.

References

1. Freedman, H.I., On a bifurcation theorem of Hopf and Friedrichs (to appear).

2. ________, The nonbifurcation of periodic solutions when the variational matrix has a zero eigenvalue (to appear).

3. Friedrechs, K.O., Advanced Ordinary Differential Equations, Gordon and Breach (1965).

4. Hopf, E., Abzueigung einer periodischen Lösung von einer stationären Lösung eines Differentialsystems, Ber.Verh.Sächs.Akad.Wiss.Leipzig. Math-Nat. Kl. 95 (1943) pp. 3-22.

N-Soliton solutions of some non-linear dispersive wave equations of physical significance

J. D. Gibbon, P. J. Caudrey, R. K. Bullough and J. C. Eilbeck

In this paper we present a derivation of an N-soliton solution of the reduced Maxwell-Bloch (RMB) equations which extends the applications of the 'inverse function method'. The RMB equations include the 'sine-Gordon' equation

$$\sigma_{xx} - \sigma_{tt} = \sin\sigma \tag{1}$$

as a special case. This equation has extensive applications in physics.[1] The RMB equations themselves are the most general form of equations governing the propagation of intense ultra-short optical pulses in certain low density dielectrics.[2] To facilitate application of the method the 'space' and 'time' variables are interchanged compared with previous work.[2] The RMB equations then take the form

$$E_{\tau}(\xi,\tau) = s(\xi,\tau) \tag{2a}$$

$$s_{\xi}(\xi,\tau) = E(\xi,\tau)\,u(\xi,\tau) + \mu\, r(\xi,\tau) \tag{2b}$$

$$r_{\xi}(\xi,\tau) = -\mu\, s(\xi,\tau) \tag{2c}$$

$$u_{\xi}(\xi,\tau) = -E(\xi,\tau)\, s(\xi,\tau) \tag{2d}$$

A distortionless solution satisfying $E,\ s,\ r \to 0,\ u \to -1$ as $|\xi| \to \infty$ is $E = E_1 \operatorname{sech}\theta_1$ with $\theta_1 \equiv \frac{1}{2}E_1\{\xi - (\frac{1}{4}E_1^2 + \mu^2)^{-1}\tau\} + \delta$. The velocity is $(\frac{1}{4}E_1^2 + \mu^2)^{-1}$ and increases with the real valued amplitude parameter E_1. This is characteristic of single soliton solutions.[3]

The inverse function method was first used[4] to find soliton solutions of the Korteweg-de Vries equation $u_t - 6uu_x + u_{xxx} = 0$. Recently it has been extended[5] to a broad class of non-linear dispersive wave equations which include the K-de V equation and sine-Gordon equations. The method we use extends this class to include the RMB equations.

Consider the linear scattering problem

$$\frac{\partial \phi_1}{\partial \xi} + i\lambda \phi_1 = V_1(\xi,\tau)\,\phi_2 \tag{3a}$$

$$\frac{\partial \phi_2}{\partial \xi} - i\lambda \phi_2 = V_2(\xi,\tau)\,\phi_1 \tag{3b}$$

The 'time' or dependence of the eigenfunctions $\phi_1(\xi,\tau)$ and $\phi_2(\xi,\tau)$ can be chosen so that

$$\frac{\partial \phi_1}{\partial \tau} = A(\xi.\tau;\lambda)\,\phi_1 + B(\xi,\tau;\lambda)\,\phi_2 \tag{4a}$$

$$\frac{\partial \phi_2}{\partial \tau} = C(\xi,\tau;\lambda)\,\phi_1 - A(\xi,\tau;\lambda)\,\phi_2 \tag{4b}$$

Conditions can be put on A, B and C so that the eigenvalues λ of (3) do not depend on τ . These are obtained by cross differentiation of (3) and (4) and prove to be

$$\frac{\partial A}{\partial \xi} = V_1\,C - V_2\,B \tag{5a}$$

$$\frac{\partial B}{\partial \xi} + 2\,i\,\lambda\,B = \frac{\partial V_1}{\partial \tau} - 2\,A\,V_1 \tag{5b}$$

$$\frac{\partial C}{\partial \xi} - 2\,i\,\lambda\,C = \frac{\partial V_2}{\partial \tau} + 2\,A\,V_2 \tag{5c}$$

We make the following choices:

$$A = \frac{i\,\lambda}{4\lambda^2 - \mu^2}\,u(\xi,\tau) \tag{6a}$$

$$B = \frac{\lambda}{4\lambda^2 - \mu^2}\left[s(\xi,\tau) + \frac{i\mu}{2\lambda}\,r(\xi,\tau)\right] \tag{6b}$$

$$C = \frac{-\lambda}{4\lambda^2 - \mu^2}\left[s(\xi,\tau) - \frac{i\mu}{2\lambda}\,r(\xi,\tau)\right] \tag{6c}$$

$$V_1 = \tfrac{1}{2} i E(\xi,\tau) = - V_2^* \tag{6d}$$

in which r, s, u, and E are real. Equations (5) with (6) are now equivalent to 3 of the RMB equations (2). Thus, the linear scattering problem equations (3) and (4), the linear equation (2c), and the condition $\lambda_\tau = 0$ are together equivalent to the non-linear RMB equations (2).

The procedure now is to solve (2) and (3) asymptotically as $|\xi| \to \infty$ using the boundary conditions $u \to -1;\ s, r, E \to 0$. Given the initial data at $\tau = 0$ these solutions determine the 'potential' functions $V_1 = V_2$ for all ξ in $-\infty < \xi < \infty$ and for each $\tau > 0$ via the linear Gel'fand-Levitan integral equations. This is the inverse function method. The initial data are determined by solving the scattering problem (2) at $\tau = 0$ using the initial data for A, B, C and the potentials $V_1 = V_2$.

The Gel'fand-Levitan integral equations for this problem take the form[6]

$$K_1(\xi,\xi') = F^*(\xi+\xi') + \int_\xi^\infty K_2^*(\xi,\xi'')\, F^*(\xi''+\xi')\, d\xi''$$

$$K_2(\xi,\xi') = - \int_\xi^\infty K_1(\xi,\xi'')\, F(\xi''+\xi')\, d\xi'' \tag{7}$$

where for each value of τ

$$F^*(\xi) = \frac{1}{2\pi} \int_{-\infty}^{\infty} R(\eta,\tau)\, e^{i\eta\xi}\, d\eta + \sum_{n=1}^{N} C_n(\tau)\, e^{i\lambda_n \xi} \tag{8}$$

The λ_n are the discrete eigenvalues of (2) which lie in the upper half η-plane: the function $R(\eta,\tau)$ plays the role of a reflexion coefficient in the continuous part of the eigenvalue spectrum. The potentials V_1 and V_2 can be found from the functions $K_1(\xi,\xi')$ and $K_2(\xi,\xi')$ by

$$|V_1(\xi,\tau)|^2 = 2 \frac{d}{d\xi} K_2(\xi,\xi)\ , \quad V_1(\xi,\tau) = -2 K_1(\xi,\xi) \tag{9}$$

For N-soliton solutions it is sufficient to discard the function $R(\eta,\tau)$ and to regard the $C_n(0)$ and λ_n as parameters. A solution of (7) for this case can be found by linear algebra and yields

$$|V_1(\xi,\tau)|^2 = \frac{d^2}{d\xi^2} \ln \det \| \underline{\underline{I}} + \underline{\underline{A}}\,\underline{\underline{A}}^* \| \tag{10a}$$

The matrix $\underline{\underline{A}}$ has elements

$$A_{nm} = (C_n C_m)^{\frac{1}{2}} \exp\left[i(\lambda_n - \lambda_m^*)\xi\right] \tag{10b}$$

and

$$C_n(\tau) = C_n(0) \exp\left[-\frac{2i\lambda_n}{\mu^2 - 4\lambda_n^2}\tau\right] \tag{10c}$$

The exponent of C_n is the exponent of $\phi_1^{-2}(\xi,\tau;\lambda_n)$ obtained from (3a) as $\xi \to \infty$. The λ_n lie in the upper half plane and either lie on the imaginary axis or form anti-hermitian pairs $\lambda_n = -\lambda_m^*$. The $C_n(0)$ determine the absolute phases of the solitons and must be chosen purely imaginary or in anti-hermitian pairs.

The solution (10) can be put in more convenient form. Define $\underline{\underline{B}}$ obtained from $\underline{\underline{I}}$ by interchanging rows (or columns) n and m for each anti-hermitian pair. Then

$$[\underline{\underline{A}}\,\underline{\underline{B}}]_{nm} = \frac{-i}{\lambda_n + \lambda_m} \exp\left[i(\lambda_n + \lambda_m)\xi - \alpha_n - \alpha_m\right] \tag{11a}$$

and

$$[\underline{\underline{B}}\,\underline{\underline{A}}^{-1}]_{nm} = \frac{-i}{\lambda_n + \lambda_m} \exp\left[-i(\lambda_n + \lambda_m)\xi + \alpha_n + \alpha_m + 2\beta_n + 2\beta_m\right] \tag{11b}$$

where

$$iC_n = \exp(-2\alpha_n) = \exp\left[-2\alpha_{no} - \frac{2i\lambda_n\tau}{\mu^2 - 4\lambda_n^2}\right] \tag{11c}$$

and

$$i\,\frac{\prod_j (\lambda_j + \lambda_n)}{\prod_{j\neq n} (\lambda_j - \lambda_n)} = \exp 2\beta_n \tag{11d}$$

Further, $\underline{\underline{B}}\,\underline{\underline{A}}^* = \underline{\underline{A}}\,\underline{\underline{B}}$ so that equations (11) yield

$$[\underline{\underline{B}}\,\underline{\underline{A}}^{-1} + \underline{\underline{B}}\,\underline{\underline{A}}]_{nm} = \exp(\beta_n + \beta_m)\, M_{nm} \tag{12a}$$

where

$$M_{nm} = \frac{2}{E_n + E_m} \cosh\tfrac{1}{2}(\theta_n + \theta_m) \tag{12b}$$

$$E_n = 4i\lambda_n \; ; \quad \theta_n = \tfrac{1}{2}E_n\left(\xi - \frac{4}{E_n^2 + 4\mu^2}\tau\right) + \delta_n \tag{12c}$$

The phases δ_n, determined by the $C_n(0)$, are arbitrary. Equations (9) and (6d) now mean

$$E^2 = 4\frac{\partial^2}{\partial\xi^2} \ln \det \| \underline{\underline{M}} \| \tag{13}$$

The 1-soliton solution is $E = E_1 \operatorname{sech} \theta_1$ quoted earlier. The N-soliton solution determined by N distinct real parameters E_n breaks up into the linear sum of 1-soliton solutions

$$E \sim \sum_{n=1}^{N} E_n \operatorname{sech}(\theta_n \pm \gamma_n) \tag{14}$$

when ξ is large and $\pm$ve The phase shifts induced by the 'collision' of these solitons are

$$2\gamma_n = \sum_{m=1}^{n-1} \ln a_{nm} - \sum_{m=n+1}^{N} \ln a_{nm} \tag{15a}$$

where

$$a_{nm} = \left(\frac{E_n - E_m}{E_n + E_m}\right)^2 \tag{15b}$$

Such solutions govern the break up of intense ultra-short optical pulses as they have so far been observed[7]. If the E_n are composed of l distinct real numbers and m complex conjugate pairs $(N = l + 2m)$ the solution consists of l single solitons and m bions (soliton-antisoliton bound states) in collision. This type of solution breaks up into l single solitons travelling at different velocities and m bions travelling at the same or different velocities. Bions have still to be observed in optical pulses. When $\mu = 0$ the equations (2) have solution $r = 0$, $s = -\sin\sigma$, $u = -\cos\sigma$

where $\sigma = \int_{-\infty}^{\xi} E(\xi',\tau)\,d\xi'$ providing $\sigma_{\xi\tau} = \sin\sigma$

This is another form of the sine-Gordon equation (1). The choice

$A = \frac{i}{4\lambda}\cos\sigma$, $B = -C = \frac{1}{4\lambda}\sin\sigma$ has, however, already been used[5] to find N soliton solutions of the sine-Gordon equation.

1. A. Barone, F. Esposita, C. J. Magee and A. C. Scott. Riv. Nuovo Cimento 1, 227 (1971).
2. J. C. Eilbeck, J. D. Gibbon, P. J. Caudrey and R. K. Bullough. J. Phys. A. 6, 1337 (1973) and some references therein.
3. N. J. Zabusky and M. D. Kruskal. Phys. Rev. Lett. 15, 240 (1965).
4. C. S. Gardner, J. M. Greene, M. D. Kruskal and R. M. Miura. Phys. Rev. Lett. 19, 1095 (1967).
5. M. J. Ablowitz, J. D. Kaup, A. C. Newell and H. Segur. Phys. Rev. Lett. 31, 125 (1973).
6. V. E. Zakharov and A. B. Shabat. Sov. Phys. JETP 34, 62 (1972).
7. H. M. Gibbs and R. E. Slusher. Phys. Rev. A 6, 2326 (1972).

Continuity of generalized solutions of the Cauchy problem for the porous medium equation

B.H. Gilding

The degenerate parabolic equation $u_t = (u^m)_{xx}$, where $m > 1$ is a constant, represents the laminar flow of a homogeneous gas through a homogeneous isotropic porous medium. (For the results of this and the following paragraph we refer to the paper [1] and the references cited there). Solutions to the Cauchy problem for this equation exist only in a generalized sense, whenever the initial data takes the value zero on some interval. However, these generalized solutions can be constructed under nominal restrictions, as the pointwise limit of a decreasing sequence of functions $u_n(x,t)$, $n = 1,2,3,\ldots$, each a positive classical solution of a first boundary value problem in the domain $R_n = (-n,n)\times(0,T]$, $n = 1,2,3,\ldots$, respectively. If then, we had some uniform estimates of Hölder continuity for this sequence of classical solutions, we should have an estimate of Hölder continuity for the generalized solution.

In the paper [1] D.G. Aronson shows that if $u(x,t)$ is a smooth positive classical solution of the equation in the domain R_{n+1}, then for any τ, $T > \tau > 0$, there exists a constant $C_1 = C_1(\sup u, m, \tau)$ such that $|(u^{m-1})_x| \leq C_1$ in $(-n,n)\times(\tau,T]$. From this it follows that for any (x_1,t), (x_2,t) in $(-n,n)\times(\tau,T]$, $|u(x_1,t) - u(x_2,t)| \leq C_2|x_1 - x_2|^{\nu}$, where $\nu = \min(1, \frac{1}{m-1})$, and C_2 depends only on $\sup u, m$ and τ. Hence, in the limit, it is seen that the constructed generalized solution of the Cauchy problem satisfies a Hölder condition with respect to x in the domain $(-\infty,\infty)\times(\tau,T]$, with exponent ν and a coefficient dependent only upon m,τ, and the supremum of the initial data. By means of an explicit example this exponent is shown to be the best possible globally obtainable.

Using the above results S.N. Kruzhkov [3] has shown that the generalized solution is also Hölder continuous with respect to t in $(-\infty,\infty)\times(\tau,T]$, with exponent $\nu/\nu+2$. It is intended to illustrate an extension of this technique by which we may show that the generalized solution is actually Hölder continuous with the optimum exponent $\nu/2$.

<u>Theorem</u>. Let $u(x,t)$ be a bounded positive classical solution of the equation

$$u_t = (u^m)_{xx} \quad , \qquad m > 1 \; ,$$

in the domain $S = (-\infty, \infty) \times [\tau, T]$, $T > \tau \geq 0$. Assume that there is a constant C_1 such that

$$|(u^{m-1})_x| \leq C_1 \qquad (1)$$

everywhere in S. Then there exists a constant K such that

$$|u(x,t_1) - u(x,t_2)| \leq K |t_1 - t_2|^{\frac{\nu}{2}}$$

for all (x,t_1), (x,t_2) in S, where $\nu = \min(1, \frac{1}{m-1})$. The constant K depends only on m, C_1, T and $M = \sup u$.

<u>Proof</u>. It follows immediately from (1) that there exists a constant $C_2 = C_2(C_1, m, M)$ such that

$$|u(x_1,t) - u(x_2,t)| \leq C_2 |x_1 - x_2|^{\nu} \qquad (2)$$

for all (x_1,t), (x_2,t) in S. It is this relation that has the greatest influence in showing that $u(x,t)$ also satisfies a Holder condition with respect to t with exponent $\nu/2$.

The principal tool in the proof is the 'strong maximum principle' ([2],p. 34), which we state in précis in the following

<u>Lemma</u>. Let $L(\,.\,) = a(x,t)(\,.\,)_{xx} + b(x,t)(\,.\,)_x - (\,.\,)_t$ be a parabolic operator with continuous coefficients defined on a domain $D \subseteq \mathbb{R}^2$. Let $u(x,t)$ be a $C^{2,1}(D)$ function. Then if $Lu \geq 0$ ($Lu \leq 0$) in D, and if u has a maximum (minimum) which is attained at some point (x_0,t_0) interior to D ; $u(x,t) = u(x_0,t_0)$ for all (x,t) in D which can be connected to (x_0,t_0) by an arc lying in D along which the t-coordinate is non-decreasing from (x,t) to (x_0,t_0).

Choose an arbitrary fixed point (x_1,t_1) in $(-\infty,\infty) \times [\tau, T)$, and an arbitrary fixed t_2 such that $T \geq t_2 > t_1$. Also, let ρ be an arbitrary positive constant.

We consider the parabolic operator

$$L(\,.\,) = mu^{m-1}(\,.\,)_{xx} + m(u^{m-1})_x(\,.\,)_x - (\,.\,)_t \; ,$$

in the domain $D = [x_1 - \rho, x_1 + \rho] \times [t_1, t_2]$. Let

$$v(x,t) = u(x,t) - u(x_1,t_1) - C_2\rho^\nu - \frac{s}{\rho^2}(x-x_1)^2 - \frac{2sm}{\rho^2}(M^{m-1} + C_1\rho)(t-t_1),$$

where

$$s = \sup_{t_1 \leq t \leq t_2} |u(x_1,t) - u(x_1,t_1)|.$$

So that

$$Lv = 0 - \frac{2sm}{\rho^2}[u^{m-1} + (u^{m-1})_x(x-x_1)] + \frac{2sm}{\rho^2}[M^{m-1} + C_1\rho] \geq 0$$

in D. Now where $t = t_1$ and $|x - x_1| \leq \rho$,

$$v(x,t) = u(x,t_1) - u(x_1,t_1) - C_2\rho^\nu - \frac{s}{\rho^2}(x-x_1)^2 \leq 0$$

by (2). Also where $|x - x_1| = \rho$ and $t_2 \geq t \geq t_1$,

$$\begin{aligned} v(x,t) &= u(x,t) - u(x_1,t_1) - C_2\rho^\nu - s - \frac{2sm}{\rho^2}(M^{m-1} + C_1\rho)(t-t_1) \\ &\leq u(x,t) - u(x_1,t) - C_2\rho^\nu + u(x_1,t) - u(x_1,t_1) - s \leq 0 \end{aligned}$$

by (2) and the definition of s.

Thus, $Lv \geq 0$ in D, and $v \leq 0$ on the parabolic boundary of D, from which it follows by the maximum principle that v cannot attain a positive maximum in D. In other words $v \leq 0$ everywhere in D. This means in particular that $v(x_1,t) \leq 0$ for all t such that $t_1 \leq t \leq t_2$, or more explicitly

$$u(x_1,t) - u(x_1,t_1) \leq C_2\rho^\nu + \frac{2sm}{\rho^2}(M^{m-1} + C_1\rho)(t-t_1)$$

for all such t. Similarly we could have considered the function

$$w(x,t) = u(x,t) - u(x_1,t_1) + C_2\rho^\nu + \frac{s}{\rho^2}(x-x_1)^2 + \frac{2sm}{\rho^2}(M^{m-1} + C_1\rho)(t-t_1)$$

in the domain D, and shown that

$$u(x_1,t) - u(x_1,t_1) \geq -C_2\rho^\nu - \frac{2sm}{\rho^2}(M^{m-1} + C_1\rho)(t-t_1),$$

for all t, $t_1 \leq t \leq t_2$. From these two inequalities we obtain the estimate

$$|u(x_1,t) - u(x_1,t_1)| \leq C_2\rho^\nu + \frac{2sm}{\rho^2}(M^{m-1} + C_1\rho)(t_2 - t_1)$$

for all t, $t_1 \leq t \leq t_2$. As t does not contribute to the last expression we can conclude that in fact

$$s \leq C_2 \rho^{\nu} + \frac{s}{2} \left[\frac{4m(M^{m-1} + C_1 \rho)(t_2 - t_1)}{\rho^2} \right].$$

To complete the proof we now have to make full use of the arbitrariness of ρ by setting it equal to

$$2m\, C_1 (t_2 - t_1) + 2\{m(t_2 - t_1)(m\, C_1^2 (t_2 - t_1) + M^{m-1})\}^{\frac{1}{2}} . \qquad (3)$$

This quantity is just the positive root of the quadratic equation obtained by equating the numerator and denominator of the expression in square brackets. This yields the result that

$$s \leq C_2 [2mC_1 (t_2 - t_1)^{\frac{1}{2}} + 2\{m^2 C_1^2 (t_2 - t_1) + m\, M^{m-1}\}^{\frac{1}{2}}]^{\nu} (t_2 - t_1)^{\frac{\nu}{2}} + \frac{s}{2},$$

or,

$$s \leq 2C_2 [2m\, C_1 T^{\frac{1}{2}} + 2\{m^2 C_1^2 T + m\, M^{m-1}\}^{\frac{1}{2}}]^{\nu} (t_2 - t_1)^{\frac{\nu}{2}}$$

$$= K (t_2 - t_1)^{\frac{\nu}{2}} .$$

Which of course means

$$|u(x_1, t_2) - u(x_1, t_1)| \leq K(t_2 - t_1)^{\frac{\nu}{2}} .$$

As (x_1, t_1) in $(-\infty, \infty) \times [\tau, T)$ and t_2, $T \geq t_2 > t_1$, were arbitrary, we have proved the theorem.

It should be noted that the theorem has been proved in an unbounded domain, for it is essential that ρ may be chosen completely arbitrarily. If we wished to prove a similar result in a bounded domain $R = (a,b) \times [\tau, T]$ we could set about it in the same way; but, for a given (x_1, t_1) in $(a,b) \times [\tau, T)$ we would have to bound the admissible values of ρ by $\min(x_1 - a,\ b - x_1)$, to ensure that the operator L is defined in the domain $D = [x_1 - \rho,\ x_1 + \rho] \times [t_1, t_2]$. However, we observe that if ρ^* is the value of ρ chosen by expression (3), then $\rho^* \to 0$ as $t_2 - t_1 \to 0$. We are able to utilize this to gain the continuity result for the generalized solution of the equation. The $u_n(x,t)$, $n = 1,2,3,\ldots$, in the approximating sequence are each defined on an increasing sequence of domains R_n. Hence for fixed (x_1, t_1), (x_1, t_2) in $(-\infty, \infty) \times [\tau, T]$, $|u_n(x_1, t_1) - u_n(x_1, t_2)| \leq K|t_2 - t_1|^{\frac{\nu}{2}}$ for large enough n. So if $y(x,t) = \lim_{n \to \infty} u_n(x,t)$ is the generalized solution of the Cauchy problem $|y(x_1,t_1) - y(x_1,t_2)| \leq K|t_2-t_1|^{\frac{\nu}{2}}$. By the arbitrary nature of (x_1,t_1), (x_1,t_2) this implies that y satisfies a Hölder condition with respect to t everywhere in $(-\infty,\infty) \times [\tau,T]$ with exponent $\nu/2$.

This result with D.G. Aronson's yields the fact that the generalized solution of the Cauchy problem for the equation $u_t = (u^m)_{xx}$ belongs to the space $C^{0+\nu,0+\frac{\nu}{2}}((-\infty,\infty)\times[\tau,T])$ for any τ, $T > \tau > 0$, and this is the best possible exponent.

The technique illustrated is not just applicable to the porous medium equation. The method can be used on other parabolic equations, to show that Hölder continuity with respect to x with exponent α, implies Hölder continuity with respect to t with exponent $\alpha/2$.

References

[1] D.G. ARONSON, Regularity properties of flows through porous media, SIAM J. Appl. Math., Vol.17, No.2, March 1969, pp.461-467.

[2] A. FRIEDMAN, Partial Differential Equations of Parabolic Type, Prentice-Hall, Inc., Englewood Cliffs, N.J., 1964.

[3] S.N. KRUZHKOV, Results concerning the nature of the continuity of solutions of parabolic equations and some of their applications, Math. Notes, Vol.6, 1969, pp.517-523.

Two Timing for Abstract Differential Equations

William S. Hall

Two-timing, known more properly as the two variable expansion procedure, has become a popular method for formally studying certain nonlinear evolution equations. Generally it is used to find expressions which are claimed to be asymptotic to the actual solutions for long time intervals. Rarely is the technique shown to be justified.

In this note we shall indicate how the two variable method can be used to give a qualitative as well as quantitative analysis of the equation

$$\dot{u} = \varepsilon f(t,u), \; u(o) = u_o \tag{1}$$

when f is T-periodic in t and ε is small. We shall see that by decomposing t into a fast time σ and a slow time τ there arises an equation

$$v' = g(\tau, v, \varepsilon) \qquad v' = dv/d\tau \tag{2}$$

whose solutions for the given initial value u_o are asymptotic to those of (1) for times comparable to its expected interval of existence. In addition, the critical points of (2) correspond to T-periodic solutions of (1). These are stable or unstable according to the stability of the corresponding equilibrium points of (2). Finally, equation (2), which cannot usually be found exactly, has a useful approximation, namely the classical equation of averaging,

$$v' = \frac{1}{T} \int_o^T f(s,v) ds \tag{3}$$

whose solutions give similar information about the behaviour of (1).

We note that many ordinary and partial differential equations can be put into form (1). For example, if the evolution equation,

$$\dot{x} = Ax + \varepsilon g(x) \tag{4}$$

has a linear part A generating a T-periodic family of operators $\{E(t)\}$, then the transformation $x(t) = E(t)u(t)$ reduces (4) to (1) with $f(t,u) = E(-t)g(E(t)u)$.

Readers desiring to see two-timing applied to ordinary differential equations should consult [3] and [7]. For its application to some wave equations see [2] or [8]. An analysis of the convergence of the method for the second order equation $x'' + x = \epsilon g(x,x')$ has been given in [9]. A systematic, rigorous approach to two-timing for abstract equations in Banach spaces can be found in [4] and [5]. Applications are also made there to the decay of wave equations with cubic damping and to the generation of self-sustained oscillations in a wave equation of Van der Pol type. Most of these notes are in fact based on the material in the last two references.

A feature of the usual two variable method is that each term in the expansions used are T-periodic in the fast time σ. We thus begin the analysis of (1) by considering the relation

$$\frac{dw}{d\sigma} = \epsilon\{f(\sigma,w) - [f(\cdot,w(\cdot))]\}, \quad w(o) = v \tag{5}$$

where [ ] is the mean value operator

$$[w] = \frac{1}{T}\int_o^T w(s)ds \tag{6}$$

and v is any initial value. Since the vector field has no mean value, every solution $w(\sigma,v,\epsilon)$ to (5) will be T-periodic. Of course it does not satisfy (1). However, our immediate concern is to insure that w forms a smooth vector field in the sense that for a given u_o in a Banach space X, there is a ball V at u_o and a neighborhood E of zero such that $w(\sigma,v,\epsilon)$ and its Frechet derivative in v, $w_v(\sigma,v,\epsilon)$ are continuous in (σ,v,ϵ) on R x V x E. This can be done by requiring, for example, that f, and f', its derivative in u, be bounded, continuous maps of (t,u) in R x X and satisfy a Lipschitz condition in u. A second, rather useful condition is to demand that f and f' satisfy Caratheodory type hypotheses of measurability in t, Lipschitz continuity in u, and with integrable bounding and Lipschitz constants. Either way, the uniform contraction mapping theorem produces a solution, possibly differentiable a.e. only, in the form

$$w(\sigma,v,\varepsilon) = v + \int_0^\sigma\{f(s,w(s,v,\varepsilon)) - [f(\cdot,w(\cdot,v,\varepsilon))]\}ds \tag{7}$$

It also follows that the derivative in v is defined by

$$w_v(\sigma,v,\varepsilon)h = h + \varepsilon \int_0^\sigma\{f'(s,w(s,v,\varepsilon))w_v(s,v,\varepsilon)h - [f'(\cdot,w(\cdot,v,\varepsilon)) w_v(\cdot,v,\varepsilon)h]\}ds \tag{8}$$

and that both have the desired continuity properties in (σ,v,ε).

A second feature of two-timing methods is that there is a "constant of integration" in each term of the expansion depending only on the slow time τ. This suggests trying for a solution to (1) in the form

$$u(t,\varepsilon) = w(\sigma,v(\tau),\varepsilon) \tag{9}$$

where σ and τ depend on t and ε. Experience indicates taking $\sigma = t$ and $\tau = \varepsilon t$. Using the chain rule and equation (5), and substituting into (1) gives the equation for v,

$$w_v(\tau/\varepsilon,v,\varepsilon)\; v' = [f(\cdot,w(\cdot,v,\varepsilon))] \tag{10}$$

Returning to (8), we see w_v has the form $I + \varepsilon B(\sigma,v,\varepsilon)$ where B is bounded on X and is well-defined for all σ in R, and all (v,ε) in $V \times E$. Thus for small ε, w_v is boundedly invertible and we arrive at the exact form for equation (2),

$$v' = w_v^{-1}(\tau/\varepsilon,v,\varepsilon)[f(\cdot,w(\cdot,v,\varepsilon))] \tag{2}$$

with the initial value $v(o) = u_o$. It is straightforward to prove the vector field for (2) is also Lipschitz continuous in v. By the standard Picard existence theorem, (2) has a solution $v(\tau,\varepsilon)$ remaining in V for ε in E, for $o \leq \tau \leq a$, for some $a > o$. Thus $u(t,\varepsilon) = w(t,v(\varepsilon t,\varepsilon),\varepsilon)$

exists on the interval $o \leq t \leq a/\varepsilon$. From the representation (7),

$$w(t,v(\varepsilon t,\varepsilon),\varepsilon) - v(\varepsilon t,\varepsilon) = O(\varepsilon)$$

since the right side of (7) is uniformly bounded providing $v(\tau,\varepsilon)$ remains in V and ε is in E. Thus, the solutions to (2) carry the asymptotic behavior of (1) for times of order $1/\varepsilon$.

Now suppose $v(\varepsilon)$ is a constant solution to (2). Then $w(t,v(\varepsilon),\varepsilon)$ is a T-periodic solution to (1). It is fairly direct to prove that the stability properties of the periodic solution are those of $v(\varepsilon)$.

Because $w_v(\sigma,v,\varepsilon)$ is T-periodic and bounded for any σ, the vector field in (2) is continuous in ε even at $\varepsilon = o$. As $\varepsilon \to o$, $w_v \to I$ and $w \to v$. Thus (2) reduces to (3), and by continuity in the parameter ε we can deduce the following result:

Theorem 1. Let $\eta > o$ be given. Then there is an $L > o$ such that the solution $v(\tau)$ to (3) with initial value u_o satisfies

$$|u(t,\varepsilon) - v(\varepsilon t)| < \eta$$

for $o \leq t \leq L/\varepsilon$.

An application of the implicit function theorem proves:

Theorem 2. Suppose there is a v_o in X such that $F(v_o) = o$ and such that $F'(v_o)$ is a linear homeomorphism on X where F and F' are defined by,

$$F(v_o) = \frac{1}{T}\int_o^T f(s,v_o)ds \text{ and } F\ (v_o)h = \frac{1}{T}\int_o^T f'(s,v_o)hds$$

Then (2) has an isolated equilibrium point $v(\varepsilon)$ and hence (1) has a T-periodic solution.

Arguments similar to those used in the Poincare-Liapunov stability theorems give:

Theorem 3. If the variational equation $\dot{z} = F'(v_0)z$ is exponentially asymptotically stable, then they same is true for the periodic solution corresponding to v_0.

Theorem 2 has been given before in [6]. Theorems 1 and 3 appear in the theory of averaging [1] . Our results, however, are obtained more directly and simply, but of course the method in its present form does not extend to the almost periodic case.

References

[1] N.N. Bogoliubov and Y.A. Mitropolsky, Asymptotic Methods in the Theory of Nonlinear Oscillations, Gordon and Breach, New York, 1961.

[2] S.C. Chikwendu and J. Kevorkian, A perturbation method for hyperbolic equations with small nonlinearities, S.I.A.M. J. Appl. Math 22 (1972), 235-258.

[3] J.D. Cole, Perturbation Methods in Applied Mathematics, Blaisdell, Waltham, Mass., 1968.

[4] J.P. Fink, W.S. Hall, and A.R. Hausrath, A convergent two-time method for periodic differential equations, J. Differential Equations, to appear.

[5] J.P. Fink, W.S. Hall, and A.R. Hausrath, Discontinuous periodic solutions for an autonomous nonlinear wave equation, preprint, University of Pittsburgh, PA., 15260.

[6] J.K. Hale, Oscillations in Nonlinear Systems, McGraw-Hill, New York, 1963.

[7] J. Kevorkian, The two variable expansion procedure for the approximate solutions of certain nonlinear differential equations, Lectures in Applied Mathematics 7, Space Mathematics III, AMS, New York, 1966.

[8] J.B. Keller and S. Kogelman, Asymptotic solutions of initial value problems for nonlinear partial differential equations, S.I.A.M. J. Appl. Math, 18 (1970), 748-758.

[9] F.W. Kollett, Two timing methods valid on expanding intervals, preprint, Bard college, N.Y.

Counterexamples in the spectral theory of singular Sturm - Liouville operators.

S.G. Halvorsen

We consider the differential operator in $L^2(0,\infty)$ associated with the differential equation

(1) $$x'' + [\lambda+f(t)]x = 0$$

on $[0,\infty)$ where $f(t)$ is real and continuous, and a linear homogeneous boundary condition

(2) $$x(0)\cos\theta + x'(0)\sin\theta = 0 \qquad (0\leq\theta<\pi).$$

This operator is self-adjoint when (1) is of limit point type at ∞ in the sense of Weyl, i.e. if for some λ (hence for every λ) at most one solution $x\in L^2(0,\infty)$. This will be assumed henceforth, without invoking any specific criterion ensuring it.

The present paper gives a theorem and some counterexamples to certain conjectures all concerning the essential spectrum S' of (1), i.e. the set of cluster points λ of the spectrum S_θ of (1) and (2). S', thus consisting of the cluster points of the point spectrum plus the continuous spectrum, is independent of the particular value of θ in (2) as observed first by Weyl [13], p.251.
About the structure of the essential spectrum it is known that S' always contains $\lambda = +\infty$, cf. [6]. With this restriction S' may be any closed real set, cf. [3].
Various conditions are known to ensure that S' covers the entire real line or that it is empty (save for $\lambda = \infty$) etc.
Here we will mainly consider the case that

(3) $f(t)$ tends nondecreasingly to ∞ as $t\to\infty$.

Hartman [4] has shown that (3) together with any one of the following conditions implies that every real λ belongs to S':

(4) $$f(t) = o(t^2)$$

or if for some $\alpha>1$ either

(5) $$\int^{\infty} f^{-\frac{\alpha}{2}}\, dt = \infty$$

or

(6) $$\int^{\infty} f^{-\frac{1}{2}}(\log f)^{-\alpha} dt = \infty$$

generally

(7) $$\int^{\infty} f^{-\frac{1}{2}}(\log f)^{-1}(\log\log f)^{-1}\cdots(\log_n f)^{-\alpha} dt = \infty$$

If "for some $\alpha>1$" is replaced by "every $\alpha<1$" in (5), (6) or (7) the conclusion concerning S' may not be drawn. It might be mentioned here that for functions f(t) of sufficiently "regular growth" (5) for $\alpha=1$ implies a purely continous spectrum over $(-\infty,\infty)$, while convergence of the integral in (5) with $\alpha=1$ implies that eg. (1) is of limit circle type at ∞, hence that the associated operators have purely discrete spectra with no finite cluster point, [12], p.123.
In Hartman's paper [4] the question is raised whether the condition $\alpha>1$ in (5) may be relaxed to $\alpha=1$ while keeping the assertion concerning S'. The question was put forward again by Hinton [10], who proved that every real λ belongs to S' provided that $f \in C^{(1)}$, $f\to\infty$ (not necessarily monotonically) and $f^{-\frac{1}{2}}f'\to 0$ as $t\to\infty$.

We shall answer this question to the negative, showing that conditions (5), (6) and (7) represent "best possible" results, in fact showing that even condition (4) cannot be improved upon, by proving

<u>THEOREM 1</u>. (i) There exist continous functions f(t) satisfying (3) and (5) with $\alpha=1$, and even (for every natural n)

(8) $$\int^{\infty} f^{-\frac{1}{2}}(\log f)^{-\frac{1}{2}}(\log\log f)^{-1}\cdots(\log_n f)^{-1} dt = \infty$$

such that the spectra of eq. (1) are purely discrete, clustering only at $\pm\infty$ (eq. (1) being of limit point type at ∞).

(ii) Continous $f(t)$ exist such that (3) holds, with $f(t) = O(t^2)$ (and hence (7) with $\alpha=1$ is satisfied for every n) such that the essential spectrum is infinite, without containing every real λ.

The last part of the theorem settles another question raised in [4], viz. whether (3) implies that S' is either empty or contains every real λ.

Two further questions concerning $S(\alpha)$ and S' that were raised by Hartman and Wintner [7] are answered through the following theorem:

THEOREM 2. Continuous $f(t)$ satisfying (3) exist such that (1) is of limit point type at infinity and for $\lambda=\lambda_0$ has all solutions bounded, while λ_0 is not in the essential spectrum.

Lastly, without any explicit assumption on the asymptotic behaviour of $f(t)$ we have

THEOREM 3. λ_0 does not belong to S' if eq. (1) with $\lambda=\lambda_0$ has two linearly independent solutions x_1, x_2 satisfying

$$(9) \qquad x_1 = O(t^{-\frac{1}{2}+k}), \quad x_2 = O(t^{-\frac{1}{2}-k})$$

for some $k>0$, or more generally

$$(10) \qquad x_1 = O(heE^{-1}), \quad x_2 = O(h^{-1}e),$$

where $e \in L^2(0,\infty)$, $E=\int_t^\infty e^2 dt$, $h>0$, abs.cont., and $\frac{h'}{h} \geq (\frac{1}{K}-\frac{1}{2})\frac{e^2}{E}$ (a.e.) for some $K>0$.

Theorems 2 and 3 (granting that (9) is compatible with the l.p. case, as will be seen) show a striking difference from the situation when $f(t)$ is bounded above, cf. [7] and [8], p. 153 where assumptions of boundedness of all solutions or the existence of a solution $x_1 \notin L^2(0,\infty)$ of (1) for $\lambda=\lambda_0$ and satisfying

(11) $\int_0^t x_1{}^2 dt = O(t^N)$ for some N

implies that λ_0 does belong to the essential spectrum.

To prove the theorems above, we shall need two important characterizations of the spectra of (1) and (2).

The first one is due to Hartman [5] p.915:

(A) The number of points λ of the spectrum satisfying $\lambda'<\lambda<\lambda''$ or $\lambda'\leq\lambda<\lambda''$ (according as (1) is oscillatory or non-oscillatory for $\lambda=\lambda''$) is exactly $n = \lim\inf\{N(T,\lambda'')-N(T,\lambda')\}$ as $T\to\infty$, where $N(T,\lambda)$ denotes the number of zeros for $0\leq t<T$ of a non-trivial solution of (1) and (2), including the possibility of $n=\infty$.

The second one is due to Hartman and Wintner [9], p.546:

(B) A real $\lambda=\lambda_0$ is in the essential spectrum of (1) if and only if there exists some real continuous $g(t)\in L^2(0,\infty)$ such that the corresponding inhomogeneous equation

(12) $y'' + [\lambda_0+f(t)]y = g(t)$

has no solution $y\in L^2(0,\infty)$.

For theorem 3 we shall also need the following

Lemma: Let $p>1$, $g\geq 0$ and $g\in L^p(0,\infty)$. Then

(13) $\int_0^\infty t^{-a}(\int_0^t g s^{\frac{a}{p}-1} ds)^p dt \leq (\frac{p}{a-1})^p \int_0^\infty g^p dt$ (a>1)

(14) $\int_0^\infty t^{-a}(\int_t^\infty g s^{\frac{a}{p}-1} ds)^p dt \leq (\frac{p}{1-a})^p \int_0^\infty g^p dt$ (a<1)

More generally, if in addition $e\in L^p(0,\infty)$, $E = \int_+^\infty e^p dt$, h>0, a.c.

(15) $$\int_0^\infty h^{-p}e^p\left(\int_0^t ghe^{p-1}E^{-1}ds\right)^p dt \le K^p \int_0^\infty g^p dt$$

if $\frac{h'}{h} \ge \left(\frac{1}{K}-\frac{1}{p}\right)\frac{e^p}{E}$ (a.e.) for some $K>0$.

(16) $$\int_0^\infty h^pe^pE^{-p}\left(\int_t^\infty gh^{-1}e^{p-1}ds\right)^p dt \le K^p\int_0^\infty g^p dt$$

if $\frac{h'}{h} \ge \left(\frac{1}{K}-\frac{p-1}{p}\right)\frac{e^p}{E}$ (a.e.) for some $K>0$.

(13) and (14) are due (mainly) to Hardy [2], p.245-6. These inequalities were generalized by Levinson [11]; (15) and (16) are further generalizations of Levinson's results. The proof is analogous to Hardy's and Levinson's proofs.

Now for the proof of theorem 3: Observing that a solution of (12) (assuming $g\in L^2(0,\infty)$) is

(17) $$y = x_1\int_t^\infty gx_2 ds + x_2\int_0^t gx_1 ds$$

when x_1, x_2 are solutions of (1) for $\lambda=\lambda_0$ such that $x_2\in L^2(0,\infty)$ and $x_1x_2'-x_1'x_2 = 1$, theorem 3 follows using characterization (B) and showing that in (17) $y\in L^2(0,\infty)$ for every $g\in L^2(0,\infty)$ by applying first (13) and (14) with $p=2$ and $a=1\pm2k$ respectively, to obtain the first part of theorem 3, and similarly applying (15) and (16) with $p=2$ to the first and second integral in (17) resp., obtaining the last part of theorem 3.

To prove theorem 2 and the last part of theorem 1 we define a nondecreasing step function $f^*(t)$ by

(18) $$f^*(t) = c_k^2 \quad \text{for} \quad t_{k-1}\le t<t_k,$$

$$c_k(t_k-t_{k-1}) = 2\pi, \quad t_0=0, \quad t_k\to\infty.$$

Then equation (1) with $\lambda=0$ and $f=f^*$ has a solution pair for $t_{k-1}\le t<t_k$

(19) $$y_1 = \cos c_k(t-t_{k-1}), \; y_2 = c_k^{-1}\sin c_k(t-t_{k-1})$$

satisfying $y_1y_2'-y_1'y_2 = 1$.

A simple and interesting choice is $c_k = Ck^{\frac{1}{2}}$, implying

$t_k = \sum_1^k 2\pi c_i^{-1} \sim \frac{4\pi}{C} k^{\frac{1}{2}} = \frac{4\pi}{C^2} c_k$,

(20) $$f^*(t) = c_k^2 \sim \frac{C^4}{16\pi^2} t_k^2 \sim \frac{C^4}{16\pi^2} t^2 \text{ for } t_{k-1} \le t < t_k, \text{ hence}$$

(21) $$y_1 = O(1), \quad y_2 = O(t^{-1}), \text{ such that}$$

according to theorem 3 (choosing $k=\frac{1}{2}$ in (9)) $\lambda=0$ is not in S', thus theorem 2 is proved if we can smoothen f^* to obtain a continuous f without perturbing the orders of the solutions. That this can be done is seen using the integral equation for a general solution of (1) for $\lambda=0$

(22) $$x = c_1y_1+c_2y_2+ \int_c^t (f-f^*)[y_1(t)y_2(s)-y_1(s)y_2(t)]x\, ds.$$

From [1], p.112 we know that $|f-f^*|^{\frac{1}{2}}y_i \in L^2(0,\infty)$ (i=1,2) implies $|f-f^*|^{\frac{1}{2}}x \in L^2(0,\infty)$, hence assuming $f-f^*$ integrable (y_1,y_2 being bounded) we may put $c=\infty$ in (22), obtaining first that x is bounded, putting next $c_1=0$, majorizing $|y_2(s)|$ by Kt^{-1} on (t,∞) and using again the integrability of $f-f^*$ and boundedness of x and y_1, we find that $c_1=0$ yields a solution $x_2 = O(t^{-1})$.

Evidently, the modification of f^* into f will not prevent that $f\sim f^*$. Lastly, Hartman's method used to establish conditions (4)-(7) when applied to the case $f(t) \le K^2t^2$ (f non-decreasing) yields that every interval $[\lambda,\lambda+2K]$ contains at least one point of S'. Hence (ii) of theorem 1 is proved.

Finally, to prove the first part of theorem 1 we observe that a general solution of eq. (1) with $f=f^*$ defined through (18) is

(23) $$y=a_n\cos[(c_n^2+\lambda)^{\frac{1}{2}}(t-t_{n-1})]+b_n\sin[(c_n^2+\lambda)^{\frac{1}{2}}(t-t_{n-1})] \quad \text{in } (t_{n-1},t_n),$$

with

(24) $$a_{n+1} = C_n a_n + S_n b_n$$

(25) $$b_{n+1} = k_n(-S_n a_n + C_n b_n)$$

where $S_n = \sin 2\pi(1+\lambda c_n^{-2})^{\frac{1}{2}}$, $C_n = \cos 2\pi(1+\lambda c_n^{-2})^{\frac{1}{2}}$, $k_n = \left(\frac{c_n^{2}+\lambda}{c_{n+1}^{2}+\lambda}\right)^{\frac{1}{2}}$.

We have $y(t_n) = a_n$, further it is clear that for large values of c_n, i.e. for large n

y(t) has 2 or 3 zeros in (t_{n-1},t_n) when $\lambda>0$

y(t) has 1 or 2 zeros in (t_{n-1},t_n) when $\lambda<0$.

If we can show that a_n remains eventually of one sign, y(t) has exactly 2 zeros on (t_{n-1},t_n) for any λ when n is sufficiently large. It will then follow from the spectral characterization (A) above that the spectrum is purely discrete, with no finite cluster point.

From (24) and (25) we get, writing $T_n = \tan 2\pi(1+\lambda c_n^{-2})^{\frac{1}{2}}$

(26) $$\frac{a_{n+1}}{a_n} = C_n\left(1+T_n \frac{b_n}{a_n}\right).$$

(27) $$\frac{b_{n+1}}{a_{n+1}} = k_n \frac{-T_n+\frac{b_n}{a_n}}{1+T_n\frac{b_n}{a_n}} .$$

Assuming now (for a suitable m and some $n_0>m$)

(28) $$k_n \le 1-|T_n| \sum_{m}^{n-1} |T_k| \qquad (n\ge n_0)$$

we may show inductively that for $n\ge n_0$

(29) $$\left|\frac{b_n}{a_n}\right| \le \sum_{m}^{n-1} |T_k|$$

hence that for $n\ge n_0$

$$\frac{a_{n+1}}{a_n} \ge C_n\left(1-|T_n|\left|\frac{b_n}{a_n}\right|\right) \ge C_n\left(1-|T_n| \sum_{m}^{n-1} |T_k|\right) \ge C_n k_n > 0$$

showing that a_n does not change sign for $n\ge n_0$.

To show (29) we choose a_n, b_n to satisfy (29) for $n=n_0$, and use (27), (28) and the induction hypothesis (29) for $n_0 \le n \le N$, obtaining

$$\left|\frac{b_{N+1}}{a_{N+1}}\right| \le k_N \frac{|T_N|+\left|\frac{b_N}{a_N}\right|}{1-|T_N|\left|\frac{b_N}{a_N}\right|} \le \frac{k_N}{1-|T_N|\sum\limits_m^{N-1}|T_k|}\sum_m^N|T_k| \le \sum_m^N|T_k|, \text{ q.e.d.}$$

This proof is a refined version of Hartman's proof in his example of an eq. (1) with a purely discrete spectrum [3], p.124-125. In his example $\int^\infty f^{-\frac{1}{2}}dt<\infty$, and he raises the question whether this is true generally. As stated in our theorem 1 (i) this is not so. To prove theorem 1 (i) it is necessary to show that (28) is satisfied for some $f^*(t)$ which satisfies the conditions of theorem 1 (i) and that we may modify $f^*(t)$ so as to obtain a continuous $f(t)$ satisfying the same conditions (which is trivial) and such that the character of the spectrum is not changed. To see this we use again the integral equation (22), valid for a general solution x of (1) when y_1,y_2 are solutions of (1) for $f=f^*$ with $y_1y_2'-y_1'y_2=1$. From (24) and (25) we get $a_n^2+b_n^2=a_{n+1}^2+k_{n+1}^{-2}b_{n+1}^2>a_{n+1}^2+b_{n+1}^2$, showing that all solutions (23) are bounded (with non-increasing sequences of consecutive extremal values, as is generally true when f is non-decreasing). Hence as before it is sufficient to make $f-f^*$ integrable, we can choose $c=\infty$ in (22) and may rewrite (22) as $x=(c_1+I_1)y_2+(c_2+I_2)y_2$, where $I_1,I_2\to 0$ when $t\to\infty$. Likewise $x'=(c_1+I_1)y_1'+(c_2+I_2)y_2'$, and we obtain $x'y_1-xy_1'=c_2+I_2\to c_2$ as $t\to\infty$. Hence for $c_2\neq 0$ we see that zeros of x and y_1 separate each other for large t-values, and we may conclude that the modification of f^* into a continuous f does not create any new cluster points of spectra.

That the spectrum clusters at both $+\infty$ and $-\infty$ is a consequence of eq. (1) being oscillatory for every λ ([6], p.313-314).

To satisfy (28) we choose

$$(30) \qquad c_n = n^{\frac{1}{2}}(\log n)^{\frac{1}{4}}\cdot d_n$$

where $d_n\to\infty$ in such a way that $\sum^n c_k^{-1} \sim 2n^{\frac{1}{2}}(\log n)^{-\frac{1}{4}}d_n^{-1}$, $\sum^n c_k^{-2}\sim 2(\log n)^{\frac{1}{2}}d_n^{-2}$

and $D_n = c_{n+1}^2 - c_n^2 \sim (\log n)^{\frac{1}{2}} d_n^2$ as $n\to\infty$.

We have $k_n \sim 1-\frac{1}{2}D_n c_n^{-2}$, $|T_n| \sim \pi|\lambda| c_n^{-2}$, and (28) for large n reduces to $d_n^4 \geq 4\pi^2\lambda^2$.

The asymptotic equalities above are valid e.g. if d_n is differentiable and satisfies $d_n^{-1}d_n' = o(n^{-1}(\log n)^{-1})$. To satisfy (8) we choose in addition $d_n = o(\log_p n)$ for every p. Finally, $f \sim f^* \sim \frac{1}{16\pi^2} t^2 \log t \cdot d_n^4$.

REFERENCES

[1] S.G. Halvorsen, On the quadratic integrability of solutions of $x''+fx = 0$. Math. Scand. 14 (1964), 111-119.

[2] G.H. Hardy, J.E. Littlewood, G. Pólya, Inequalities. Cambridge, 1934.

[3] P. Hartman, Some examples in the theory of singular boundary value problems. Amer. J. Math. 74 (1952), 107-126.

[4] P. Hartman, On the essential spectra of ordinary differential operators. Amer. J. Math. 76 (1954), 831-838.

[5] P. Hartman, A characterization of the spectra of one-dimensional wave equations. Amer. J. Math. 71 (1949), 915-920.

[6] P. Hartman and A. Wintner, On the orientation of unilateral spectra. Amer. J. Math. 70 (1948), 309-316.

[7] P. Hartman and A. Wintner, On the location of spectra of wave equations. Amer. J. Math. 71(1949), 214-217.

[8] P. Hartman and A. Wintner, On the derivatives of the solutions of one-dimensional wave equations. Amer. J. Math. 72 (1950), 148-156.

[9] P. Hartman and A. Wintner, On the essential spectra of singular eigenvalue problems. Amer. J. Math. 72 (1950), 545-552.

[10] D.B. Hinton, Continuous spectra of second-order differential operators. Pac. J. Math. 33 (1970), 641-643.

[11] N. Levinson, Generalizations of an inequality of Hardy. Duke Math. J. 31 (1964), 389-394.

[12] E.C. Titchmarsh, Eigenfunction Expansions, part I, 2nd ed., Oxford 1962.

[13] H. Weyl, Über gewöhnliche Differentialgleichungen mit Singularitäten und die zugehörigen Entwicklungen willkürlicher Funktionen. Math. Ann. 68 (1910), 220-269.

Perturbation of a Singular Boundary Value Problem arising from Torsional Vibration of an Inhomogeneous Cylinder

D.J. Heath and A.D. Wood

The Physical Problem

This problem arises from the propagation of torsional waves in a semi-infinite inhomogeneous cylinder of radius a. The central axis of the cylinder coincides with the z-axis and one end is in the z=0 plane. We assume that torsional waves have been excited (see Kolsky [1]: 95). The resulting motion of the cylinder is symmetrical about the z-axis. A transverse wave is propagated in the positive z-direction. Each cross-section remains in its own plane and rotates about the z-axis, which remains undisturbed. The amplitude of the torsional waves at any point is a function only of the radial coordinate r.

The type of inhomogeneity that is usually considered is when μ, the Lamé shear coefficient, and the density ρ, vary radially with r. We shall assume that ρ is constant: the more general case has been considered by Heath in his dissertation [2] but presents no new mathematical features.

The angular displacement $u_\theta(r,z,t)$ satisfies the equation of motion

$$\mu(r)\left\{\frac{\partial^2 u_\theta}{\partial r^2}+\frac{\partial^2 u_\theta}{\partial z^2}+\frac{1}{r}\frac{\partial u_\theta}{\partial r}-\frac{u_\theta}{r^2}\right\}+\mu'(r)\left\{\frac{\partial u_\theta}{\partial r}-\frac{u_\theta}{r}\right\}=\rho\frac{\partial^2 u_\theta}{\partial t^2}$$

where $\mu'(r)=\frac{d\mu}{dr}$. Because there is no displacement of the axis, we have $u_\theta(0,z,t)=0$. Because the lateral surfaces of the cylinder are stress free, we have at r=a the second boundary condition

$$\sigma_{r\theta}\equiv\mu(r)\left\{\frac{\partial u_\theta}{\partial r}-\frac{u_\theta}{r}\right\}=0,\qquad \text{for all } z,t\ .$$

To find sinusoidal solutions we put $u_\theta(r,z,t)=v(r)\exp\left[i(\omega t-\alpha z)\right]$ where the angular frequency ω is constant, and α is the reduced wave number. We obtain the singular boundary-value problem

$$v''+\left(\frac{1}{r}+\frac{\mu'}{\mu}\right)v'+\left(\frac{\rho}{\mu}\omega^2-\alpha^2-\frac{1}{r^2}-\frac{\mu'}{r\mu}\right)v=0, \qquad (1a)$$

$$v(0) = 0, \qquad (1b) \qquad v'(a) = \frac{1}{a} v(a) . \qquad (1c)$$

Achenbach [3] treats (1a) over an interval (b,a) where $0 < b < a$: this is the hollow cylinder problem. His methods break down when $b = 0$.

2. The Perturbation Problem

We assume that $\mu(r) = \mu_o(1 + \varepsilon\phi(r))$ where $\phi \in C^2[a,b]$: (1a) may then be written

$$v'' + \left(\frac{1}{r} + \varepsilon\theta(r,\varepsilon)\right) v' + \left(\lambda - \varepsilon\chi(r,\varepsilon) - \frac{1}{r^2} - \frac{\varepsilon\theta(r,\varepsilon)}{r}\right) v = 0 \qquad (2)$$

where $\theta(r,\varepsilon) \equiv \dfrac{\phi'(r)}{1 + \varepsilon\phi(r)}$ and $\chi(r,\varepsilon) \equiv \dfrac{\rho\omega^2\phi(r)}{\mu_o(1 + \varepsilon\phi(r))}$ are analytic functions of ε for small $|\varepsilon|$, and $\lambda \equiv \dfrac{\rho\omega^2}{\mu_o} - \alpha^2$.

We now have a perturbation of Bessel's equation of order 1 by a first-order differential expression. Neither the perturbed or unperturbed differential expression is formally self-adjoint, and both have a singularity at $r=0$. The standard methods of analytic perturbation theory are not immediately applicable. Heath deals with the problem of perturbation by first or second order differential expressions in [2]. Because these give rise to unbounded operators when considered in $L^2(0,a)$, it seems preferable to transform (2) into Liouville normal form by setting

$$v(r) = \exp\left[-\frac{1}{2}\int_0^r \left(\frac{1}{t} + \varepsilon\theta(t,\varepsilon)\right) dt\right] u(r) . \qquad (3)$$

We obtain

$$u'' + \{\lambda - \frac{3}{4r^2} - \varepsilon\psi(r,\varepsilon)\}u = 0 \qquad (4a)$$

where $\psi(r,\varepsilon) \equiv \dfrac{3\theta(r,\varepsilon)}{2r} + \dfrac{1}{2}\theta'(r,\varepsilon) + \chi(r,\varepsilon) + \dfrac{\varepsilon}{4}\theta^2(r,\varepsilon)$ is analytic in ε for small $|\varepsilon|$. When $\varepsilon=0$ (4a) is the Fourier-Bessel equation with solutions $r^{\frac{1}{2}}J_1(\kappa r)$, $r^{\frac{1}{2}}Y_1(\kappa r)$, where κ is the square root of λ which is real and positive when λ is real and positive (Titchmarsh [4]: 81).

The transformation (3) removes the first order term from (2) only at the cost of introducing the perturbation parameter ε into the boundary condition at $r=a$ which becomes $u'(a) = \left(\dfrac{3}{2a} + \dfrac{1}{2}\varepsilon\theta(a,\varepsilon)\right) u(a)$. (4c)

The boundary condition at r=0 remains unchanged as $u(0) = 0$. (4b)

The problem (4a,b,c) with $\varepsilon=0$ is self-adjoint with one singular end-point at $r=0$ where it is in the limit-point case ([4]: 82). It has a countable number of real, simple eigenvalues $\lambda_0, \lambda_1, \lambda_2, \ldots$ where $\lambda_0 = 0$ and for $n = 1,2,3, \ldots, \kappa_n = \lambda_n^{\frac{1}{2}}$ satisfies $J_2(\kappa_n a) = 0$. The associated eigenfunctions are $u_0(r) = A_0 r^{3/2}$, $u_n(r) = A_n r^{\frac{1}{2}} J_1(\kappa_n r)$, $n > 1$. With a suitable choice of constants A_n, $\{u_n\}_{n=0}^{\infty}$ is an orthonormal set in $L^2(0,a)$. The vanishing of $u_n(r)$ at $r=0$ is consistent with the boundary condition (4b), although this boundary condition is not used in the construction of eigenfunctions by the methods of [4].

It is not immediately obvious that the problem (4a,b,c) with $\varepsilon \neq 0$ is self-adjoint, but this is a routine verification using the results given in Coddington and Levinson ([5]:189).

We now appeal to the perturbation theory of Kato [6], Ch. VII. Let $L_\varepsilon u \equiv -u'' + \left(\frac{3}{4r^2} + \varepsilon\psi(r,\varepsilon)\right) u$. Consider the quadratic form

$$\mathfrak{t}_\varepsilon[u] = \int_0^a \left(|u'(r)|^2 + \left(\frac{3}{4r^2} + \psi(r,\varepsilon)\right)|u(r)|^2\right)dr + \left(\frac{3}{2a} + \frac{1}{2}\varepsilon\theta(a,\varepsilon)\right)|u(a)|^2 .$$

By ([6]: 398) $\{\mathfrak{t}_\varepsilon\}$ is a self-adjoint holomorphic family of forms of type (a). The representation theorem associates with $\mathfrak{t}_\varepsilon$ a self-adjoint operator $T_\varepsilon = L_\varepsilon$ in $L^2(0,a)$ given by $u \in D(T_\varepsilon)$ if

(i) u and $u' \in AC[\delta,a]$ for all $\delta \in (0,a)$, (ii) $u'(a) = \left(\frac{3}{2a} + \frac{1}{2}\theta(a,\varepsilon)\right)u(a)$,

(iii) $\int_0^a \frac{3}{4r^2}|u(r)|^2dr < \infty$, $\int_0^a \psi(r,\varepsilon)\ |u(r)|^2dr < \infty$, (iv) $L_\varepsilon u \in L^2(0,a)$.

It can be shown ([6], 409) that $\{T_\varepsilon\}$ is a holomorphic family of type (B). This implies that (4) has real eigenvalues $\Lambda_0, \Lambda_1, \Lambda_2, \ldots$ with associated orthonormal eigenfunctions $U_0(r), U_1(r), U_2(r), \ldots$ and that

$$\Lambda_n = \lambda_n + \varepsilon\lambda_n^{(1)} + \varepsilon^2\lambda_n^{(2)} + \ldots \tag{5}$$

$$U_n(r) = u_n(r) + \varepsilon\tilde{u}_n^{(1)}(r) + \varepsilon^2\tilde{u}_n^{(2)}(r) + \ldots$$

where λ_n, $u_n(r)$ are the nth eigenvalue and eigenfunction of the unperturbed problem and the coefficients of $\varepsilon, \varepsilon^2, \ldots$ are to be determined. Both series

(5) are convergent in some neighbourhood of $\varepsilon = 0$ whose radius may be estimated.

3. The coefficients in the perturbation series

The authors are unaware of any method in the literature for finding the coefficients $\lambda_n^{(k)}$ and $\tilde{u}_n^{(k)}(r)$ for problems where the perturbation appears in the boundary conditions. We give a modification of the method of Titchmarsh ([7]: Section 19.3). The original method depends on expanding the unknown functions $\tilde{u}_n^{(k)}(r)$ in the eigenfunctions $\{u_n(r)\}$ of the unperturbed problem. But here the functions $\tilde{u}_n^{(k)}(r)$ satisfy a different boundary condition and such an expansion is impossible. We overcome this by transforming problem (4) by $w(r) = u(r) \exp\left[-\frac{1}{2}\varepsilon \int_0^r \theta(s,\varepsilon)ds\right]$ to obtain

$$w'' + 2\theta(r,\varepsilon)w' + \{\lambda - \frac{3}{4r^2} + \varepsilon(\tfrac{1}{2}\theta'(r,\varepsilon) - \psi(r,\varepsilon) + \tfrac{1}{4}\varepsilon\theta^2(r,\varepsilon))\}w = 0 , \quad (6a)$$

$$w(0) = 0 , \qquad w'(a) = \frac{3}{2a} w(a) . \quad (6b,c)$$

The eigenvalues of (6) are those of (4) and the eigenfunctions $W_n(r)$ of (6) are related to the eigenfunctions $U_n(r)$ of (4) by

$$W_n(r) = U_n(r) \exp\left[-\tfrac{1}{2}\varepsilon \int_0^r \theta(s,\varepsilon)ds\right] \quad (7)$$

and the orthonormality of $\{U_n\}$ implies that

$$\int_0^a W_n(r)\, W_m(r) \exp\left[\varepsilon \int_0^r \theta(s,\varepsilon)ds\right] dr = \delta_{nm} . \quad (8)$$

The boundary conditions of both perturbed and unperturbed problems are now the same and we may write $W_n(r) = u_n(r) + \varepsilon u_n^{(1)}(r) + \varepsilon^2 u_n^{(2)}(r) + \ldots$, where $u_n^{(1)}(r) = \sum_{p=0}^{\infty} \alpha_{np}\, u_p(r)$, $u_n^{(2)}(r) = \sum_{p=0}^{\infty} \beta_{np}\, u_p(r)$, ... and α_{np}, β_{np} are the Fourier coefficients of $u_n^{(1)}$, $u_n^{(2)}$ respectively with respect to $\{u_n\}$. We may proceed exactly as in [7] to find the coefficients α_{np}, β_{np} and hence $u_n^{(1)}$, $u_n^{(2)}$ and $\lambda_n^{(1)}$, $\lambda_n^{(2)}$, except that (8) is used where Titchmarsh uses the orthogonality relation. The expressions for $W_n(r)$ may be transformed back to those for $U_n(r)$ by (7) and thence by (3) to obtain expressions for the eigenfunctions of the original problem (1). The eigenvalues of all three problems remain unchanged.

Physical implications

It is now possible to express the natural frequencies and normal modes for torsional vibrations of the unhomogeneous elastic cylinder as perturbations of those for the homogeneous cylinder. Using these, Heath has gone on to show in [2], Section 4,5, that the fundamental mode of vibration for torsional waves in an inhomogeneous cylinder is non-dispersive, as is the case for a homogeneous cylinder.

REFERENCES

1. Kolsky, H., Stress Waves in Solids, New York, Dover Publications Inc.

2. Heath, D.J., M.Sc. Dissertation, Cranfield Institute of Technology, Bedford, England, 1973.

3. Achenbach, J.D., Forced torsional motion of an inhomogeneous hollow cylinder, Revue Roumaine des Sciences technique: Serie de mecanique applique, 1966.

4. Titchmarsh, E.C., Eigenfunction Expansions associated with second order differential equations, Part I, Oxford 1962.

5. Coddington, A.E. and Levinson, N., Theory of Ordinary Differential Equations, New York, McGraw-Hill, 1955.

6. Kato, T., Perturbation Theory for Linear Operators, New York, Springer-Verlag, 1966.

7. Titchmarsh, E.C., Eigenfunction Expansions associated with second order differential equations, Part II, Oxford, 1958.

Some function theoretical aspects in the theory of partial differential equations

G. Jank

1 Introduction

We consider the differential equation

$$(1) \qquad w_{z\bar{z}} - \frac{m+1}{z-\bar{z}} w_z + \frac{n+1}{z-\bar{z}} w_{\bar{z}} = 0, \qquad n, m \in \mathbb{N}_0 ,$$

which is a generalisation of an equation firstly treated by E. Peschl and K. W. Bauer [1] with the aid of differential operators. From [3] one can see that all solutions of (1) defined in a simply connected domain of the upper half plane can be represented by a certain differential operator. This representation theorem can be extended to solutions of (1) having isolated singularities (theorem 2). It is also possible to carry over Riemann's theorem about locally bounded functions (theorem 3). If we consider the solutions of (1), defined in the whole upper half plane except at a finite set of singular points, and bounded as z approach to the real axis, we also have for this class a representation theorem (theorem 4). The functions in this class allow an analytic continuation over the real axis. If in (1) n = m one can also consider real solutions. In that case one gets the results of St. Ruscheweyh [5].

2 Representation theorems, Riemann's theorem

For our simple differential equation (1) we get the following

theorem 1: (i) *To each solution w of (1) defined in a simply connected domain G of the upper half plane, there are two in*

G holomorphic functions f(z), g(z), so that

(2) $$w = \frac{\partial^{n+m}}{\partial z^n \partial \bar{z}^m} \left(\frac{f(z)+\overline{g(z)}}{z-\bar{z}}\right) .$$

This functions f, g will be called "generating functions" of w.

(ii) Conversely (2) represents a solution of (1) defined in G for each pair of in G holomorphic functions.

(iii) Given a solution (defined in G) of (1) the functions $\frac{d^{n+m+1}}{dz^{n+m+1}} f(z)$, $\frac{d^{n+m+1}}{dz^{n+m+1}} g(z)$ are determined unambiguously according to

$$f^{(n+m+1)}(z) = \frac{d^{m+1}[(z-\bar{z})^{n+1} w]}{m!(z-\bar{z})^{n+m+2}}$$

$$g^{(n+m+1)}(z) = -\frac{d^{n+1}[(z-\bar{z})^{m+1} \bar{w}]}{n!(z-\bar{z})^{n+m+2}}$$

with $d := (z-\bar{z})^2 \frac{\partial}{\partial z}$.

(iv) The solution $w \equiv 0$ has the generating functions

$$f_o = \sum_{k=0}^{n+m} a_k z^k, \qquad g_o = -\sum_{k=0}^{n+m} \overline{a_k}\, z^k, \qquad a_k \in \mathbb{C} .$$

This result one can get from [2], [4]. Later we will also need the representation of solutions of (1) with isolated singularities. We have

theorem 2: If w is a solution of (1) defined for $0 < |z-z_o| < r$, with z_o and z in the upper half plane, then w has a representation (2). The generating functions are according to

$$f(z) = f_1(z) + P_{n+m}(z) \log(z-z_o)$$

$$g(z) = g_1(z) + \overline{P_{n+m}(\bar{z})} \log(z-z_o),$$

where the functions f_1 and g_1 are Laurent series in $0 < |z-z_0| < r$, and $P_{n+m}(z)$ is a polynomial of degree not greater then n+m.

The proof follows from the fact that for a solution with an isolated singularity in z_0 the $(n+m+1)^{th}$ derivative of a generating functions is defined uniquely in a puctured neighbourhood of z_0. Easily we can get now a generalisation of a theorem of the classical theory of functions (Riemann's theorem).

Theorem 3: If w is a solution of (1), defined in the punctured neighbourhood $\dot{U}(z_0)$ of z_0 in the upper half plane with the generating functions

$$f(z) = \sum_{-\infty}^{\infty} a_k (z-z_0)^k + P_{n+m}(z)\log(z-z_0)$$

$$g(z) = \sum_{-\infty}^{\infty} b_k (z-z_0)^k + \overline{P_{n+m}(\bar{z})}\log(z-z_0) ,$$

and if there is a constant c, so that $|w| < c$ in $\dot{U}(z_0)$, then there is a holomorphic continuation of f and g in $U(z_0)$. That means $a_k = b_k = 0, \quad k = -1, -2, \ldots, \quad P_{n+m}(z) \equiv 0$.

For the proof one has to notice that under the hypothesis of theorem 3 there exists the real analytic continuation of w in $U(z_0)$. It follows from theorem 1 that for example the function $f^{(n+m+1)}(z)$ is bounded in $\dot{U}(z_0)$ and therefore the holomorphic continuation of f in $U(z_0)$ exists. Now let us prove a lemma we need for the next theorem. As an abbreviation we write

$$E_{nm}f(z) = \frac{\partial^{n+m}}{\partial z^n \partial \bar{z}^m} \frac{f(z)}{z-\bar{z}} .$$

<u>Lemma 1 : If $z_o \in \mathbb{R}$ and f(z) holomorphic in z_o, then with n+m>0</u>

$$\lim_{\substack{z\to z_o \\ \mathrm{Im}\, z>0}} (E_{nm}\, f(z) \; - E_{nm}\, f(\bar{z})) = 0.$$

It can be proved that the function

$$h(z) := \begin{cases} \dfrac{f(z) - f(\bar{z})}{z - \bar{z}} & z \neq z_o \\ f'(z_o) & \end{cases}$$

is real analytic in a neighbourhood of z_o and therefore limit and differentiation may be changed.

3 A representation theorem for solutions of (1) bounded on the real axis

We define B^{nm} as the set of solutions of (1) defined in the upper half plane except at a finite set of points. The functions in B^{nm} should remain bounded as z (with Im z > 0) approach the real axis. H^{nm} denotes the set of holomorphic functions of the form

$$f(z) = g(z) + \sum_{l=1}^{k} S_l(z) \log(z - z_l),$$

where: g(z) is holomorphic in the extended complex plane except at some isolated singularities which are not situated on the real axis. $S_l(z)$ is a polynomial whose degree does not exceed n+m. z_l is a fixed point of the upper half plane.

<u>Theorem 4 : To each $w \in B^{nm}$ there is a function $f \in H^{nm}$ so that</u> (n+m > 0)

$$w = E_{nm}(f(z) - f(\bar{z})). \tag{3}$$

<u>Conversely</u> (3) <u>represents a function of</u> B^{nm} <u>for each</u> $f \in H^{nm}$.

To prove the second assertion, from theorem 2 we can see that (3) represents a solution of (1) in the upper half plane except at a finite set of points. From lemma 1 we deduce, that w remains bounded as z (Im z > 0) approach the real axis. For the first assertion of theorem 4 we need a

<u>lemma 2 : If $w \in B^{nm}$ does not have any singularities in the upper</u>

<u>half plane, then</u> $w \equiv 0$.

If we expand w in a double series in a point z_o ($\operatorname{Im} z_o > 0$) with the coefficients a_{rs} and convergent in $|z-z_o| < \operatorname{Im} z_o < 1$, the series

$$\emptyset(z,\bar z) = \sum_{r=0}^{\infty}\sum_{s=0}^{\infty} \frac{a_{rs}\,(z-z_o)^{r+n}(\bar z-\bar z_o)^{s+m}}{(r+1)\dots(r+n)(s+1)\dots(s+m)}$$

is absolute uniformly convergent in $|z-z_o| < \operatorname{Im} z_o$ and we have $w = \frac{\partial^{n+m}}{\partial z^n \partial \bar z^m}\emptyset$. $\emptyset$ also remains bounded as z approach the real axis. On the other hand because of theorem 1 there are two in the upper half plane holomorphic functions $f(z)$ and $g(z)$, so that $\emptyset$ is the expansion of $\frac{f(z)+\overline{g(z)}}{z-\bar z}$ in z_o. So we get $(f(z)+\overline{g(z)})\big|_{z\in\mathbb{R}} = 0$. As the real and imaginary part of $f(z)+\overline{g(z)}$ are harmonic functions in the upper half plane vanishing on the real axis we conclude that $f(z) + \overline{g(z)}$ must vanish in the upper half plane and the lemma is proved.

Now we complete the proof of theorem 4. If $w \in B^{nm}$ and z_l, $l=1,\dots,k$ denotes the singularities of w we see from theorem 2 that in $\dot U(z_l)$ exists a representation in the form

$$w = E_{nm}\left[\, f_l + {}_lP_{n+m}\log(z-z_l) + \overline{g_l} + \overline{{}_lP_{n+m}\log(z-z_l)}\,\right].$$

We denote with f_l^{*} and g_l^{*} the principal parts of the Laurent series f_l and g_l respectively, and form the function $w_1 = w - E_{nm}[\alpha(z)-\alpha(\bar z)]$, with $\alpha(z) := \sum_{l=1}^{k}\left[f_l^{*}(z) - \overline{g_l^{*}(\bar z)} + {}_lP_{n+m}\log(z-z_l) + \overline{{}_lP_{n+m}\log(\bar z-z_l)}\right] \in H^{nm}$.

We conclude that $w_1 \in B^{nm}$ and has no singularities in the upper half plane. Lemma 2 gives $w_1 \equiv 0$ and theorem 4 is proved.

References

[1] Bauer K.W., Über eine der Differentialgleichung $(1 \pm z\bar z)^2 w_{z\bar z} \pm n(n+1)w = 0$ zugeordnete Funktionentheorie, Bonner Math. Schr. <u>23</u> (1965)

[2] Bauer K.W., Florian H., Bergman-Operatoren mit Polynomerzeugenden Applicable Analysis (im Druck)

[3] Bauer K.W., Jank G., Differentialoperatoren bei einer inhomogenen elliptischen Differentialgleichung, Rend. Ist. Mat. Univ. Trieste, Heft II, 1-29 (1971).

[4] Bauer K.W., Ruscheweyh St., Polynomoperatoren in der Theorie partieller Differentialgleichungen (erscheint demnächst)

[5] Ruscheweyh St., Über den Rand des Einheitskreises fortsetzbare Lösungen der Differentialgleichung von Peschl und Bauer, Ber. d. Gesellsch. f. Math. u. Datenverarb., Bonn 57, 29-36 (1972)

A MULTI-PARAMETER STURM-LIOUVILLE PROBLEM

by

A. Källström and B. D. Sleeman

§1. Introduction.

Consider the system of ordinary differential equations

$$\frac{d^2 y_r}{dx_r^2} + \left\{ \sum_{s=1}^{k} a_{rs}(x_r)\lambda_s - q_r(x_r) \right\} y_r = 0, \qquad (1.)$$

$0 \leq x_r \leq 1, \quad r = 1,2,\ldots,k,$

in which

$$a_{rs}(x_r) \in C'[0,1] \text{ and real valued,}$$

$$q_r(x_r) \in C'[0,1] \text{ and real valued,}$$

$r,s = 1,2,\ldots,k.$

We formulate a Sturm-Liouville problem for (1.1) by seeking non-trivial solutions $y_r(x_r;\lambda_1,\ldots,\lambda_k)$ together with corresponding eigenvalues $\lambda = (\lambda_1,\ldots,\lambda_k)$ satisfying the boundary conditions

$$y_r(0) = y_r(1) = 0, \quad r = 1,2,\ldots,k, \qquad (1.2a)$$

or

$$\frac{dy_r(0)}{dx_r} = \frac{dy_r(1)}{dx_r} = 0, \quad r = 1,2,\ldots,k. \qquad (1.2b)$$

When discussing questions of existence of eigenfunctions to the problems defined by (1.1) (1.2a) or (1.1) (1.2b) it is clear that further conditions must be imposed on the coefficients a_{rs}, q_r in addition to the continuity requirements given above. It turns out that there are two possible hypotheses which can be introduced. One is to make the hypothesis

(A) $$\Delta_k = \det\{a_{rs}(x_r)\}_{r,s=1}^{k} > 0$$

for all $\underset{\sim}{x} = (x_1,\ldots,x_k) \in I_k$,

where I_k denotes the Cartesian product of the k intervals [0,1].

Alternatively we could suppose

(B):
$$\begin{vmatrix} \alpha_1 & \alpha_2 \cdots\cdots & \alpha_k \\ a_{21} & a_{22} & a_{2k} \\ \vdots & & \\ a_{k1} & a_{k2} & a_{kk} \end{vmatrix} > 0, \ldots \quad \begin{vmatrix} a_{11} & \cdots\cdots & a_{1k} \\ \vdots & & \\ a_{r-1,1} & a_{r-1,2} & a_{r-1,k} \\ \alpha_1 & \alpha_2 & \alpha_k \\ a_{r+1,1} & a_{r+1,2} \cdots & a_{r+1,k} \\ \vdots & & \\ a_{k,1} & a_{k2} & a_{kk} \end{vmatrix} > 0$$

$$\cdots\cdots \quad \begin{vmatrix} a_{11} & \cdots\cdots & a_{1k} \\ \vdots & & \\ a_{k-1,1} & \cdots\cdots & a_{k-1,k} \\ \alpha_1 & \cdots\cdots & \alpha_k \end{vmatrix} > 0$$

for some real k-tuple of numbers $\alpha_1,\ldots,\alpha_k$ not identically zero. The inequalities holding for all $\underset{\sim}{x} \in I_k$.

In the case k = 1 hypothesis (A) reduces to the positivity requirement demanded by the coefficient of λ_1 in the familiar one-parameter Sturm-Liouville problem. When k = 2 it is not difficult to establish that hypothesis (A) implies hypothesis (B). But the converse is not true in general. For $k \geq 3$ the problem defined by (1.1) (1.2), (A) and that by (1.1) (1.2) and (B) are distinct. We note, for further reference, that hypothesis (B) can be expressed in the more convenient form

$$h_s = \sum_{r=1}^{k} \alpha_r \, a_{sr}^* > 0 \qquad (1.3)$$

for all $\underset{\sim}{x} \in I_k$ where a_{sr}^* denotes the cofactor of a_{sr} in the determinant $\Delta_k = \det\{a_{rs}\}_{e,s=1}^{k}$.

The Sturm-Liouville problem for (1.1) (1.2) under the hypothesis (A) has been treated with varying degrees of generality in [1], [2], [3] and [5]. However the situation in which hypothesis (B) is invoked remains, to the best of our knowledge, uninvestigated and it is to this problem we address ourselves.

As a preliminary we can state that if eigenfunctions exist satisfying (1.1) (1.2) and (B) then the associated eigenvalues are real and the eigenfunctions form an orthogonal set with respect to the inner product defined in (2.5) below, as well as the inner product defined by

$$K(u,v) = \int_{I_k} u\bar{v}\ \Delta_k\ d\underset{\sim}{x}.$$

These results will be discussed in depth elsewhere.

§2. An elliptic boundary value problem.

Using a procedure similar to that in [6] it may be shown that the problem defined by (1.1) (1.2) can be formally replaced by the following system of eigenvalue problems

$$\sum_{s=1}^{k} (a_{sr}^{*} \frac{\partial^2 Y}{\partial x_s^2} - q_s a_{sr}^{*} Y) = - \lambda_r \Delta_k Y, \tag{2.1}$$

$$r = 1,2,\ldots,k,$$

where Y is a non-trivial solution satisfying boundary conditions of the form (1.2) on the sides of the k-dimensional cube I_k. Multiplication of (2.1) by α_r and summation over r = 1,2,...,k leads to

$$\sum_{s=1}^{k} (h_s \frac{\partial^2 Y}{\partial x_s^2} - h_s q_s Y) = - \Lambda \Delta_k Y, \tag{2.2}$$

where

$$\Lambda = \sum_{r=1}^{k} \alpha_r \lambda_r . \tag{2.3}$$

Because of (1.3) we see that we now have an elliptic eigenvalue problem for Y in which the boundary conditions on the cube I_k are of either Dirichlet or Neumann type. However such a problem is not in the usual form since the coefficient of the spectral parameter Λ, is not necessarily definite and so the standard theory is not applicable. If we had formulated (2.2) under the hypothesis (A) then although Δ_k is positive, the left hand side of (2.2) would not in general be elliptic. The exception of course being the case $k = 2$.

In [4] Pleijel has developed a spectral theory for problems of the type governed by (2.2) and Dirichlet or Neumann conditions. In such a theory a suitable Hilbert space must be defined. The usual space consisting of those functions which are Lebesque measurable in $L^2_{\Delta_k}[I_k]$ is not a Hilbert space since Δ_k is not positive definite. Instead a positive definite Dirichlet integral is associated with (2.2) and the spectral theory is developed in the Hilbert space H which, in the case of Dirichlet boundary conditions, is the completion of $C_0'(I_k)$ with respect to the inner product

$$D(u,v) = \int_{I_k} \sum_{s=1}^{k} (h_s \frac{\partial u}{\partial x_s} \frac{\partial \bar{v}}{\partial x_s} + h_s q_s u\bar{v}) d\underline{x} . \tag{2.4}$$

For the Neumann problem H is the completion of $C'(I_k)$ with respect to the inner product (2.4). In order that the norm associated with (2.4) be positive definite we need the additional hypothesis

$$\sum_{s=1}^{k} h_s q_s \geq 0 \tag{2.5}$$

for all $\underline{x} \in I_k$. Such a condition is realised if in addition to hypothesis (B) we suppose

(C): $$q_r(x_r) \geq 0 \quad \text{for all } x_r \in [0,1]. \quad r = 1,2,\dots,k. \tag{2.6}$$

<u>Remark 1</u>

It may be shown, using a result of Atkinson [1] applied to any one of the determinants in (B) and "shifting" the origin of the spectrum for (1.1) that it is sufficient to demand that only one of the $q_r(x_r)$ be positive.

Remark 2

Although Pleijel's analysis specifically deals with the equation

$$\Delta u - qu + \lambda k u = 0$$

in which q is non-negative and k indefinite, it can be easily extended to the situation in hand.

For a spectral analysis of the problem outlined above we may prove that (2.2) has a discrete spectrum which can be ordered in magnitude as

$$\cdots \leq \Lambda_{-2} \leq \Lambda_{-1} \ (< 0) < \Lambda_1 \leq \Lambda_2 \leq \cdots \tag{2.7}$$

In addition, if $\phi_n(\underset{\sim}{x};\ \Lambda_n)$ is an eigenfunction of (2.2) corresponding to the eigenvalue Λ_n then we have the Parseval equality

$$D(u) = \sum_{-\infty}^{\infty} |\Lambda_n| \{K(u,\phi_n)\}^2 \tag{2.8}$$

for all $u \in H \ominus H(\infty)$, where $H(\infty)$ is the set $\{u \in H;\ \Delta_k u = 0$ and u satisfies the boundary conditions$\}$. In (2.8) K(u,v) is as defined in Section 1.

§3. Decomposition of $\phi_n(\underset{\sim}{x};\ \Lambda_n)$

In this section we relate the eigenfunctions $\phi_n(\underset{\sim}{x};\ \Lambda_n)$ to those of the original problem defined by (1.1) (1.2). Since not all the α_r, $r = 1,2,\ldots,k$ in (B) are zero we may suppose for example that $\alpha_k \neq 0$; then from (2.3) we can solve for λ_k to give

$$\lambda_k = \frac{\Lambda_n}{\alpha_k} - \sum_{r=1}^{k-1} \frac{\alpha_r}{\alpha_k} \lambda_r. \tag{3.1}$$

Substituting this into the first k - 1 equations of (1.1) say, gives the new system

$$\frac{d^2 y_r}{dx_r^2} + \left\{ \sum_{s=1}^{k-1} a_{rs}(x_r)\lambda_s + a_{rk}\frac{\Lambda_n}{\alpha_k} - a_{rk}\sum_{t=1}^{k-1}\frac{\alpha_t}{\alpha_k}\lambda_t - q_r(x_r)\right\} y_r = 0$$

or

$$\frac{d^2y_r}{dx_r^2} + \left\{ \sum_{s=1}^{k-1} (a_{rs} - a_{rk}\frac{\alpha_s}{\alpha_k})\lambda_s + a_{rk}\frac{\Lambda_n}{\alpha_k} - q_r \right\} y_r = 0, \quad r = 1,2,\dots,k-1. \tag{3.2}$$

Thus we have generated a (k - 1)-parameter eigenvalue problem.

Consider the determinant

$$\Delta_{k-1} = \det\left\{ a_{rs} - \frac{\alpha_s}{\alpha_k} a_{rk} \right\}_{r,s=1}^{k-1}$$

$$= \det\left\{ a_{rs} \right\}_{r,s=1}^{k-1} - \sum_{s=1}^{k-1} \frac{\alpha_s}{\alpha_k} \det_s \left\{ a_{rs} \right\}_{r=1}^{k-1} \tag{3.3}$$

in which $\det_s$ denotes the $(k - 1) \times (k - 1)$ determinant $\det\{a_{rs}\}_{r,s=1}^{k-1}$ with the s-th column replaced by the entries a_{rk}, $r = 1,2,\dots,k-1$. In terms of the cofactors of Δ_k we have, using (1.3)

$$\Delta_{k-1} = a^*_{kk} + \sum_{s=1}^{k-1} \frac{\alpha_s}{\alpha_k} a^*_{ks} = \frac{1}{\alpha_k} h_k, \tag{3.4}$$

which is of fixed sign and non zero for all $\underset{\sim}{x} \in I_{k-1}$.

Thus the problem defined by (3.2) and (1.2) constitutes a (k - 1)-parameter eigenvalue problem for which the equivalent of hypothesis (A) holds. This problem has been studied by Browne [2] who has proved that the eigenfunctions are complete in the space $\tilde{K} \equiv L^2_{\Delta_{k-1}}[I_{k-1}]$. Let the eigenfunctions be denoted by

$$E^m_{k-1} \equiv y^m_1(x_1;\ \tilde{\lambda}^m) \otimes y^m_2(x_2;\ \tilde{\lambda}^m) \otimes \dots \otimes y^m_{k-1}(x_{k-1};\ \tilde{\lambda}^m), \tag{3.5}$$

where $\tilde{\lambda}^m = (\lambda^m_1, \lambda^m_2, \dots, \lambda^m_{k-1})$ and $\lambda^m_r = \lambda^m_r(\Lambda_n)$, $r = 1,2,\dots,k-1$.

Clearly the eigenfunction ϕ_n of (2.2) considered as a function of $x_1, x_2, \dots, x_{k-1}$ is an element of $\tilde{K}$ and so ϕ_n can be expanded as a convergent series

$$\phi_n(x_1,\dots,x_k;\ \Lambda_n) = \sum_{\tilde{\lambda}^m} c_m(x_k)\ E^m_{k-1}, \tag{3.6}$$

convergence of course being in the sense of the norm induced by $\tilde{K}$.

Using the orthogonality property of the eigenfunctions E^m_{k-1} we have

$$C_m(x_k) \|E^m_{k-1}\|^2_{\Delta_{k-1}} = \int_{I_{k-1}} \phi_n E^m_{k-1} \Delta_{k-1} dx_1 \dots dx_{k-1}, \tag{3.7}$$

where the norm on the left hand side of (3.7) is the norm induced by the space $\tilde{K}$.

By differentiating (3.7) twice with respect to x_k and using the fact that ϕ_n satisfies (2.2) we prove after succesive integrations by parts that $C_m(x_k)$ satisfies the equation

$$C''_m(x_k) + (\sum_{r=1}^{k} \lambda_r a_{kr} - q_r) C_m(x_k) = 0. \tag{3.8}$$

That is $C_m(x_k)$ is an eigenfunction of the k-th member of the system (1.1) (1.2) with the same eigenvalues $\lambda^m_1, \dots, \lambda^m_k$, ($\lambda^m_k$ being determined from (3.1)) as the previous k - 1 members of the system. We also note that the series (3.6) is a finite sum. This follows from the fact that the eigenvalues Λ_n have finite multiplicity which implies that only a finite number of the eigenfunctions $(C_m E^m_{k-1})^{\infty}_{m=1}$ are linearly independent.

Thus from (3.6) and the fact that $C_m(x_k)$ satisfies (3.8) we have the Parseval equality (2.8) expressible in terms of the eigenfunctions of our original problem (1.1) (1.2).

§4. Concluding remarks.

It is hoped to generalise the analysis in [4] to include more general boundary conditions of Sturm-Liouville type and thus to extend the completeness result discussed in this paper. Further, it is possible to generalise the arguments discussed here to the abstract multi-parameter spectral problem for both bounded and unbounded operator equations under hypotheses analagous to (A) and (B). These and other problems will be taken up elsewhere.

The research reported in this paper was supported by a grant from the Science Research Council and is gratefully acknowledged.

References

[1] F. V. Atkinson. Multiparameter eigenvalue problems, matrices and compact operators. Academic Press 1972.

[2] P. J. Browne A multiparameter eigenvalue problem. J. Math. Analysis Applic., 38 553-568 (1972).

[3] M. Faierman. The completeness and expansion theorems associated with the multiparameter eigenvalue problem in ordinary differential equations. J. Diff. Equations., 5 197-213 (1969).

[4] Å. Pleijel. Le problème spectral de certaines equations aux dérivées partielles. Ark. Mat. Astr. Fysik. 30A 21. 1-47 (1944).

[5] B. D. Sleeman. Completeness and expansion theorems for a two-parameter eigenvalue problem in ordinary differential equations using variational principles. J. Lond. Math. Soc. 2 (6) 705-712 (1973).

[6] B. D. Sleeman. Singular linear differential operators with many parameters. Proc. Roy. Soc. Edin. (A) 71 18. 199-232 (1973).

Periodic Solutions of Holomorphic Differential Equations

by

N.G.Lloyd

Let $\omega \in \mathbb{R}$ be fixed and let $t \in \mathbb{R}$, $z \in \mathbb{C}^N$; define $\mathfrak{F}$ to be the collection of differential equations

$$\dot{z} = f(t,z), \qquad (1)$$

where (i) f is continuous on $\mathbb{R} \times \mathbb{C}^N$, (ii) f is holomorphic in $\mathbb{C}^N$ for every fixed t, and (iii) $f(t+\omega, z) = f(t,z)$ for all t,z. The equation (1) is identified with the function f, and $\mathfrak{F}$ is then given the structure of a Fréchet space by means of the seminorms

$$p_k(f) = \max [\ |f^i(t,z)|;\ 1 \leq i \leq N,\ |z| \leq k,\ 0 \leq t \leq \omega\ ]$$

(k=1,2,...); here $|z| = \max [\ |z^i|:\ 1 \leq i \leq N\ ]$.

The solution of (1) satisfying $z(t_o) = z_o$ is written $z_f(t;t_o,z_o)$. Define, for positive integers m,

$$q_m^{(\tau)}(f,c) = z_f(\tau+m\omega;\tau,c) - c;$$

it will be convenient to write $q_m^{(0)} = q_m$. The solutions of (1) of period $m\omega$ (abbreviated $m\omega$-p.s.) correspond to the zeros of $q_m(f,.)$. Clearly $q_m^{(\tau)}$ is continuous in a neighbourhood of (f,c) if $z_f(\tau+m\omega;\tau,c)$ is defined.

<u>Definition 1</u> An $m\omega$-p.s. $\psi(t)$ of (1) is an <u>m-harmonic</u> if $\psi(0)$ is an isolated zero of q_m, and $\psi(t)$ does not admit a period $m_1\omega$, $0 < m_1 < m$, m_1 integral.

Suppose $U \subset \mathbb{C}^N$ and $g : U \to \mathbb{C}^N$ is holomorphic: if z_o is an isolated p-point of g in U, the multiplicity of z_o as a p-point is defined to be $i(g,z_o,p)$, the degree of g at p relative to sufficiently small neighbourhoods of z_o.

Definition 2 Let $\psi(t)$ be an isolated $m\omega$-p.s. of (1); the multiplicity k of ψ as an $m\omega$-p.s. is $i(q_m(f,.), \psi(0),0)$; ψ is simple if $k = 1$.

Theorem 1 Let D be a bounded domain in $\mathbb{C}^N$ and f, g holomorphic maps of $\overline{D}$ into $\mathbb{C}^N$ such that $|g(z)| < |f(z)|$ for $z \in \partial D$. If f has finitely many zeros in D, then f and f+g have the same number of zeros in D, counting multiplicity.

The proof of Theorem 1 depends on the fact that compact analytic sets in $\mathbb{C}^N$ are finite; the result is used to prove the following (note that multiplicity is always taken into account):

Theorem 2 Suppose the equation f has M isolated $m\omega$-p.s., of which M_1 are m-harmonics. Then there is a neighbourhood U of 0 in $\mathfrak{F}$ such that if $g \in f+U$, g has at least M isolated $m\omega$-p.s., at least M_1 of which are m-harmonics.

Using the fact that holomorphic maps are open, it may be shown that ψ is a simple $m\omega$-p.s. of (1) if and only if the variational equations of (1) relative to ψ have no non-trivial solution of period $m\omega$. It is then proved, using Theorem 2, that a periodic solution of multiplicity k can be separated into k simple periodic solutions by means of a small change in f.

Definition 3 Let $\mathcal{S}$ be a closed subset of $\mathfrak{F}$, with the induced topology. Suppose $f_n \to f$ in $\mathcal{S}$ and $c_n \to c$ in $\mathbb{C}^N$; if, for some τ, $q_m^{(\tau)}(f_n, c_n) = 0$ $(n=1,2,\ldots)$, but $q_m^{(\tau)}(f,c)$ is not defined, $z_f(t;\tau,c)$ is said to be a singular $m\omega$-periodic solution of f with respect to $\mathcal{S}$ (abbreviated $m\omega$-s.p.s.).

Henceforth $\mathcal{S}$ is regarded as fixed.

<u>Definition 4</u> (i) $\mathcal{B}_m = \{f \in \mathcal{S} :$ f has an $m\omega$ -s.p.s.$\}$

(ii) $\mathcal{I}_m = \{f \in \mathcal{S} :$ f has infinitely many $m\omega$ -p.s.$\}$

(iii) $\mathcal{A}_m = \mathcal{S} \setminus (\mathcal{B}_m \cup \mathcal{I}_m)$

Singular periodic solutions are characterised by the property of becoming infinite both as t increases and decreases:

<u>Theorem 3</u> (1) An $m\omega$ -s.p.s. is defined for a t-interval of length at most $m\omega$.

(2) If f has an $m\omega$ -s.p.s., there are sequences (f_n), (α_n) and $\tau \in \mathbb{R}$ such that $q_m^{(\tau)}(f_n, \alpha_n) = 0$ and $\alpha_n \to \infty$.

Singular periodic solutions are unusual occurrences :

<u>Theorem 4</u> For all m, $\mathcal{A}_m \cup \mathcal{I}_m$ is dense in $\mathcal{S}$.

A restriction must now be placed on $\mathcal{S}$; it is henceforth supposed that $\mathcal{S}$ satisfies the following hypothesis.

<u>Hypothesis A</u> There is a function $\rho : \mathcal{S} \to \mathbb{R}$ such that every $m\omega$ -p.s. of f intersects the ball $B(0, \rho(f))$, and every f has a neighbourhood U_f such that $\rho[U_f]$ is bounded.

The converse of Theorem 3, (2), now holds, and the following results may be proved:

<u>Theorem 5</u> For each m, (1) $\mathcal{I}_m \subset \mathcal{B}_m$, and (2) $\mathcal{A}_m$ is open in $\mathcal{S}$.

The result that was sought strengthens Theorem 2 :

<u>Theorem 6</u> Suppose that $\mathcal{S}$ satisfies Hypothesis A and that $f \in \mathcal{A}_m$. If f has exactly M $m\omega$ -p.s., M_1 of which are m-harmonics, then there is a neighbourhood f+U of f such that every equation $g \in f+U$ has exactly M $m\omega$ -p.s., at least M_1 of which are m-harmonics.

<u>Corollary</u> All equations in the same component of $\mathcal{A}_m$ have the same number of $\lambda\omega$ -p.s. for all divisors λ of m.

Remark. In [1] the set $\mathcal{S}$ was taken to be the collection of equations

$$\dot{z} = z^n + p_1(t)z^{n-1} + \dots + p_n(t), \qquad (2)$$

with $z \in \mathbb{C}$, n fixed and $p_i(t)$ continuous, periodic and real-valued. A number of results on the periodic solutions of these equations were proved, many showing that equations of certain kinds have exactly n periodic solutions. It was shown that when n=2, (2) can have two periodic solutions, infinitely many, or none. It now transpires that $\mathcal{Q}_1$ has countably many components even when n = 2.

Reference

[1] N.G.Lloyd, 'The number of periodic solutions of the equation $\dot{z} = z^N + p_1(t)z^{N-1} + \dots + p_N(t)$'; Proc. London Math. Soc. 27 (1973), 667–700.

VOLTERRA EQUATIONS ON A HILBERT SPACE

Stig-Olof Londen

The asymptotic behavior as $t \to \infty$ of the solutions of the nonlinear Volterra equation

$$(1) \qquad x(t) + \int_0^t A(t-\tau)g(x(\tau))d\tau = f(t), \quad 0 \leq t < \infty,$$

(where $A(t)$, $f(t)$, $g(x)$ are prescribed real functions and $x(t)$ is the unknown) has recently been extensively investigated, see for example [1-3,7]. A typical result obtained in these works is the following [3]: Assume $A(t) \geq 0$ and nonincreasing on $[0,\infty)$, $A(0) < \infty$, $A(t) \in L_1[0,\infty)$, $f(t) \in C[0,\infty) \cap BV[0,\infty)$, $g(x) \in C(-\infty,\infty)$. Let $x(t)$ be a solution of (1) on $[0,\infty)$ such that $\sup_{0\leq t<\infty} |x(t)| < \infty$. Then

$$(2) \qquad \lim_{t\to\infty} [x(t) + g(x(t)) \int_0^\infty a(\tau)d\tau] = f(\infty) .$$

It is sometimes of interest to study (1), not on R, but on an arbitrary real Hilbert space H. This may be motivated by the fact that certain nonlinear partial differential equations can be put into this form, see for example [6]. In this case $x(t)$, $f(t)$: $R^+ \to H$, and $A(t)$ denotes a oneparameter family of linear symmetric mappings of $H \to H$. The integral is now to be taken as a Bochner integral. Recent work in this direction include [6].

When working in R one often divides the analysis in two parts; existence and boundedness on one hand and a more detailed asymptotic analysis on the other. Suppose we keep this distinction even when working in a Hilbert space and consider the second part. Thus assume the existence of a solution $x(t)$ of (1) such that $\sup_{0\leq t<\infty} \|g(x(t))\| < \infty$ and consider the problem of demonstrating the existence of limiting values when $t \to \infty$.

If one tries to extend to an arbitrary Hilbert space the proofs originally constructed to yield asymptotic information on (1) for $H = R$ one runs into certain difficulties centering around how to handle $g(x(t))$. When working in R one assumes $g(x) \in C(-\infty,\infty)$ which then combined with reasonable hypotheses on A and f is shown to imply uniform continuity of $g(x(t))$ with respect to t on $[0,\infty)$. This last fact often forms the starting point for the detailed asymptotic analysis, see [3]. In an arbitrary Hilbert space one may of course a priori postulate uniform continuity, [5,6]. However, if g is a nonlinear partial differential operator, then this uniform continuity may turn out to be rather hard to verify. A somewhat different approach for studying (1) on R is that created by Levin [1]. His proof is not so heavily dependent on the continuity of $g(x(t))$; it is however dependent on the ordering of the real line and may as such be hard to extend to an arbitrary Hilbert space.

The point made in the preceeding paragraph motivated recent work by the author. The aim was thus to obtain asymptotic results on (1) on a Hilbert space, postulating the existence of a solution such that $\sup_{0 \le t < \infty} \|g(x(t))\| < \infty$, but without a priori imposing any continuity of $g(x(t))$ with respect to t, and without demonstrating it in the course of the proof.

The main result we obtained is

THEOREM 1. Let, in (1), the following hold:

$x(t),\ g(x(t)) \in H$, $x(t)$ satisfies (1), $0 \le t < \infty$,

$\langle g(x(t)), x(t)\rangle \ge 0$, $0 \le t < \infty$,

$\sup_{0 \le t < \infty} \|g(x(t))\| < \infty$,

$A(t)$ is strongly continuously differentiable on R^+,

$$\int_0^\infty \|A^{(k)}(t)\|dt < \infty, \quad k = 0,1,$$

$$\langle y, \int_0^\infty A(\tau)d\tau\, y\rangle \geq \alpha\|y\|^2, \quad y \in H,$$

for some constant $\alpha > 0$. Assume there exists a nonincreasing function $\beta(t) > 0$, $0 \leq t < \infty$, such that

$$\langle y, A'(t)y\rangle \leq -\beta(t)\|y\|^2, \quad y \in H, \quad t \in R^+, \tag{3}$$

and let $\lim_{t\to\infty} \|f(t)\| = 0$. Let x,f be strongly differentiable a.e. on R^+ and suppose $\int_0^\infty \|f'(\tau)\|d\tau < \infty$.

Finally assume that there exists a locally absolutely continuous nonnegative function $G(x(t))$ such that

$$\frac{d}{dt} G(x(t)) = \langle g(x(t)), x'(t)\rangle ,$$

a.e. on R^+ and that given any $\varepsilon > 0$ there exists $\rho > 0$ such that if $G(x(t)) \geq \varepsilon$, then $\langle g(x(t)), x(t)\rangle \geq \rho$ follows. Under these conditions

$$\lim_{t\to\infty} G(x(t)) = 0 .$$

The proof of Theorem 1 works briefly as follows. Using above all (3) and borrowing certain ideas from [3] we show at first that if $\lim_{t\to\infty} G(x(t)) = 0$ does not hold, then there exist $t_n, \hat{t}_n \to \infty$ such that

(4) $\qquad G(x(t)) \geq \varepsilon, \quad \hat{t}_n \leq t \leq t_n, \quad t_n - \hat{t}_n \to \infty .$

Having (4) and observing that by the present hypothesis $\int_0^\infty \sin(\omega t)A(t)dt$ is positive definite, we apply certain Fourier transform techniques used in [4] to complete the proof.

Note that (3) - if applied to $H = R$ - of course is stronger than corresponding assumptions in recent papers. For example the result quoted above after (1) only assumed $A(t)$ nonincreasing. It is thus reasonable to expect that (3) can be weakened. Work aiming at this and at establishing applications to partial differential equations is currently in progress.

REFERENCES

[1] J.J. Levin, On a nonlinear Volterra equations, J.Math.Anal. Appl., 39 (1972), 458-476.

[2] J.J. Levin and D.F. Shea, On the asymptotic behavior of the bounded solutions of some integral equations, I, II, III, J.Math.Anal.Appl. 37 (1972), 42-82, 288-326, 537-575.

[3] S-O. Londen, On a nonlinear Volterra integral equation, J. Differential Eqs., 14 (1973), 106-120.

[4] S-O. Londen, On some integrodifferential equations in a Hilbert space, Report A 45, Institute of Mathematics, Helsinki University of Technology, 1974.

[5] R.C. MacCamy and J.S.W. Wong, Stability theorems for some functional equations, Trans. A.M.S., 164 (1972), 1-37.

[6] R.C. MacCamy, Nonlinear Volterra equations on a Hilbert space, Report 73-14, Department of Mathematics, Carnegie-Mellon University, 1973.

[7] C.C. Shilepsky, The asymptotic behavior of an integral equation with an application to Volterra's population equation , to appear.

How to get numerical convergence out of instable algorithms

K. Nickel

The following theorem is well known: consistency and global stability imply numerical convergence. Unfortunately, many algorithms which arise in the numerical solution of (ordinary or partial) differential equations are globally instable, but locally stable. Using appropriate termination criteria one finds that consistency and local stability are already sufficient for convergence. The rate of convergence is investigated. The connection to the Dahlquist theory is explained. The numerical solution of a simple initial value problem by using a highly unstable difference formula is treated as an example. These results can be found in:

K. Nickel: Stability and convergence of numeric algorithms, Parts I and II. Technical Notes BN-785 February 1974 and BN-788 March 1974. Institute for Fluid Dynamics and Applied Mathematics; University of Maryland, College Park/ Maryland, USA.

On the existence of an interface in nonlinear diffusion processes

L. A. Peletier

Let u be the solution of the heat equation

$$u_t = u_{xx}$$

in the strip $S = (-\infty, \infty) \times (0, T]$ which satisfies the initial condition

$$u(x, 0) = u_0(x) \qquad -\infty < x < \infty \tag{1}$$

in which u_0 is a given bounded, continuous function on the real line.

Suppose $u_0 \geqslant 0$ on $(-\infty, \infty)$, but $u_0 \not\equiv 0$. Then it is well known that

$$u(x, t) > 0 \qquad \text{in } S.$$

In particular, if $u(x,t)$ has compact support as a function of x at $t = 0$, then it ceases to have this property for any time $t > 0$. Thus, the heat equation describes a process in which disturbances propagate at infinite speed.

For the porous media equation

$$u_t = (u^m)_{xx} \qquad (x, t) \in S \tag{2}$$

in which $m > 1$, the situation is different. Let u be the (weak) solution of this equation which satisfies the initial condition (1). Suppose again that $u_0 \geqslant 0$ on $(-\infty, \infty)$, and that $u_0 \not\equiv 0$. Then we can only conclude that

$$u(x, t) \geqslant 0 \qquad \text{in } S.$$

Moreover, if $u(x,t)$ has compact support as a function of x at $t = 0$, then $u(x,t)$ continues to have this property for any time $t \in (0, T]$. Evidently the porous media equation describes a process in which disturbances may propagate at finite speed. In general, the transition from a region where $u > 0$ to a region where $u = 0$ is not smooth, and one can speak of an interface between such regions. For a more detailed discussion of these results, we refer to [7] .

In this note we consider the Cauchy problem

$$u_t = (D(u)u_x)_x \qquad (x,t) \in S \tag{3}$$

$$u(x,0) = u_0(x) \qquad x \in (-\infty, \infty) \tag{4}$$

in which D is a smooth, non-negative function, defined on $[0,\infty)$ and positive on $(0,\infty)$; and u_0 a smooth non-negative function with compact support in $(-\infty,\infty)$, which is not identically zero. Under these conditions a solution u of problem (3), (4) exists, and is unique [6]. By a solution, we mean here a weak solution, since, as was the case with the porous media equation, we are interested in solutions which may not be everywhere smooth.

Our object is to obtain a condition on the diffusion coefficient D which is both necessary and sufficient for equation (3) to describe a process in which the speed of propagation is finite, and hence to have solutions which exhibit an interface.

A sufficient condition for the existence of an interface has been known for some time. Oleinik, Kalashnikov and Yui-Lin [6] showed that if

$$\int_0 \frac{D(s)}{s}\, ds < \infty \tag{A}$$

then, if $u_0(x)$ has compact support, so has $u(x,t)$ for all $t \in [0,T]$, and there exists an interface. In her seminar at Moscow University, Oleinik made the conjecture that condition (A) is also a necessary one.

Recently Kalashnikov partly proved this conjecture [4]. He showed that condition (A) is indeed necessary, although he needed the additional assumption

$$\delta D(s) - s D'(s) \geq 0 \quad (s > 0) \qquad \text{for some } \delta \in (0, \tfrac{1}{3}), \tag{B}$$

in which the prime denotes differentiation. In this note we shall improve on this result, and establish the necessity of condition (A) without demanding condition (B).

Thus, we shall have completed the proof of the following result.

THEOREM. Let u be a weak solution of the Cauchy problem (3), (4) in which the function u_0 has compact support. Then $u(x,t)$ has compact support as a function of x for any $t \in [0,T]$ if and only if condition (A) is satisfied.

It is clear that the diffusion coefficient in the porous media equation, $D(s) = m s^{m-1}$ $(m > 1)$ satisfies condition (A), and that the one in the heat equation, $D(s) \equiv 1$, does not.

In proving that condition (A) is necessary for the existence of an interface we make use of the construction devised in [6] to prove existence of a weak solution $u(x,t)$. In [6] it was shown that $u(x,t)$ is the pointwise limit of a decreasing sequence $\{u_n(x,t)\}$ of classical solutions of equation (3) in a series of expanding cylindrical domains $\bar{Q}_n = [-n,n] \times [0,T]$. We recall that for each $n \geq 1$, (i) $u_n \in C^{2,1}(\bar{Q}_n)$, (ii) $u_n(x,0)$ is bounded away from zero on $[-n,n]$ and (iii) $u_n(\pm n,t) = M+\delta$ on $[0,T]$ for some $\delta > 0$, where $M = \max\{u_0(x) : -\infty < x < \infty\}$.

To prove that $u(x,t) > 0$ in S, we shall construct a sequence of functions $w_n(x,t)$, defined in $\bar{Q}_n$ with the properties:

P1. $w_n(x,t) \leq u_n(x,t)$ in $\bar{Q}_n$ for each $n \geq 1$,

P2. $w(x,t) = \lim_{n\to\infty} w_n(x,t) > 0$ for each $(x,t) \in S$.

Assuming for the moment that such a sequence has been constructed, it follows that for any $(x,t) \in S$,

$$u(x,t) = \lim_{n\to\infty} u_n(x,t) \geq \lim_{n\to\infty} w_n(x,t) = w(x,t) > 0 .$$

We now turn to the construction of the sequence $\{w_n\}$. Because $u_0 \geq 0$ on $(-\infty,\infty)$ and $u_0 \not\equiv 0$, there exists a point $x_0 \in (-\infty,\infty)$ such that $u_0(x_0) > 0$.

Without loss of generality we may assume that $x_0 = 0$. Thus $u_0(0) > 0$ and hence $u(0,t) > 0$ for all $t \in [0,T]$ [3] . Let

$$\min \{u(0,t) : \; 0 \leq t \leq T\} = 2\mu \quad .$$

Then, because $u \in C(\bar{S})$, μ is a positive number.

The construction of the functions w_n is a modification of a construction developed in an earlier paper which dealt with the mixed problem for equation (3) [5] . It exploits the properties of a class of similarity solutions of equation (3). Set $\eta = x(t+\tau)^{-\frac{1}{2}}$, where $x \geq 0$ and $\tau > 0$. Then the function $u(x,t) = f(\eta)$ is a solution of (3) if it satisfies the equation

$$(D(f)f')' + \tfrac{1}{2}\eta f' = 0 \qquad 0 < \eta < \infty \tag{5}$$

in which primes denote differentiation. As boundary values we choose

$$f(0) = \mu \,, \qquad \lim_{\eta \to \infty} f(\eta) = 0 \,. \tag{6}$$

In [1] and [2] it has been shown that problem (5), (6) has a unique positive solution on $[0,\infty)$ if and only if condition (A) is violated. We now define

$$w_n(x,t) = \begin{cases} f\{x(t+\tau_n)^{-\frac{1}{2}}\} & x \geq 0,\; 0 \leq t \leq T\,, \\ f\{-x(t+\tau_n)^{-\frac{1}{2}}\} & x \leq 0,\; 0 \leq t \leq T\,, \end{cases}$$

where τ_n is so chosen that

$$w_n(x,0) \leq u_n(x,0) \,.$$

It is shown in [5] that this is possible, and that

$$\lim_{n \to \infty} \tau_n = 0 \,. \tag{7}$$

Applying the maximum principle in the domains $(-\infty,0) \times (0,T)$ and $(0,\infty) \times (0,T)$, remembering that $w_n(0,t) = \mu < 2\mu \leq u(0,t)$, we establish property P1.

Property P2 follows from the fact that, in view of (7),

$$w(x,t) = \begin{cases} f(x\,t^{-\frac{1}{2}}) & x \geq 0, \quad 0 < t \leq T \\ f(-x\,t^{-\frac{1}{2}}) & x \leq 0, \quad 0 < t \leq T \end{cases}$$

and that $f(\eta) > 0$ for $0 \leq \eta < \infty$.

REFERENCES

1. ATKINSON, F.V. and L.A. PELETIER, Similarity profiles of flows through porous media. Arch. Rational Mech. Anal. 42 (1971) 369-379.

2. ATKINSON, F.V. and L.A. PELETIER, Similarity solutions of the nonlinear diffusion equation. Arch. Rational Mech. Anal. To appear.

3. KALASHNIKOV, A.S., The occurrence of singularities in solutions of the non-steady seepage equation. USSR Computational Math. and Math. Phys. 7 (1967) 269-275.

4. KALASHNIKOV, A.S., On equations of the nonstationary-filtration type in which the perturbation is propagated at infinite velocity. Vestnik Moskovskogo Universiteta. Matematika, 27 (1972) 45-49.

5. PELETIER, L.A., A necessary and sufficient condition for the existence of an interface in flows through porous media. Arch. Rational Mech. Anal. To appear.

6. OLEINIK, O.A., A.S. KALASHNIKOV and CHZOU YUI-LIN, The Cauchy problem and boundary problems for equations of the type of nonstationary filtration. Izv. Akad. Nauk. SSSR, ser. mat. 22 (1958) 667-704.

7. OLEINIK, O.A., On some degenerate quasilinear parabolic equations. Seminari dell' Instituto Nazionale di Alta Mathematica 1962-63, Odcrisi, Gubbio, 1964, 355-371.

Hölder Continuous Nonlinear Product Integrals

A.T.Plant

In this paper $A(t)$ ($0 \leqslant t \leqslant T$) is a set valued nonlinear accretive operator on Banach space X. Then (see for example [2] or [4]) $J_\lambda(t) = (I + \lambda A(t))^{-1}$ is single valued and nonexpansive on its domain for $\lambda > 0$. In [2] solutions of the evolution equation

$$x'(t) + A(t)x(t) \ni 0 \quad , \quad x(s) = x \tag{1}$$

are shown to have product integral representation

$$U(t,s)x = \lim_{n\to\infty} \prod_{i=1}^{n} J_{(t-s)/n}(s + i(t-s)/n)x \tag{2}$$

if the following holds:

A) D = closure domain $A(t)$ is t-independent.

B) $D \subset$ domain $J_\lambda(t)$ $(\lambda > 0)$

C) $\|J_\lambda(t)x - J_\lambda(s)x\| \leqslant \lambda\,\|f(t) - f(s)\|\, L(\|x\|)$

where L is continuous and f is X-valued and continuous.

In [4] we weakened the time dependence in C) by showing f need only be Riemann integrable. Here we show λ-dependence can be weakened * if t-dependence is strengthened. The following condition D) replaces C).

D) $\|J_\lambda(t)x - J_\lambda(s)x\| < \lambda^p\, |t-s|^q\, L(\|x\|)$ $(p < 1)$

Since $q > 1$ implies $J_\lambda(t) = J_\lambda(s)$, we assume $q \leqslant 1$. Anticipating what is to follow we also assume $r = p + q/2 - 1 > 0$, so $p > 1/2$, $q > 0$.

With $p = 1$ it is shown in [2] that $|A(t)x| = \sup_{\lambda>0} \lambda^{-1}\|x-J_\lambda(t)x\|$ is either $+\infty$ for all t or bounded for all t , the latter being the case if $x \in$ domain $A(t)$ for some t. When $p < 1$ we cannot draw

* I should like to thank Professor Tosio Kato for suggesting this possibility to me.

this conclusion, and we make the following rather weak assumption

E) There exists $x_o \in X$ such that $M(x_o) = \sup_t |A(t)x_o| < \infty$

THEOREM. Suppose A),B),D),E) hold and $r > 0$. Let x be such that $N(x) = \inf_t |A(t)x| < \infty$ (the set of such x is dense in D). Then the limit in (2) exists,

$$||U(t',s')x - U(t,s)x|| < K_x(|t'-t|^{2r} + |s'-s|^{2r/q}) \tag{3}$$

and $U(t,s)$ has unique extension to a nonexpansive evolution operator on D.

Sketch Proof. Assume $0 < \mu \leqslant \lambda$, and let $\alpha = \mu/\lambda$, $\beta = 1-\alpha$ and K be a generic constant bounded if λ, r^{-1}, T, $||x||$, $N(x)$ are all bounded. Define for $x \in D$

$$P_{\lambda,k}(s)x = \begin{cases} \prod_{i=1}^{k} J_\lambda(s+i\lambda)x & k = 1,2,\ldots\ldots \quad s + k\lambda \leqslant T \\ x & k = 0 \end{cases}$$

$$a_{k,l} = ||P_{\lambda,k}(s)x - P_{\mu,l}(t)x|| \tag{4}$$

Now using E) (this is the only place we need E)) we obtain

$$L(||P_{\mu,l}(t)x||) < K \tag{5}$$

The proof of [2;Lemma 2.2] can be modified to give

$$a_{0,l} \leqslant l\mu^p[\mu^{1-p}N(x) + T^qL(||x||)] < Kl\mu^p \quad \text{similarly} \quad a_{k,0} < Kk\lambda^p \tag{6}$$

Precisely as in [2] or [4] we use the nonlinear resolvent formula together with D) and (5) to obtain

$$a_{k,l} \leqslant \alpha a_{k-1,l-1} + \beta a_{k,l-1} + b_{kl}, \quad b_{kl} = K\mu^p|k\lambda - l\mu + s - t|^q \quad k,l > 0$$

Now suppose $(c_{k,l})$ is the solution (obviously unique) of

$$\left.\begin{aligned} c_{k,l} &= \alpha c_{k-1,l-1} + \beta c_{k,l-1} + b_{kl} \qquad k,l > 0 \\ c_{k,0} &= b_{k0} = K\mu^{p-1}k\lambda \ (\geqslant Kk\lambda^p), \quad c_{0,l} = b_{0l} = K\mu^{p-1}l\mu \end{aligned}\right\} \tag{7}$$

then (since $\alpha, \beta > 0$, $\mu \leqslant \lambda$) $a_{m,n} \leqslant c_{m,n}$.

The solution of (7) can be expressed very simply as follows. Consider the random walk on the integer lattice of the positive quadrant of the plane. The walk has initial point (m,n), and each step of the walk goes from (k,l) to either $(k-1,l-1)$ or $(k,l-1)$ with probability α, β respectively. The axes are absorbing barriers. Let p_{kl} be the probability the walk hits (k,l). Then by inspection

$$c_{m,n} = \sum_{i,j} p_{ij} b_{ij} \tag{8}$$

Let X^j be the random variable giving the number of successes in j Bernoulli trials each with probability α of success. Similarly Y^m the number of trials required to achieve m successes. (X^j has binomial distribution, Y^m negative binomial distribution [3]). Note that $E(X^j) = j\alpha$, $\sigma^2(X^j) = \mathrm{Var}(X^j) = j\alpha\beta$. Similar expressions can be found for the mean and variance of Y^m, but we don't need them.

Now
$$p_{ij} = \begin{cases} P[\, X^{n-j} = m-i \,] & i > 0 \\ P[\, Y^m = n-j \,] & i = 0 \end{cases} \tag{9}$$

We substitute (9) into (8) and make the following simplification. First $b_{0j} = K\mu^p j = K\mu^{p-1}\lambda E(X^j)$ so

$$\sum_{j>0} P[\, Y^m = n-j \,] b_{0j} = K\mu^{p-1} \sum_{j>0} P[\, Y^m = n-j \,] \sum_{i<0} P[\, X^j = -i \,] \, |\lambda i|$$
$$= K\mu^{p-1} \sum_{i<0} P[\, X^n = m-i \,] \, |\lambda i|$$

$$c_{m,n} = K\mu^{p-1} \sum_{i} P[\, X^n = m-i \,] \, |\lambda i| + \sum_{i,j>0} P[\, X^{n-j} = m-i \,] \, b_{ij}$$
$$\leqslant K\mu^{p-1} E(|\lambda(m-X^n)|) + K\mu^p \sum_{j=1}^{n} E(|\lambda(m-X^{n-j}) - j\mu + s - t|^q)$$

The estimate $E(|X|^q) \leqslant [\, \sigma^2(X) + E(X)^2 \,]^{q/2}$ $(0 < q \leqslant 2)$ (X any random variable with finite mean and variance) then gives

$$c_{m,n} \leqslant K\mu^{p-1}[n\mu(\lambda-\mu)+(m\lambda-n\mu)^2]^{1/2} + K\mu^{p-1} n\mu [n\mu(\lambda-\mu)+(m\lambda-n\mu+s-t)^2]^{q/2}$$

which simplifies to

$$||P_{\lambda,m}(s)x - P_{\mu,n}(t)x|| \leqslant c_{m,n} \leqslant K\mu^{p-1}[\ |\lambda-\mu|^{q/2}+|m\lambda-n\mu|+|m\lambda-n\mu+s-t|^q\ ] \tag{10}$$

Unfortunately we cannot use this estimate directly since μ^{p-1} blows up to $+\infty$ as $\mu\downarrow 0$. However it is not too difficult to show that (10) in fact implies (possibly with a different K)

$$||P_{\lambda,m}(s')x - P_{\mu,n}(s)x|| < K\lambda^{p-1}[\ \lambda^{q/2} + |m\lambda-n\mu| + |m\lambda-n\mu+s'-s|^q\ ] \tag{11}$$

From (11) it follows at once that $P_{\lambda,m}(s)x$ is Cauchy if $|m\lambda-t+s| = o(\ \lambda^{(1-p)/q}\ )$ as $\lambda\downarrow 0$. Since D is complete this gives (2) as a special case and also shows U(t,s) is nonexpansive and can be extended to an evolution operator on D.

It now follows from (11) that

$$||P_{\lambda,m}(s')x - U(t,s)x|| \leqslant K\lambda^{p-1}[\ \lambda^{q/2} + |m\lambda+s-t| + |m\lambda+s'-t|^q\ ]$$

$$||P_{\lambda,m}(s')x - U(t',s')x|| \leqslant K\lambda^{p-1}[\ \lambda^{q/2} + |m\lambda+s'-t'| + |m\lambda+s'-t'|^q\ ] \tag{12}$$

Now choose m so that $|m\lambda+s'-t'| \leqslant \lambda$, then

$$\begin{aligned} ||U(t',s')x - U(t,s)x|| &\leqslant K\lambda^{p-1}[\ \lambda^{q/2}+|t'-s'+s-t|+\lambda+(|t'-t|+\lambda)^q] \\ &\quad +K\lambda^{p-1}[\ \lambda^{q/2} + \lambda + \lambda^q\ ] \\ &\leqslant K\lambda^{p-1}[\ \lambda^{q/2} + |s'-s| + |t'-t|^q\ ] \end{aligned}$$

and set $\lambda = (\ |s'-s| + |t'-t|^q\ )^{2/q}$ to obtain

$$||U(t',s')x - U(t,s)x|| < K(\ |s'-s| + |t'-t|^q\ )^{2r/q}$$

which gives (3) and completes the sketch proof.

COROLLARY. <u>The rate of convergence of</u> (2) <u>is given by</u> (12) <u>as</u>

$$||P_{(t-s)/n,n}(s)x - U(t,s)x|| = O((\tfrac{t-s}{n})^r) \qquad \text{if } N(x) < \infty.$$

Remarks. Clearly most of the estimates made in the proof of the Theorem can be tightened, and it is possible to obtain an exact expression for the quantity $K = K_x$. The Theorem also generalises in the usual way to the case $A(t)$ is ω-accretive (i.e. $J_\lambda(t)$ has Lipschitz norm $(1-\lambda\omega)^{-1}$ for $0<\lambda$, $\lambda\omega<1$).

If problem (1) has a strong solution $x(t)$ then the proof of [2;Theorem 3.1] needs only slight modification to show $U(t,s)x = x(t)$, so we may regard $U(t,s)x$ in the general case to be a weak or generalised solution of (1).

REFERENCES

[1] M.G. Crandall and T.M. Liggett, "Generation of semi-groups of nonlinear transformations on general Banach spaces", Amer. J. Math. 93 (1971) 265-298.

[2] M.G. Crandall and A. Pazy, "Nonlinear evolution equations in Banach spaces", Israel J. Math. 11 (1972) 57-94.

[3] W. Feller, "An introduction to probability theory and its applications", Vol.1 (third edition), John Wiley & Sons.

[4] A.T. Plant, "The product integral method for nonlinear evolution equations", Technical Report No. 23, Control Theory Centre, University of Warwick, England.

CONTINUOUS SPECTRA AND LINEAR OPERATOR EQUATIONS

G. F. Roach

1. INTRODUCTION

Let L be a linear partial differential expression of order 2m, where m is a positive integer. Consider the boundary value problem

$$Lu - \lambda u = f, \quad \lambda \in \mathbb{C} \tag{1.1}$$

$$u \in (bc) \tag{1.2}$$

where (1.1) is defined in an open, bounded, connected region $G \subset E^n$ and (1.2) indicates that solutions, u, of (1.1) must satisfy certain boundary conditions, denoted typically by (bc), on the closed, smooth boundary ∂G of G.

Let $H_o \equiv H_o(G)$ denote a complex Hilbert space, whose elements are functions defined over G, and introduce a linear operator $A : H_o \to H_o$ defined by

$$Au = Lu, \quad u \in D(A) \tag{1.3}$$

$$D(A) = \{u \in H_o : Au \in H_o, \quad u \in (bc)\} \tag{1.4}$$

The boundary value problem (1.1), (1.2) has the following realisation as an operator equation in H_o

$$(A-\lambda)u = f \in H_o, \quad u \in D(A) \subset H_o . \tag{1.5}$$

The theory of such equations is well developed if A is assumed to be compact but not so otherwise.

2. DEFINITIONS AND ASSUMPTIONS

Let H_o be a complex Hilbert space with structure $(.,.)_o$, $||.||_o$. Given a linear operator $A : H_o \to H_o$ a scalar $\lambda \in \mathbb{C}$ belongs to $\rho(A)$, the resolvent set of A, if $\overline{R(A-\lambda)} = H_o$ and there exists a constant $c > 0$ such that $||(A-\lambda)u||_o \geq c||u||_o$, $u \in D(A)$. The spectrum of A is then $\sigma(A) = \mathbb{C}\setminus\rho(A)$. When $A \in K(H_o)$, the set of all compact linear operators mapping H_o into itself, then $\sigma(A)$ consists of a countable number of isolated eigenvalues of finite multiplicity. When $A \notin K(H_o)$ the spectrum of A contains more than eigenvalues. In particular it can contain a subset, $\sigma_c(A)$, the continuous spectrum of A. A scalar $\lambda \in \sigma_c(A)$ if (i) $\overline{R(A-\lambda)} = H_o$ and (ii) $R_\lambda \equiv (A-\lambda)^{-1}$ exists as an unbounded operator on $R(A-\lambda)$. Consequently, for $A \notin K(H_o)$, $\lambda \in \sigma_c(A)$ equation (1.5) has a solution $u = R_\lambda f$ provided $f \in R(A-\lambda)$. However, this characterisation of the class of given functions is inadequate if it is only known that $f \in H_o$: denseness arguments being invalid due to the unboundedness of R_λ.

In the analysis of (1.5) the following assumptions are made:

A1 $\overline{D(A)} = H_o$

A2 A is self adjoint with respect to the structure of H_o

A3 $A \notin K(H_o)$ and $\sigma_c(A)$ is non empty

A4 $\lambda \in \sigma_c(A)$

A5 The Hilbert space H_o has a dense subspace H_+ which is a Hilbert space with respect to the structure $(.,.)_+$, $||.||_+$, where $||u||_o \leq ||u||_+$, $u \in H_+$ and $H_+ \subseteq H_o$, the inclusion being dense.

As a consequence of A5 it is possible to construct a Hilbert space H_- with structure $(.,.)_-$, $||.||_-$ such that

$$H_+ \subseteq H_o \subseteq H_- \tag{2.1}$$

$$||u||_+ \geq ||u||_o \geq ||u||_-, \quad u \in H_+$$

the inclusions being dense. In this case H_o is said to be equipped with a positive space H_+ and a negative space H_-.

3. SELF-ADJOINT CASE

The following Theorem provides a basis for much of the subsequent analysis.

Theorem 3.1. Let H_o be a Hilbert space equipped as in (2.1). Let T be a linear operator, with domain $D(T) \subseteq H_+$, acting in H_o and satisfying

$$|(Tu,u)_o| \geq c_1||u||_+^2, \quad c_1 > 0, \ u \in D(T) \tag{3.1}$$

$$|(Tu,v)_o| \leq c_2||u||_+ \, ||v||_+, \quad c_2 > 0, \quad u,v \in D(T). \tag{3.2}$$

Assume that D(T), in general not dense in H_+, is dense in H_o. Let $H'_+ \subseteq H_+$ denote the closure of D(T) in H_+ and form the negative space H'_- with respect to H_o and H'_+. The operator T, considered as an operator from H'_+ to H'_- admits a closure Λ which is a homeomorphism between all of H'_+ and all of H'_-.

To be able to use this Theorem in its present form an additional assumption must be made.

A6 There exists a real scalar $\lambda_1 \in \rho(A)$ such that the self-adjoint operator $T \equiv (A-\lambda_1)$ satisfies (3.1) and (3.2).

On the basis of A6 take Λ to be the closure of T, considered as acting from H'_+ to H'_- and let $B : H_o \to H_o$ denote the restriction of Λ to D(B) where

$$D(B) = \{u \in H'_+ : \Lambda u \in H_o\} \tag{3.3}$$

Introduce the imbedding maps

$$O_1 : H_o \to H'_- \quad \text{and} \quad O_2 : H'_+ \to H_o \tag{3.4}$$

and notice that

$$B^{-1} = O_2 \Lambda^{-1} O_1 \tag{3.5}$$

Since Λ^{-1} is a bounded linear operator it follows that if O_1 and O_2 are compact then so also is B^{-1}.

Recall that a linear operator $F : H_1 \to H_2$, where H_1, H_2 are Hilbert spaces, is a Fredholm operator if: (i) $\overline{D(F)} = H_1$; (ii) $\alpha(F) = \dim N(F) < \infty$; (iii) $\beta(F) = \dim N(F^*) < \infty$; (iv) R(F) closed in H_2, where N(F) denotes the null space of F. The set of all such operators is denoted by $\Phi\ (H_1, H_2)$ with $\Phi\ (H_1, H_1) \equiv \Phi\ (H_1)$. Furthermore, i(F), the index of an operator $F \in \Phi\ (H_1, H_2)$ is defined by

$$i(F) = \alpha(F) - \beta(F).$$

The Riesz-Schauder theory of compact operators demonstrates that if $B^{-1} \in K(H_o)$ then $(\mu - B^{-1}) \in \Phi\ (H_o)$ and $i(\mu - B^{-1}) = 0$.

With this notation equation (1.5) can be written in the form

$$(B-\eta)u = f \in H_o, \quad \eta = \lambda - \lambda_1, \quad u \in D(B) \subseteq H'_+ \tag{3.6}$$

Furthermore, since B^{-1} exists and is a left inverse of B, equation (3.6) can be reduced to the equivalent form

$$(\mu - B^{-1})u = \mu B^{-1} f \in H_o, \quad \mu = \eta^{-1}, \quad u \in D(B) \subseteq H'_+ \tag{3.7}$$

Notice that since $D(A) \subseteq D(B)$ the operator B is an extension of the operator A. Furthermore, since D(A) is dense in H_o it follows that D(B) is dense in H_o. The above results can be summarised as follows.

Theorem 3.2. Let H_o be a Hilbert space equipped as in (2.1). Let $A : H_o \to H_o$ be a linear operator with $D(A) \subseteq H_+$. Assume that there exists a real scalar $\lambda_1 \in \rho(A)$ such that the operator $T \equiv (A - \lambda_1)$ satisfies (3.1) and (3.2) for all $u, v \in D(A) = D(T)$. Assume that D(T), in general not dense in H_+, is dense in

H_o. Let $H'_+ \subseteq H_+$ denote the closure of D(T) in H_+ and construct the negative space H'_- with respect to H_o and H'_+. If the imbedding operators $O_1 : H_o \to H'_-$ and $O_2 : H'_+ \to H_o$ are compact then T admits an extension $B : H_o \to H_o$, defined as a restriction of a homeomorphism $A : H'_+ \to H'_-$ to $D(B) = \{u \in H'_+ : Au \in H_o\}$, with the properties

(i) $B^{-1} = O_2 A^{-1} O_1 \in K(H_o)$

(ii) $(\mu - B^{-1}) \in \Phi(H_o)$, $(\bar{\mu} - (B^{-1})^*) \in \Phi(H_o)$

Hence the equations

$$(B-\eta)u = f \in H_o, \quad (B^*-\bar{\eta})v = g \in H_o$$

form a Fredholm pair and B is a Fredholm extension of T.

It should be remarked that A3 precludes the possibility of A^{-1} being compact. Indeed, since A is densely defined and self-adjoint it must be a closed operator and the assumption $A^{-1} = (A-\lambda_1)^{-1} \in K(H'_-, H'_+)$ implies that A is a closed operator with compact resolvent. This in turn implies that A has a spectrum consisting entirely of isolated eigenvalues at finite multiplicity which contradicts A3.

4. NON-SELF-ADJOINT CASE

Although A has been assumed self-adjoint, in general the operator $T = A-\lambda$, $\lambda \in \rho(A) \subset \mathbb{C}$ is not self-adjoint. A partial resolution of this difficulty is effected by introducing the operator T^+, the restriction to H_+ of T^* the Hilbert space adjoint of T in H_o and assuming

A7 $D(T^+)$ is dense in H_o

A8 $D(T)$ and $D(T^+)$ are "close" in sense that they have a common closure in H_+

As a consequence of A8 the inequalities (3.1) and (3.2) are equivalent to

$$|(v,T^+v)_o| \geq c_3 \, ||v||_+^2, \quad c_3 > 0, \quad v \in D(T^+) \tag{4.1}$$

$$|(u,T^+v)_o| \leq c_4 \, ||u||_+ \, ||v||_+, \quad c_4 > 0, \quad u \in D(T), \; v \in D(T^+) \tag{4.2}$$

Therefore Theorems 3.1 and 3.2 continue to hold in this more general situation and similar results to those already obtained can be established.

5. UNBOUNDED REGIONS

When the region $G \subset E^n$ is unbounded problems relating to the imbedding of function spaces arise. To illustrate these consider the particular case when the Hilbert spaces H_o, H_+, H_- are respectively the Sobolev spaces $W_2^o(G) = L_2(G)$, $W_2^m(G)$, $W_2^{-m}(G)$. If $m > n/2$ then the functions in $W_2^m(G)$ are continuous in $\bar{G} = G \cup \partial G$ and by virtue of the Imbedding Theorem the estimate

$$|u(x)| \leq C(x) \, ||u||_m, \quad u \in W_2^m(G), \quad x \in G$$

can be obtained, where $C(x) \geq \delta > 0$, $x \in G$ is a function which grows large as $x \to \infty$. When G is a conical region it can be shown that

$$C(x) = K(1+|x|^{m-n/2}), \quad K > 0, \quad x \in G.$$

An operator $T : H_1 \to H_2$ is a Hilbert-Schmidt operator if for some, and hence all, orthonormal bases $\{e_k\}_{k=1}^{\infty}$ of H_1 the series $\sum_{k=1}^{\infty} ||Te_k||_{H_2}^2$ converges. If $H_1 \subseteq H_2$ the inclusion is quasi-nuclear (q.n) if the inclusion operator is Hilbert-Schmidt.

It can be shown that a Hilbert-Schmidt operator is compact. Furthermore, for H_o equipped as in (2.1) it can be established that the inclusion $H_+ \to H_o$ is q.n iff the inclusion $H_o \to H_-$ is q.n.

When G is unbounded the inclusion $W_2^m(G) \to L_2(G)$, $m > n/2$, is not q.n. To overcome this a positive space is introduced in a different way. Let $q \in C^m(\bar{G})$ be a fixed function satisfying $q(x) \geq 1$ for all $x \in \bar{G}$. Define $W_2^{\sigma}(G)$, $\sigma = (m,q)$, as the closure, with respect to the inner product $(u,v)_{\sigma} = (qu,qv)_m$, of all functions in $W_2^m(G)$ which vanish in a neighbourhood of infinity. Since $||u||_o \leq ||u||_{\sigma}$ the space $W_2^{\sigma}(G)$ can be taken as a positive space relative to a zero space $L_2(G)$; the associated negative space being denoted by $W_2^{-\sigma}(G)$. Compact imbeddings can now be obtained as a consequence of

<u>Theorem 5.1.</u> Assume $m > n/2$. If the function q satisfies

$$\int_G \frac{C^2(x)}{q^2(x)} \, dx < \infty \tag{5.1}$$

then the inclusion

$$W_2^{\sigma}(G) \to L_2(G), \quad \sigma = (m,q)$$

is quasi-nuclear.

For a conical region it is easily seen that the integral (5.1) converges with $q(x) = (1+|x|^{m+\varepsilon})$, $m > n/2$, $\varepsilon > 0$. Consequently results similar to those obtained for bounded regions can be obtained for unbounded regions provided weighted spaces of the form $W_2^{\sigma}(G)$, $\sigma = (m,q)$ are used.

6. EXAMPLE

Consider the differential expression L defined by

$$(Lu)(x) = -u''(x) + p(x)u(x), \quad 0 < x < \infty \tag{6.1}$$

where p is a real valued function, summable on every interval $(0,\alpha)$, $0 < \alpha < \infty$ and decreasing sufficiently rapidly as $x \to \infty$.

Let G denote the interval $(0,\infty) \subset \mathbb{R}$ and set

$$H_o = L_2(G), \quad H_+ = W_2^2(G), \quad H_- = W_2^{-2}(G).$$

Further, let F denote the set of all finite functions defined on G, that is the set of all functions which vanish in a neighbourhood of $x = 0$ and $x = \infty$. Associated with (6.1) and the boundary condition

$$u(0) = 0 \tag{6.2}$$

an operator $A_1 : H_o \to H_o$ defined by

$$A_1 u = Lu, \quad u \in D(A_1)$$

$$D(A_1) = \{u \in H_+ \cap F : Lu \in H_o\}.$$

The following results can be established [2].

1. L is a self-adjoint differential expression with respect to the structure of H_o
2. $\overline{D(A_1)} = H_o$
3. A_1 is a linear, self-adjoint operator in H_o
4. A_1 has no positive eigenvalues.
5. The set of eigenvalues of A_1 is bounded, at most countable and its limit points can only lie on the semi-axis $\lambda \geq 0$.
6. If for some $\gamma > 0$ the integral
$$\int_x^\infty \xi^{\gamma+1} |p(\xi)| d\xi$$
converges then $\lambda = 0$ is not an eigenvalue of A_1.
7. Every $\lambda > 0$ (and if 6 holds, $\lambda \geq 0$) belongs to the continuous spectrum of A_1.

Now consider the operator equation

$$(A_1 - \lambda)u = f \in H_o, \quad u \in D(A_1), \quad \lambda \in \sigma_c(A_1).$$

Direct calculation reveals that A_1 is a positive operator on H_o and H_+ and furthermore the operator

$$T_1 = A_1 - \lambda_1, \quad \lambda_1 = -|\lambda_1| \in \rho(A_1)$$

satisfies the inequalities (3.1) and (3.2). However, when G is unbounded the imbeddings $H_+ \to H_0$, $H_0 \to H_-$ are not compact and Theorem 3.2 is not available. This situation can be improved by specifying more precisely the behaviour of the solutions at infinity. This will mean dealing with a subset of $D(A_1)$ and consequently a different operator realisation of boundary value problems associated with (6.1) and (6.2) will be obtained. To this end let q be a sufficiently smooth, fixed function satisfying $q(x) \geq 1$ for all $x \in \bar{G}$. Define an operator

$$A : H_0 \to H_0 \quad \text{by}$$

$$Au = Lu, \quad u \in D(A)$$

$$D(A) = \{u \in D(A_1) : qu \in H_+\}.$$

It is clear that the results obtained above for the operator A_1 also apply to the operator A. However, the advantage in this case is that H_+', the closure of D(A) with respect to the norm induced by the inner product $(u, v)_+ = (u, v)_2$, $u,v \in D(A)$, is the same as the closure of $D(A_1)$ with respect to the norm induced by the inner product $(qu, qv)_2$, $u,v \in D(A_1)$. Therefore H_+' is a proper subset of $W_2^\sigma(G)$, $\sigma = (2,q)$, and furthermore it is dense in H_0. Compactness of the appropriate embeddings can be obtained by ensuring that q satisfies (5.1)

The original problem now reduces to the form (3.7) where the operator B^{-1}, being the resolvent R_{λ_1} of $(A-\lambda_1)$, exists and has the form

$$R_{\lambda_1} f(x) = \int_0^\infty G(x,y\,;\,\lambda_1)\, f(y)dy$$

where $G(x,y\,;\,\lambda_1)$ is the appropriate Green's function.

REFERENCES

1. I.M. Gelfand and N. Ja. Vilenkin, Generalised Functions Part IV Academic Press, New York 1964.

2. M.A. Naimark, Linear Differential Operators Part II Ungar, New York, 1968.

ASYMPTOTIC LINEARITY AND A CLASS OF NONLINEAR STURM-LIOUVILLE PROBLEMS ON THE HALF LINE

by

J.F. Toland

Let X be a Banach space, and $\mathbb{R}$ the real line. We consider the eigenvalue problem

$$u = \lambda G(\lambda,u) \qquad , (\lambda,u) \in \mathbb{R} \times X \tag{1}$$

where $G : \mathbb{R} \times X \to X$ is a continuous mapping such that

(H1) $G(\lambda,u) = \lambda Au + R(\lambda,u)$ for all $(\lambda,u) \in \mathbb{R} \times X$,

(H2) $A : X \to X$ is a bounded linear operator

(H3) $R : \mathbb{R} \times X \to X$ is such that $\lim_{||u|| \to \infty} ||R(\lambda,u)||/||u|| = 0$ uniformly on bounded λ - intervals.

(H4) $G(\lambda\cdot)$ is a k-set contraction for each $\lambda \in \mathbb{R}$.

In [1] the author considered (1) when $G : \mathbb{R} \times X \to X$ is compact (i.e. $k = 0$ in (H4)). In this note we extend the abstract results of [1] to the case of $k = 0$, and we shall merely indicate how these results can find application to the theory of nonlinear ordinary differential equations on $[0,\infty)$.

Our results depend on the following observation. If G satisfies (H1) - (H4) above we can define a mapping $\tilde{G} : \mathbb{R} \times X \to X$ as follows

$\tilde{G}(\lambda,u) = ||u||^2 G(\lambda,u/||u||^2)$ for all $u(\neq \theta)$ in $X, \lambda \in \mathbb{R}$

$\tilde{G}(\lambda,\theta) = \theta$, for all $\lambda \in \mathbb{R}$.

<u>Theorem 1.</u> If $G : \mathbb{R} \times X \to X$ satisfies (H1)-(H4) above, then G is a continuous map from $\mathbb{R} \times X \to X$ such that (i) λA is the Fréchet derivative, at θ, of $G(\lambda,)$ for each $\lambda \in \mathbb{R}$, (ii) (λ,u) is a non-trivial $(u \neq \theta)$ solution of (1) if and only if (λ,ν) is a non-trivial solution of

$$\nu = G(\lambda,\nu) \tag{2}$$

where $\nu = u/||u||^2$, (iii) $G(\lambda,)$ is a k-set contraction with the same k as in (H4), for each $\lambda \in \mathbb{R}$.

(i) and (ii) of Theorem 1 are immediate from (H1)-(H4). To prove (iii) we we shall need the following result from the theory of k-set contractions, (e.g. see [4]).

<u>Lemma 1</u> Let $T : D(X) \to X$ be a k-set contraction and let $f : D \to R^+$ be a continuous function such that $\sup\{f(x) : x \in D\} = \lambda < \infty$. Define $T^1 : D \to X$, by $T^1(x) = f(x)\, Tx$ for all $x \in D$. Then T^1 is a k -set contraction.

<u>Proof of Theorem 1 (iii)</u> ($\alpha(\Omega)$ shall denote the measure of noncompactness of a bounded subset Ω of X, see [3]). If Ω is a closed bounded subset of X with $\alpha(\Omega) = d$ it will suffice to show that $G(\Omega)$ can be covered by a finite number of sets of diameter less than, or equal to $kd + \varepsilon$, for arbitary $\varepsilon > 0$.

Given $\varepsilon > 0$ let $B_{(kd+\varepsilon)/2}$ denote that ball, centred at Θ, and the radius $(kd + \varepsilon)/2$. Let $\Omega^1 \equiv \Omega \setminus \{\tilde{G}^{-1}(B_{(kd+\varepsilon)/2} \cap G)$. So it suffices to show that $\tilde{G}(\Omega^1)$ can be covered by a finite number of sets of diameter less than, or equal to $kd + \varepsilon$. Since Ω^1 is a closed, bounded subset of X which does not contain Θ, $I = \inf\{||u|| : u \in \Omega^1\} > 0$.

Choose $\eta > 0$ such that $\eta < \varepsilon I/2d$

Then $$I + \eta < I + I\varepsilon/2d$$

Hence, for any non-negative integer m

$$\frac{I + (m+1)\eta}{I + m\eta} \leqslant 1 + \varepsilon/2d$$

For each $m \geqslant 0$, put

$$E_m = \{u \in X : I + m\eta < ||u||^2 < I + (m+1)\eta\}.$$

Clearly there exists $M > 0$ such that $\Omega^1 \subset \bigcup^{M} E_m$, and so it suffices to show that $\tilde{G}(\Omega^1 \cap E_m)$ can be covered by a finite number of sets of diameter less than, or equal to $k\,d + \varepsilon$, for $m = 0, \ldots, M$.

For $m \geqslant 0$, $\tilde{G}(\Omega^1 \cap E_m) = \{||u||^2 G(u/||u||^2) : u \in \Omega^1 \cap E_m\}$. Hence, by Lemma 1,

$$\alpha G(\Omega^1 \cap E_m) \leqslant (I + (m+1)\eta)\, \alpha\{G(u/||u||^2) : u \in \Omega^1 \cap E_m\}.$$

But $\{u/||u||^2 : u \in \Omega^1 \cap E_m\}$ is a bounded subset of X, and so

Therefore $\alpha\tilde{G}(\Omega' \cap E_m) \leq \dfrac{k(I+(m+1)\eta}{I+m\eta}\,\alpha(\Omega' \cap E_m)$

$$\leq k(1+\varepsilon/2d)d = kd + \varepsilon.$$

This completes the proof of the theorem.

Definition: A point $\alpha \in \mathbb{R}$ is an asymptotic bifurcation point of (1) if and only if there exists a sequence $\{(\lambda_n, u_n)\} \subset \mathbb{R} \times X$ of solutions of (1) with $\lambda_n \to \alpha$, $||u_n|| \to \infty$ as $n \to \infty$.

Theorem 1 implies that α is an asymptotic bifurcation point for (1) if and only if α is a bifurcation point for (2). We adopt the following convention in which to state our results.

If Ω_1 and $\Omega_2 \subseteq \mathbb{R} \times X$ are connected sets which consist of non-trivial solutions of (1) such that there exists sequences $\{(\alpha_n, u_n)\}$ and $\{(\beta_n, v_n)\}$ in Ω_1 and Ω_2 respectively, with $\alpha_n \to \gamma$, $\beta_n \to \gamma$ and $||u_n|| \to \infty$, $||v_n|| \to \infty$ as $n \to \infty$ we shall say that $\Omega_1 \cup \Omega_2$ is a continuum of solutions of (1) in $\mathbb{R} \times X$ which bifurcates asymptotically near γ.

The main result is now immediate.

Theorem 2 Let $G : \mathbb{R} \times X \to X$ satisfy (H1)-(H5). Suppose $|\mu| < k^{-1}$, and μ is a characteristic value of odd multiplicity of A, then there exists a continuum $\mathcal{C}_\mu$ of solutions of (1) which bifurcates asymptotically near μ and for which at least one of the following holds.

(i) $\mathcal{C}_\mu$ bifurcates asymptotically near $\hat{\mu}$, where $\hat{\mu} \neq \mu$, and $\hat{\mu}$ is a characteristic value of A

(ii) $\inf\{||u|| : (\lambda, u) \in \mathcal{C}_\mu \cap ([-k^{-1}, k^{-1}] \times X\} = 0$

(iii) $\inf\{|\lambda \pm k^{-1}| : (\lambda, u) \in \mathcal{C}_\mu\} = 0.$

Proof: We apply the main theorem of Stuart [2] to equation (2) and interpret the result in the light of Theorem 1 above.

Applications.

Consider
$$\left.\begin{aligned} -(p(x)u'(x))' + q(x)u(x) &= \lambda\{u(x) + g(x)f(u(x))\}, \quad x \in (0,\infty) \\ u(0) &= 0 \end{aligned}\right\} \quad (3)$$

where (I) $p : [0,\infty) \to R$ is continuously differentiable in $[0,\infty)$ with p' bounded and $0 < P_1 \leq p(x) \leq P_2 < \infty$ for all $x \in [0, \infty)$

(II) $q : [0,\infty) \to R$ is continuous, with $0 < Q_1 \leq q(x) \leq Q_2 < \infty$ for all $x \in [0,\infty)$.

(III) f is continuously differentiable from $\mathbf{R} \to \mathbf{R}$, and there exists $P > 0$ and $K > 0$ such that $|f(p)| < K|p|^r$ for all $p > P$ and some $r < 1$.

(IV) $g \in H_0^{\,1}\ (0,\infty)$.

Remark: These are far from the most general hypotheses under which the following results hold but they serve to illustrate the type of application which is possible.

Result A. If $\mathcal{A}$ denotes the differential operator, whose domain $D(\mathcal{A})$ is the set of twice continuously differentiable functions with compact support in $(0,\infty)$, which is given by $\mathcal{A}u = -(p(x)u'(x))' + q(x)u(x)$, for all $x \in (0,\infty)$ and such a functions $u \in D(\mathcal{A})$, then $\mathcal{A}$ has self-adjoint extension A in L^2 with $D(A) = H_0^1 \cap W^{2,2}$ and $\sigma_e(A) \subset [Q,\infty)$ where $Q = \liminf_{x\to\infty} q(x)$ and $\sigma_e(A)$ denotes the essential spectrum of the closed, densely defined linear operator A. If $\mu \in \sigma(A)$ and $\mu < Q$ then μ is an eigenvalue of A of multiplicity 1. A^{-1} exists and is a bounded linear operator from all of L^2 into itself with $\gamma(A^{-1}) \leq Q^{-1}$. A is a positive self-adjoint operator in L^2. Furthermore $A^{\frac{1}{2}}$ is a linear homeomorphism of H_0^1 onto L^2 and $k_1||u|| < |||A^{-\frac{1}{2}}u||| < k_2||u||$ for all $u \in L^2$. There exists a constant $C > 0$ such that, for $\xi > C$, the operator $(A+\xi)^{-1}$ is defined on all of L^2 and is a linear homeomorphism from L^2 onto $H_0^1 \cap W^{2,2}$. Also $(A^{\frac{1}{2}})^{-1}$ is a bounded linear operator from all of L^2 into itself and $\gamma((A^{\frac{1}{2}})^{-1}) \leq Q^{-\frac{1}{2}}$.

For $u : [0,\infty) \to \mathbf{R}$ and $x \in [0,\infty)$ let us define H as follows

$$(Hu)\ (x) = g(x)\ f(u(x))$$

We state, without proof the following result

Lemma 2. Let f and g satisfy (III) and (IV), then the operator $u \quad A^{-\frac{1}{2}}HA^{-\frac{1}{2}}u$ takes L^2 into itself, is compact and continuous.

Furthermore, $\lim_{\|u\|\to\infty} ||A^{-\frac{1}{2}}\ HA^{-\frac{1}{2}}u||_2/||u||_2 = 0.$

It is easy to see that (λ,u) is a classical solution of (3) if and only if $u = A^{-\frac{1}{2}} \nu$ where $(\lambda,\nu) \in \mathbb{R} \times L^2(0,\infty)$ and (λ,ν) is a solution of

$$\nu = \lambda A^{-1}\nu + \lambda A^{-\frac{1}{2}} H A^{-\frac{1}{2}} \nu .$$

Thus we see that we have reduced problem (3) to an abstract equation of type (1). More complete results on the application of the results of section (1) are in preparation.

REFERENCES

1. J.F. Toland "Asymptotic Linearity and nonlinear Eigenvalue Problems" Quart. Jour. Math. 24.(94) 1973.

2. C.A. Stuart "Some Bifurcation theory for k-set contractions" Proc. Lond. Math. Soc (3) 27 (1973).

3. R.D. Nussbaum "The Fixed Point Index for Local Gondensing maps" Annol di Math (IV) vol LXXXIX.

4. M.A. Krasnasel'skii "Topological Methods in the Theory of nonlinear integral equations" Pergamon 1964.

5. A.J.B. Potter "A Fixed Point Theorem for k-set contractions" (to appear).

An Elementary Calculus for Green's Functions Associated with Singular Boundary Value Problems

by

Ulrich Trottenberg

1. Introduction

Proving non-negativity or total positivity of Green's functions associated with regular linear ordinary differential operators of order n is in general difficult. (As for methods concerning these subjects, see [3] and [2], for example.) Only in very special cases can these properties be proved by representing the considered operator - together with the boundary conditions - as a product of regular differential operators of lower order. A much wider class of operators can however be covered if one admits representations with singular differential operators. With the calculus of singular differential operators presented here (section 3), one can easily derive sufficient conditions for the desired properties. In illustration, we consider the following very simple but characteristic example.

Example. For the Green's function G of $M = L|_R$, with

$$Lu = u^{iv} \quad \text{on} \quad [0,1], \qquad R = \{u \in C_4[0,1] : u(0) = u'(0) = u(1) = u'(1) = 0\},$$

a representation

$$(1) \qquad G(x,y) = \int_0^1 G_1(x,t)\, G_2(t,y)\, dt \qquad (0 < x,y < 1)$$

with Green's functions G_1, G_2 of second order differential operators $M_1 = L_1|_{R_1}$, $M_2 = L_2|_{R_2}$ is only possible if one allows L_1, L_2 to be singular formal differential operators. For instance, this representation - as an improper integral - and the relation $(Lu)(x) = (L_2L_1u)(x)$ $(0<x<1;\ u \in C_4(0,1))$ hold with

$$(2) \qquad G_1(x,y) = \begin{cases} u_1(x)\, v_1(y) \,/\, w_1(y) & (0 < x \leq y < 1) \\ v_1(x)\, u_1(y) \,/\, w_1(y) & (0 < y \leq x < 1), \end{cases}$$

$$(3) \qquad G_2(x,y) = \begin{cases} \hat{u}_2(x)\, \hat{v}_2(y) \,/\, \hat{w}_2(y) & (0 < x \leq y < 1) \\ \hat{v}_2(x)\, \hat{u}_2(y) \,/\, \hat{w}_2(y) & (0 < y \leq x < 1), \end{cases}$$

$(L_1u)(x) = W(u_1,v_1,u)(x)/w_1(x)$, $(L_2u)(x) = W(\hat{u}_2,\hat{v}_2,u)(x)/\hat{w}_2(x)$ $(0<x<1)$, [1]

where

(4) $w_1 = W(u_1,v_1)$, $\hat{u}_2 = L_1u_2$, $\hat{v}_2 = L_1v_2$, $\hat{w}_2 = W(\hat{u}_2,\hat{v}_2)$

and $u_1(x) = x^3$, $v_1(x) = (1-x)^3$, $u_2(x) = x^2$, $v_2(x) = (1-x)^2$. (Also see section 4, where we treat a some more general case. For the explicit form of L_1 and R_1 see section 3.) Because $-G_1$ and $-G_2$ can easily be proved to be totally positive, in account of (1), G is totally positive too.

2. The Calculus in the Regular Case

Let $M = L|_R$ be an invertable differential operator on a compact interval $I = [a,b]$ $(-\infty < a < b < \infty)$, where

(5) $$(Lu)(x) = u^{(n)} + a_{n-1}(x)\, u^{(n-1)} + \dots + a_0(x)\, u \qquad (x \in I)$$

with

(6) $$a_k \in C(I) \qquad (k = 0,1,\dots,n-1)$$

and $R = R^a \cap R^b$ is characterized by $n = p + q$ separate boundary conditions at $x = a$ and $x = b$:

$$R^a = \left\{u \in C_n(I):\ \sum_{k=0}^{n-1} \alpha_{ik}\, u^{(k)}(a) = 0 \quad (i = 1,\dots,p)\right\},$$

$$R^b = \left\{u \in C_n(I):\ \sum_{k=0}^{n-1} \beta_{jk}\, u^{(k)}(b) = 0 \quad (j = 1,\dots,q)\right\}.$$

Because M is invertable there exist $n = q + p$ linear independent solutions $u_1,\dots,u_q,v_1,\dots,v_p$ of the equation $(Lu)(x) = 0$ $(x \in I)$ such that

$$u_j \in R^a \quad (j = 1,\dots,q), \qquad v_i \in R^b \quad (i = 1,\dots,p).$$

Proposition. With such functions u_j, v_i $(j = 1,\dots,q;\ i = 1,\dots,p)$ and

(7) $$w(x) := W(u_1,\dots,u_q,v_1,\dots,v_p)(x) \neq 0 \qquad (x \in I)$$

one can represent the formal differential operator L, the spaces R^a and R^b and the associated Green's function G in the following special forms:

(8) $$(Lu)(x) = W(u_1,\dots,u_q,v_1,\dots,v_p,u)(x) \,/\, w(x) \qquad (x \in I),$$

---

[1]) $W(f_1,\dots,f_k)$ denotes the Wronskian of $f_1,\dots,f_k$.

$$R^a = \{u \in C_n(I) : W(u_1,\dots,u_q,v_1,\dots,v_{i-1},u,v_{i+1},\dots,v_p)(a) = 0 \quad (i = 1,\dots,p)\},$$

$$R^b = \{u \in C_n(I) : W(u_1,\dots,u_{j-1},u,u_{j+1},\dots,u_q,v_1,\dots,v_p)(b) = 0 \quad (j = 1,\dots,q)\},$$

$$(9) \qquad G(x,y) = \begin{cases} u_1(x)\, v_1^*(y) + \dots + u_q(x)\, v_q^*(y) & (x \leq y) \\ v_1(x)\, u_1^*(y) + \dots + v_p(x)\, u_p^*(y) & (y < x) \end{cases} \qquad (x,y \in I),$$

with

$$(10) \qquad \begin{aligned} v_j^* &= (-1)^{n-j+1}\, W(u_1,\dots,u_{j-1},u_{j+1},\dots,u_q,v_1,\dots,v_p) \,/\, w \quad (j = 1,\dots,q), \\ u_i^* &= (-1)^{p-i}\, W(u_1,\dots,u_q,v_1,\dots,v_{i-1},v_{i+1},\dots,v_p) \,/\, w \quad (i = 1,\dots,p). \end{aligned}$$

Some of these relations are well-known. As for the rest see the next section.

3. The Calculus in the General Case

Let I now denote an arbitrary interval with endpoints a and b $(-\infty \leq a < b \leq \infty)$. We consider a formal differential operator L on I of the form (5) with (6). Then the inequality in (7) and the identity (8) hold with any chosen $n = q + p$ linear independent solutions

$$(11) \qquad u_1,\dots,u_q,v_1,\dots,v_p \in C_n(I)$$

of the equation $(Lu)(x) = 0$ $(x \in I)$. With these functions one can shape a function G according to (9), (10). Moreover, let U_i, V_j denote the following "boundary operators" (for $x \in I$ and $u \in C_n(I)$)

$$(12) \qquad \begin{aligned} (U_i u)(x) &= W(u_1,\dots,u_q,v_1,\dots,v_{i-1},u,v_{i+1},\dots,v_p)(x) \,/\, w(x) \quad (i = 1,\dots,p), \\ (V_j u)(x) &= W(u_1,\dots,u_{j-1},u,u_{j+1},\dots,u_q,v_1,\dots,v_p)(x) \,/\, w(x) \quad (j = 1,\dots,q). \end{aligned}$$

Theorem. *With these quantities the relation*

$$u(x) = \int_a^b G(x,y)\,(Lu)(y)\,dy \qquad (x \in I)$$

holds (as an improper integral) for all $u \in R = R^a \cap R^b$, *with*

$$R^a = \{u \in C_n(I) : (U_i u)(x) \to 0 \quad (x \to a) \qquad (i = 1,\dots,p)\},$$

$$R^b = \{u \in C_n(I) : (V_j u)(x) \to 0 \quad (x \to b) \qquad (j = 1,\dots,q)\}.$$

Proof. Using (8), (10), (12) and well-known facts on Wronskian determinants one gets $u_i^* \cdot Lu = (U_i u)'$. Therefore (for $a < a_1 < x < b$)

$$\int_{a_1}^{x} G(x,y)\,(Lu)(y)\,dy = \sum_{i=1}^{p} v_i(x)\left[U_i u\right]_{a_1}^{x} \longrightarrow \sum_{i=1}^{p} v_i(x)\,(U_i u)(x) \qquad (a_1 \to a)$$

for $u \in R^a$. By the same treatment of the remaining integral $\int_x^b$ we get for $u \in R$

$$\int_a^b G(x,y)\,(Lu)(y)\,dy = \sum_{i=1}^{p} v_i(x)\,(U_i u)(x) + \sum_{j=1}^{q} u_j(x)\,(V_j u)(x) = u(x) \quad (x \in I)$$

where the last equation is a simple determinant identity.

<u>Remark</u>. The spaces R^a and R^b, the Green's function G and the formal differential operator L don't change if the functions (11) are replaced by linear independent functions $\tilde{u}_1,\dots,\tilde{u}_q,\tilde{v}_1,\dots,\tilde{v}_p$, where every $\tilde{u}_j$ is a linear combination of $u_1,\dots,u_q$ and every $\tilde{v}_i$ is a linear combination of $v_1,\dots,v_p$.

Instead of proceeding from a given differential operator L on I and choosing special solutions (11), one can of course start with fixed functions (11) with (7) and then construct L according to (8). (Also see the next section.)

<u>Example</u>. For $u_1(x) = x$, $v_1(x) = 1$, one gets on $[0,\infty)$:

$$Lu = u'', \qquad R = \{u \in C_2[0,\infty) : u(0) = 0,\ u'(x) \to 0\ (x \to \infty)\},$$

$$G(x,y) = -x \quad (0 \leq x \leq y < \infty), \qquad G(x,y) = -y \quad (0 \leq y \leq x < \infty).$$

<u>Example</u>. The function G_1 in (2) with $u_1(x) = x^3$, $v_1(x) = (1-x)^3$ turns out to be the Green's function of the operator $M_1 = L_1|_{R_1}$:

$$(L_1 u)(x) = u'' + \frac{4x-2}{x\,(1-x)}\,u' - \frac{6}{x\,(1-x)}\,u \qquad (0<x<1),$$

$$R_1 = \{u \in C_2(0,1) : xu'(x) - 3u(x) \to 0\ (x \to 0),\ (1-x)u'(x) + 3u(x) \to 0\ (x \to 1)\}.$$

4. <u>An Application</u>

As a simple application, we consider an invertable differential operator $M = L|_R$ of order $n = 4$ on $[0,1]$, where L is given by (5) with $a_k \in C_k[0,1]$ $(k = 0,1,2,3)$ and $R = R^0 \cap R^1$,

$$R^0 = \{u \in C_4[0,1] : u(0) = \alpha_{21}\,u'(0) + \alpha_{22}\,u''(0) = 0\},$$

$$R^1 = \{u \in C_4[0,1] : u(1) = \beta_{21}\,u'(1) + \beta_{22}\,u''(1) = 0\},$$

with arbitrary coefficients α_{21}, α_{22}, β_{21}, β_{22}. We assume that the equation $(Lu)(x) = 0$ $(0 \leq x \leq 1)$ is disconjugate on $[0,1]$ (see [1]).

Then let u_1, v_1 be the solutions of $(Lu)(x) = 0$ $(0 \leq x \leq 1)$ satisfying the boundary conditions $u_1(0) = u_1'(0) = u_1''(0) = v_1(1) = v_1'(1) = v_1''(1) = 0$, $u_1(1) = 1$, $v_1(0) = 1$. The assumed disconjugacy implies their existence and that $u_1(x) > 0$, $v_1(x) > 0$, $w_1(x) = W(u_1, v_1)(x) < 0$ $(0 < x < 1)$ (while $w_1 = w_1' = 0$, $w_1'' < 0$ at $x=0$ and $x=1$). Thus one can shape G_1 as in (2), and $-G_1$ turns out to be totally positive. Moreover, with arbitrary solutions $u_2 \in R^0$, $v_2 \in R^1$ of $(Lu)(x) = 0$ $(0 \leq x \leq 1)$, such that u_1, v_1, u_2, v_2 are linear independent, one can construct G_2 according to (3) and (4), because $\hat{\omega}_2(x) = W(u_1, v_1, u_2, v_2)(x)/w_1(x) \neq 0$ $(0 < x < 1)$.

With these G_1 and G_2, which according to the above theorem are the Green's functions of some singular differential operators M_1 and M_2, and with the Green's function G of M the relation (1) is true. For the proof, use the representation (9) of G, some Wronskian identities and the fact that $u_2 \in R^0 \subset R_1^0$, $v_2 \in R^1 \subset R_1^1$, with the spaces R_1^0, R_1^1 belonging to M_1. (These last inclusions, for their part, hold because of the special form of R and because $w_1'' < 0$ at $x=0$ and $x=1$.)

The relation (1) yields the non-negativity (total positivity) of G if $-G_2$ is non-negative (totally positive). Moreover, it can be proved (see [4]) that $\hat{u}_2/\hat{\omega}_2 = c\psi^*$, $\hat{v}_2/\hat{\omega}_2 = c'\varphi^*$, with the adjoint limit functions ψ^*, φ^* of M defined in [3], Chapter 6, and unessential constants c, c'. These functions ψ^*, φ^* are non-negative if G is non-negative (see [3]). Together with (1), this fact implies, that for the case in question the non-negativity of ψ^* and φ^* is a necessary and sufficient condition for G to be non-negative.

References

1. Coppel, W.A.: Disconjugacy. Lecture notes in Math., vol. 220. Berlin - Heidelberg - New York: Springer 1971.

2. Karlin, S.: Total positivity. Stanford, California: Stanford University Press 1968.

3. Schröder, J.: On linear differential inequalities. J. Math. Anal. Appl. 22, 188-216 (1968).

4. Trottenberg, U.: Zur Inverspositivität linearer gewöhnlicher Differentialoperatoren höherer Ordnung. Dissertation Köln 1972.

Gewöhnliche Differentialungleichungen mit quasimonoton wachsenden Funktionen in Banachräumen P. Volkman

1. Ein grundlegender Satz. Sei E ein reeller Banachraum, geordnet durch einen abgeschlossenen, konvexen Kegel $K \subseteq E$, d.h. für $x,y \in E$ wird gesetzt

$$x \leqslant y \iff y-x \in K, \quad x \ll y \iff y-x \in \operatorname{Int} K \;^{1)}.$$

K^* sei die Menge der auf E linearen, stetigen Funktionale φ mit $\varphi(x) \geqslant 0$ für $x \in K$.

Definition. $f: D \longrightarrow E$ (wobei $D \subseteq E$) heißt quasimonoton wachsend (kurz q.-m.w.), wenn aus $x,y \in D$, $x \leqslant y$, $\varphi \in K^*$, $\varphi(x)=\varphi(y)$ stets $\varphi(f(x)) \leqslant \varphi(f(y))$ folgt.

Hierdurch wird der im Falle $E = \mathbb{R}^n$ geläufige Quasimonotonie-Begriff (vgl. z.B. WALTER [6]) verallgemeinert, und im folgenden soll die Übertragung einiger für $E = \mathbb{R}^n$ bekannter Ergebnisse auf beliebige Banachräume diskutiert werden. Zunächst gilt nach [3]

Satz A. Sei $f(t,x): (0,T] \times D \longrightarrow E$, $D \subseteq E$, q.-m.w. in x, und seien $v,w: [0,T] \longrightarrow D$ stetig. Dann folgt aus

$$v(0) \ll w(0), \quad v_l'(t) - f(t,v(t)) \ll w_l'(t) - f(t,w(t)) \;^{2)}$$

$(0 < t \leqslant T)$ die Ungleichung $v(t) \ll w(t)$ auf $[0,T]$.

(Satz A ist in [3] mit v', w' statt v_l', w_l' formuliert, im Beweis werden aber nur die linksseitigen Ableitungen benötigt.)

---

1) Int K bezeichnet das Innere von K.

2) v_l', w_l' bezeichnen die linksseitigen Ableitungen von v,w.

2. Existenz von Maximalintegralen. Ist $f: [0,T] \times D \longrightarrow E$ vollstetig, D offen und $u_0 \in D$, so ist das Anfangswertproblem

$$u(0) = u_0, \quad u'(t) = f(t,u(t)) \qquad (0 \leq t \leq T_0) \tag{1}$$

für kleine $T_0 > 0$ lösbar. Ist außerdem $\operatorname{Int} K \neq \emptyset$ und $f(t,x)$ monoton wachsend in x (d.h. $x \leq y \Longrightarrow f(t,x) \leq f(t,y)$), so existiert nach MLAK [1] ein Maximalintegral für (1).

Das ist wegen Satz A auch richtig, wenn $f(t,x)$ nur q.-m.w. in x ist. Ein Maximalintegral $\bar{u}$ für (1) erhält man in diesem Falle analog zu [1] und zum Vorgehen im $\mathbb{R}^n$ durch eine konvergente Folge von Oberfunktionen, d.h. man wählt $v_0 \gg \theta$ [3]), löst für kleine $\varepsilon > 0$ die Anfangswertprobleme

$$u_\varepsilon(0) = u_0 + \varepsilon v_0, \quad u_\varepsilon'(t) = f(t,u_\varepsilon(t)) + \varepsilon v_0 \qquad (0 \leq t \leq T_0) \tag{2}$$

und findet eine Folge $\varepsilon_n \searrow 0$, so daß $u_{\varepsilon_n} \longrightarrow \bar{u}$ gilt. (Nach Satz A folgt aus (1),(2) die Ungleichung $u(t) \leq u_\varepsilon(t)$ auf $[0,T_0]$.)

3. Über die $\leq$-Fassung des Satzes A. Ist in Satz A D offen und $f(t,x)$ Lipschitz-stetig in x,

$$\| f(t,x) - f(t,y) \| \leq L \| x - y \| \qquad (0 < t \leq T;\ x,y \in D), \tag{3}$$

so bleibt der Satz richtig, wenn man (an allen drei Stellen) $\ll$ durch $\leq$ ersetzt.

3.1. Ist $\operatorname{Int} K \neq \emptyset$, so kann das durch Zurückführung auf Satz A bewiesen werden (vgl. [3]).

3.2. Ist aber $\operatorname{Int} K = \emptyset$, so hilft das folgende, in [4] entwickelte Verfahren:

Man bildet den Banachraum $\tilde{E} = E \oplus \mathbb{R}$. Mit der Bezeichnung $p = (\theta,1)$ gilt also

$$\tilde{E} = E \oplus \mathbb{R}p = \{x + \xi p \mid x \in E,\ \xi \in \mathbb{R}\}.$$

---

[3]) θ bezeichnet das Nullelement von E.

Es sei noch $C = \{x + \xi p \mid \|x\| \leq \xi\}$ und $\tilde{K} = \overline{K + C}$. Dann ist $\tilde{K}$ ein abgeschlossener, konvexer Kegel in $\tilde{E}$ mit der Eigenschaft $\operatorname{Int} \tilde{K} \neq \emptyset$.

Neben der von K in E erzeugten Ordnung wird im folgenden auch die von $\tilde{K}$ in $\tilde{E}$ erzeugte Ordnung benötigt; es wird daher $\leq_K$ bzw. $\leq_{\tilde{K}}$ geschrieben und zwischen K-q.-m.w. und $\tilde{K}$-q.-m.w. Funktionen unterschieden.

Ist nun $f(t,x):(0,T] \times E \longrightarrow E$ [4]) K-q.-m.w. in x und gilt (3), so ist die durch

$$\tilde{f}(t,x + \xi p) = f(t,x) + L\xi p$$

definierte Funktion $\tilde{f}:(0,T] \times \tilde{E} \longrightarrow \tilde{E}$ in $x + \xi p$ lipschitzstetig und $\tilde{K}$-q.-m.w. (vgl. Nr.4).

Setzt man für die stetigen Funktionen $v,w: [0,T] \longrightarrow E$ die Ungleichungen

$$v(0) \leq_K w(0), \quad v_1'(t) - f(t,v(t)) \leq_K w_1'(t) - f(t,w(t)) \tag{4}$$

$(0 < t \leq T)$ voraus, so gelten diese auch mit $\leq_{\tilde{K}}$ statt $\leq_K$ und $\tilde{f}$ statt f. Wegen $\operatorname{Int} \tilde{K} \neq \emptyset$ folgt aus dem erledigten Fall 3.1 die Ungleichung $v(t) \leq_{\tilde{K}} w(t)$, und da $v(t)$, $w(t)$ in E liegen, ergibt sich $v(t) \leq_K w(t)$ auf $[0,T]$.

4. Über Fortsetzung von q.-m.w. Funktionen. Die Hauptschwierigkeit in 3.2 besteht im Nachweis der $\tilde{K}$-Quasimonotonie von $\tilde{f}$. In [4] konnte nur die $\tilde{K}$-Quasimonotonie von $\tilde{f}(x + \xi p) = f(x) + 4L\xi p$ [5]) gezeigt werden. Durch Verbesserung der Beweistechnik erweist sich der Faktor 4 als überflüssig, und man hat damit ein natürlicheres Ergebnis.

---

[4]) Der Einfachheit halber wird $D = E$ vorausgesetzt.

[5]) Die Veränderliche t wird in dieser Nr. weggelassen.

Es sei also $f: E \longrightarrow E$ K-q.-m.w. Das bedeutet nach einer von REDHEFFER und WALTER [2] gegebenen Charakterisierung der q.-m.w. Funktionen

$$\lim_{h \searrow 0} \frac{1}{h} \operatorname{dist}(K,\ y - x + h[f(y) - f(x)]) = 0 \,{}^{6)} \text{ für } x \leq_K y. \quad (5)$$

Setzt man für $\xi \geq 0$

$$K_\xi = \{q \in E \mid \operatorname{dist}(K,q) \leq \xi\},$$

so folgt nach [5] aus (3),(5)

$$\overline{\lim_{h \searrow 0}} \frac{1}{h} \operatorname{dist}(K_\xi, y - x + h[f(y) - f(x)]) \leq L\xi \quad \text{für } y-x \in K_\xi. \quad (6)$$

(In den Beweis dieser Formel geht ein Satz von BISHOP und PHELPS über die Stützpunkte konvexer Mengen ein.)

Wird $\tilde{E}$ durch $\|x + \xi p\| = \|x\| + |\xi|$ normiert, so ist es leicht, aus (6) für $\tilde{f}(x + \xi p) = f(x) + L\xi p$ die Formel

$$\lim_{h \searrow 0} \frac{1}{h} \operatorname{dist}(\tilde{K}, y+\eta p - (x+\xi p) + h[\tilde{f}(y+\eta p) - \tilde{f}(x+\xi p)]) = 0 \quad \text{für } x+\xi p \leq_{\tilde{K}} y+\eta p \quad (7)$$

herzuleiten. (7) bedeutet nach dem oben erwähnten Ergebnis aus [2], daß $\tilde{f}$ $\tilde{K}$-q.-m.w. ist.

5. Zusammenhänge mit Invarianzsätzen. (5) liefert einen Zusammenhang zwischen Quasimonotonie- und Invarianz-Bedingungen; das ist von REDHEFFER und WALTER [2] ausführlich erörtert worden. Zwei Konsequenzen seien erwähnt: Einerseits konnte das in 3.2. dargestellte Verfahren in [4] ausgenutzt werden, um einen Invarianzsatz für konvexe Mengen zu beweisen. Andererseits gelten nach [2],[5] $\leq$-Formen des Satzes A mit rechtsseitigen an Stelle von linksseitigen Ableitungen in (4), wobei statt (3) Bedingungen der Form $\|f(t,x) - f(t,y)\|$

---

6) $\operatorname{dist}(M,q)$ bezeichnet den Abstand zwischen der Menge M und dem Punkte q.

$\leqslant \omega(t, \|x - y\|)$ mit recht allgemeinen Eindeutigkeitsfunktionen ω gefordert werden. Es ist zu hoffen, daß ähnlich allgemeine Ergebnisse für Differentialungleichungen mit linksseitigen Ableitungen gefunden werden können.

Literatur.

[1] W.MLAK, Differential inequalities in linear spaces, Ann. Polon. math. 5, 95 - 101 (1958).

[2] R.M.REDHEFFER und W.WALTER, Flow-invariant sets and differential inequalities in normed spaces, Applicable Analysis (erscheint demnächst).

[3] P.VOLKMANN, Gewöhnliche Differentialungleichungen mit quasimonoton wachsenden Funktionen in topologischen Vektorräumen, Math. Z. 127, 157 - 164 (1972).

[4] – , Über die Invarianz konvexer Mengen und Differentialungleichungen in einem normierten Raume, Math. Ann. 203, 201 - 210 (1973).

[5] – , Über die positive Invarianz einer abgeschlossenen Teilmenge eines Banachschen Raumes bezüglich der Differentialgleichung $u' = f(t,u)$, J. reine angew. Math. (erscheint demnächst).

[6] W.WALTER, Differential and integral inequalities, Berlin-Heidelberg-New York: Springer 1970. [Übersetzung aus dem Deutschen.]

On the Integral Equation Method for Boundary-Value Problems

L. J. Walpole

It is wellknown that much theoretical and practical advantage can be gained by reducing a given boundary-value problem to the solution of an integral equation, especially when the domain in question is the infinite one exterior to a given boundary. We show how an appeal to the variation principles associated with the boundary-value problem enables various integral equations to be derived in a way that reflects the inherent fundamental properties of 'symmetry' (or 'self-adjointness') and (if extremum principles are available) 'positive definiteness'. These properties may not be revealed as naturally and usefully in other approaches.

For illustration, consider the exterior potential problems for which a function $u(\underline{r})$ is to satisfy Laplace's equation

$$\nabla^2 u = 0 \tag{1}$$

in some three-dimensional exterior domain, while subject to prescribed boundary conditions on the smooth closed surface B which forms the internal boundary, and while vanishing at infinity. For the Dirichlet problem where $u(\underline{r})$ coincides on B with a given function $f(\underline{r})$ and for the Neumann problem where $\partial u/\partial n$, the derivative of u along the outward normal on B, coincides with a given function $g(\underline{r})$, we recall in turn the extremum principles of Thomson and Dirichlet which imply that

$$-\int_B f \frac{\partial u}{\partial n} dS > -2\int_B f \frac{\partial u^*}{\partial n} dS + \int_B u^* \frac{\partial u^*}{\partial n} dS, \tag{2}$$

$$-\int_B u\, g\, dS > -2\int_B u^* g\, dS + \int_B u^* \frac{\partial u^*}{\partial n} dS, \tag{3}$$

where u^* is any solution of (1), not subject to boundary conditions on B but vanishing at infinity. For instance, suppose that in the Dirichlet problem u^* is represented as the single-layer potential

$$u^*(\underline{r}) = \int_B \frac{\sigma^*(\underline{r}')}{|\underline{r}-\underline{r}'|} dS'.$$

Then $u^*(\underline{r})$ coincides everywhere with $u(\underline{r})$ when the arbitrary source distribution $\sigma^*(\underline{r}')$ takes the precise value $\sigma(\underline{r}')$ determined by the wellknown integral equation

$$\int_B \frac{\sigma(\underline{r}')}{|\underline{r}-\underline{r}'|} dS' = f(\underline{r}), \quad \underline{r} \text{ on } B, \tag{4}$$

given, for instance, by Stakgold (1968, p. 125) and by Burton and Miller (1971). On the other hand, after some manipulation in the inequality (2), to allow for the discontinuity properties of the single-layer potential and to drop out a further second-order positive quantity, it is found that

$$\int_B f\,\sigma\,dS > 2\int_B f\,\sigma^*\,dS - \int_B \sigma^*(\underline{r})dS \int_B \frac{\sigma^*(\underline{r}')}{|\underline{r}-\underline{r}'|} dS'. \tag{5}$$

The integral equation (4) is recovered at once by seeking the choice of σ^* that makes this inequality sharpest. It is evident now that its kernel is positive definite (as well as symmetric) since equality would be restored in (5) by adding to the right-hand side the quantity

$$\int_B \left[\sigma(\underline{r}) - \sigma^*(\underline{r})\right]dS \int_B \frac{\sigma(\underline{r}') - \sigma^*(\underline{r}')}{|\underline{r}-\underline{r}'|} dS',$$

which therefore must be positive for all σ^*.

We recall next the Green's formula

$$\frac{1}{4\pi}\int_B \{u(\underline{r})\frac{\partial}{\partial n}\frac{1}{|\underline{r}'-\underline{r}|} - \frac{1}{|\underline{r}'-\underline{r}|}\frac{\partial}{\partial n}u(\underline{r})\}dS = u(\underline{r}'), \text{ if } \underline{r}' \text{ outside B}, \tag{6}$$

$$= \tfrac{1}{2}u(\underline{r}'), \text{ if } \underline{r}' \text{ on B}, \tag{7}$$

$$= 0, \text{ if } \underline{r}' \text{ inside B}. \tag{8}$$

The equalities (7) and (8) both provide integral equations to determine on B whichever is unknown of u or $\partial u/\partial n$, and then (6) specifies u everywhere. Equation (8) is helpful especially in axially-symmetric circumstances where $u(\underline{r})$ depends only on $(x^2+y^2)^{\frac{1}{2}}$ and z. For then {cf. Copley (1967) for a related discussion of acoustical problems} $\underline{r}'$ may be confined to points (0,0,z') on L, the interior axis of symmetry of B, to simplify the integral equations to

$$\int_B \frac{\partial u/\partial n}{\{x^2+y^2+(z-z')^2\}^{\frac{1}{2}}}\, dS = \int_B f \frac{\partial}{\partial n} \frac{1}{\{x^2+y^2+(z-z')^2\}^{\frac{1}{2}}}\, dS,$$

$$\int_B u \frac{\partial}{\partial n} \frac{1}{\{x^2+y^2+(z-z')^2\}^{\frac{1}{2}}}\, dS = \int_B \frac{1}{\{x^2+y^2+(z-z')^2\}^{\frac{1}{2}}}\, g\, dS, \tag{9}$$

for the Dirichlet and Neumann problems respectively. An alternative approach to these axi-symmetric problems expresses $u^*(\underline{r})$ as

$$u^*(\underline{r}) = \int_L \frac{\sigma^*(z')}{\{x^2+y^2+(z-z')^2\}^{\frac{1}{2}}}\, dz',$$

that is, as the potential due to sources lying simply along L. The integral equations

$$u^*(\underline{r}) = f(\underline{r}), \quad \partial u^*/\partial n = g(\underline{r}), \quad \underline{r} \text{ on } B, \tag{10}$$

then determine (when B is sufficiently smooth) the required exact value of σ^*. {A form of the second of these equations is solved when B is 'slender' by Handelsman and Keller (1967).} On the other hand, the inequalities (2) and (3) are made sharpest of all when σ^* satisfies the integral equations

$$\int_L K(z',z'')\,\sigma^*(z'')dz'' = -\int_B f \frac{\partial}{\partial n} \frac{1}{\{x^2+y^2+(z-z')^2\}^{\frac{1}{2}}}\, dS,$$

$$\int_L K(z',z'')\,\sigma^*(z'')dz'' = -\int_B \frac{1}{\{x^2+y^2+(z-z')^2\}^{\frac{1}{2}}}\, g\, dS, \tag{11}$$

respectively, where

$$K(z',z'') = -\int_B \frac{1}{\{x^2+y^2+(z-z')^2\}^{\frac{1}{2}}} \frac{\partial}{\partial n} \frac{1}{\{x^2+y^2+(z-z'')^2\}^{\frac{1}{2}}} dS$$

$$= -\int_B \frac{1}{\{x^2+y^2+(z-z'')^2\}^{\frac{1}{2}}} \frac{\partial}{\partial n} \frac{1}{\{x^2+y^2+(z-z')^2\}^{\frac{1}{2}}} dS.$$

These two expressions of the kernel allow both pairs of integral equations (9) and (10) to be extracted from the pair (11), at the expense of losing the properties of symmetry and positive definiteness manifest evidently {on consideration of the inequalities (2) and (3)} in equations (11). Though the three pairs of equations (9), (10) and (11) are equivalent in essence, each will suggest its own particular means of solution.

References

Burton, A. J. and Miller, G. F. 1971 Integral equation methods for exterior problems, Proc. Roy. Soc. Lond. A 323, 201-210.

Copley, L. G. 1967 Integral equation method for radiation from vibrating bodies, J. Acoust. Soc. Am. 41, 807-816.

Handelsman, R. A. and Keller, J. B. 1967 Axially symmetric potential flow around a slender body, J. Fluid Mech. 28, 131-147.

Stakgold, I. 1968 Boundary value problems of mathematical physics, volume II, The Macmillan Company, New York.

Vol. 247: Lectures on Operator Algebras. Tulane University Ring and Operator Theory Year, 1970–1971. Volume II. XI, 786 pages. 1972. DM 40,–

Vol. 248: Lectures on the Applications of Sheaves to Ring Theory. Tulane University Ring and Operator Theory Year, 1970–1971. Volume III. VIII, 315 pages. 1971. DM 26,–

Vol. 249: Symposium on Algebraic Topology. Edited by P. J. Hilton. VII, 111 pages. 1971. DM 16,–

Vol. 250: B. Jónsson, Topics in Universal Algebra. VI, 220 pages. 1972. DM 20,–

Vol. 251: The Theory of Arithmetic Functions. Edited by A. A. Gioia and D. L. Goldsmith VI, 287 pages. 1972. DM 24,–

Vol. 252: D. A. Stone, Stratified Polyhedra. IX, 193 pages. 1972. DM 18,–

Vol. 253: V. Komkov, Optimal Control Theory for the Damping of Vibrations of Simple Elastic Systems. V, 240 pages. 1972. DM 20,–

Vol. 254: C. U. Jensen, Les Foncteurs Dérivés de $\varprojlim$ et leurs Applications en Théorie des Modules. V, 103 pages. 1972. DM 16,–

Vol. 255: Conference in Mathematical Logic – London '70. Edited by W. Hodges. VIII, 351 pages. 1972. DM 26,–

Vol. 256: C. A. Berenstein and M. A. Dostal, Analytically Uniform Spaces and their Applications to Convolution Equations. VII, 130 pages. 1972. DM 16,–

Vol. 257: R. B. Holmes, A Course on Optimization and Best Approximation. VIII, 233 pages. 1972. DM 20,–

Vol. 258: Séminaire de Probabilités VI. Edited by P. A. Meyer. VI, 253 pages. 1972. DM 22,–

Vol. 259: N. Moulis, Structures de Fredholm sur les Variétés Hilbertiennes. V, 123 pages. 1972. DM 16,–

Vol. 260: R. Godement and H. Jacquet, Zeta Functions of Simple Algebras. IX, 188 pages. 1972. DM 18,–

Vol. 261: A. Guichardet, Symmetric Hilbert Spaces and Related Topics. V, 197 pages. 1972. DM 18,–

Vol. 262: H. G. Zimmer, Computational Problems, Methods, and Results in Algebraic Number Theory. V, 103 pages. 1972. DM 16,–

Vol. 263: T. Parthasarathy, Selection Theorems and their Applications. VII, 101 pages. 1972. DM 16,–

Vol. 264: W. Messing, The Crystals Associated to Barsotti-Tate Groups: With Applications to Abelian Schemes. III, 190 pages. 1972. DM 18,–

Vol. 265: N. Saavedra Rivano, Catégories Tannakiennes. II, 418 pages. 1972. DM 26,–

Vol. 266: Conference on Harmonic Analysis. Edited by D. Gulick and R. L. Lipsman. VI, 323 pages. 1972. DM 24,–

Vol. 267: Numerische Lösung nichtlinearer partieller Differential- und Integro-Differentialgleichungen. Herausgegeben von R. Ansorge und W. Törnig. VI, 339 Seiten. 1972. DM 26,–

Vol. 268: C. G. Simader, On Dirichlet's Boundary Value Problem. IV, 238 pages. 1972. DM 20,–

Vol. 269: Théorie des Topos et Cohomologie Etale des Schémas. (SGA 4). Dirigé par M. Artin, A. Grothendieck et J. L. Verdier. XIX, 525 pages. 1972. DM 50,–

Vol. 270: Théorie des Topos et Cohomologie Etale des Schémas. Tome 2. (SGA 4). Dirigé par M. Artin, A. Grothendieck et J. L. Verdier. V, 418 pages. 1972. DM 50,–

Vol. 271: J. P. May, The Geometry of Iterated Loop Spaces. IX, 175 pages. 1972. DM 18,–

Vol. 272: K. R. Parthasarathy and K. Schmidt, Positive Definite Kernels, Continuous Tensor Products, and Central Limit Theorems of Probability Theory. VI, 107 pages. 1972. DM 16,–

Vol. 273: U. Seip, Kompakt erzeugte Vektorräume und Analysis. IX, 119 Seiten. 1972. DM 16,–

Vol. 274: Toposes, Algebraic Geometry and Logic. Edited by. F. W. Lawvere. VI, 189 pages. 1972. DM 18,–

Vol. 275: Séminaire Pierre Lelong (Analyse) Année 1970–1971. VI, 181 pages. 1972. DM 18,–

Vol. 276: A. Borel, Représentations de Groupes Localement Compacts. V, 98 pages. 1972. DM 16,–

Vol. 277: Séminaire Banach. Edité par C. Houzel. VII, 229 pages. 1972. DM 20,–

Vol. 278: H. Jacquet, Automorphic Forms on GL(2). Part II. XIII, 142 pages. 1972. DM 16,–

Vol. 279: R. Bott, S. Gitler and I. M. James, Lectures on Algebraic and Differential Topology. V, 174 pages. 1972. DM 18,–

Vol. 280: Conference on the Theory of Ordinary and Partial Differential Equations. Edited by W. N. Everitt and B. D. Sleeman. XV, 367 pages. 1972. DM 26,–

Vol. 281: Coherence in Categories. Edited by S. Mac Lane. VII, 235 pages. 1972. DM 20,–

Vol. 282: W. Klingenberg und P. Flaschel, Riemannsche Hilbertmannigfaltigkeiten. Periodische Geodätische. VII, 211 Seiten. 1972. DM 20,–

Vol. 283: L. Illusie, Complexe Cotangent et Déformations II. VII, 304 pages. 1972. DM 24,–

Vol. 284: P. A. Meyer, Martingales and Stochastic Integrals I. VI, 89 pages. 1972. DM 16,–

Vol. 285: P. de la Harpe, Classical Banach-Lie Algebras and Banach-Lie Groups of Operators in Hilbert Space. III, 160 pages. 1972. DM 16,–

Vol. 286: S. Murakami, On Automorphisms of Siegel Domains. V, 95 pages. 1972. DM 16,–

Vol. 287: Hyperfunctions and Pseudo-Differential Equations. Edited by H. Komatsu. VII, 529 pages. 1973. DM 36,–

Vol. 288: Groupes de Monodromie en Géométrie Algébrique. (SGA 7 I). Dirigé par A. Grothendieck. IX, 523 pages. 1972. DM 50,–

Vol. 289: B. Fuglede, Finely Harmonic Functions. III, 188. 1972. DM 18,–

Vol. 290: D. B. Zagier, Equivariant Pontrjagin Classes and Applications to Orbit Spaces. IX, 130 pages. 1972. DM 16,–

Vol. 291: P. Orlik, Seifert Manifolds. VIII, 155 pages. 1972. DM 16,–

Vol. 292: W. D. Wallis, A. P. Street and J. S. Wallis, Combinatorics: Room Squares, Sum-Free Sets, Hadamard Matrices. V, 508 pages. 1972. DM 50,–

Vol. 293: R. A. DeVore, The Approximation of Continuous Functions by Positive Linear Operators. VIII, 289 pages. 1972. DM 24,–

Vol. 294: Stability of Stochastic Dynamical Systems. Edited by R. F. Curtain. IX, 332 pages. 1972. DM 26,–

Vol. 295: C. Dellacherie, Ensembles Analytiques, Capacités, Mesures de Hausdorff. XII, 123 pages. 1972. DM 16,–

Vol. 296: Probability and Information Theory II. Edited by M. Behara, K. Krickeberg and J. Wolfowitz. V, 223 pages. 1973. DM 20,–

Vol. 297: J. Garnett, Analytic Capacity and Measure. IV, 138 pages. 1972. DM 16,–

Vol. 298: Proceedings of the Second Conference on Compact Transformation Groups. Part 1. XIII, 453 pages. 1972. DM 32,–

Vol. 299: Proceedings of the Second Conference on Compact Transformation Groups. Part 2. XIV, 327 pages. 1972. DM 26,–

Vol. 300: P. Eymard, Moyennes Invariantes et Représentations Unitaires. II. 113 pages. 1972. DM 16,–

Vol. 301: F. Pittnauer, Vorlesungen über asymptotische Reihen. VI, 186 Seiten. 1972. DM 18,–

Vol. 302: M. Demazure, Lectures on p-Divisible Groups. V, 98 pages. 1972. DM 16,–

Vol. 303: Graph Theory and Applications. Edited by Y. Alavi, D. R. Lick and A. T. White. IX, 329 pages. 1972. DM 26,–

Vol. 304: A. K. Bousfield and D. M. Kan, Homotopy Limits, Completions and Localizations. V, 348 pages. 1972. DM 26,–

Vol. 305: Théorie des Topos et Cohomologie Etale des Schémas. Tome 3. (SGA 4). Dirigé par M. Artin, A. Grothendieck et J. L. Verdier. VI, 640 pages. 1973. DM 50,–

Vol. 306: H. Luckhardt, Extensional Gödel Functional Interpretation. VI, 161 pages. 1973. DM 18,–

Vol. 307: J. L. Bretagnolle, S. D. Chatterji et P.-A. Meyer, Ecole d'été de Probabilités: Processus Stochastiques. VI, 198 pages. 1973. DM 20,–

Vol. 308: D. Knutson, λ-Rings and the Representation Theory of the Symmetric Group. IV, 203 pages. 1973. DM 20,–

Vol. 309: D. H. Sattinger, Topics in Stability and Bifurcation Theory. VI, 190 pages. 1973. DM 18,–

Vol. 310: B. Iversen, Generic Local Structure of the Morphisms in Commutative Algebra. IV, 108 pages. 1973. DM 16,–

Vol. 311: Conference on Commutative Algebra. Edited by J. W. Brewer and E. A. Rutter. VII, 251 pages. 1973. DM 22,–

Vol. 312: Symposium on Ordinary Differential Equations. Edited by W. A. Harris, Jr. and Y. Sibuya. VIII, 204 pages. 1973. DM 22,–

Vol. 313: K. Jörgens and J. Weidmann, Spectral Properties of Hamiltonian Operators. III, 140 pages. 1973. DM 16,–

Vol. 314: M. Deuring, Lectures on the Theory of Algebraic Functions of One Variable. VI, 151 pages. 1973. DM 16,–

Vol. 315: K. Bichteler, Integration Theory (with Special Attention to Vector Measures). VI, 357 pages. 1973. DM 26,–

Vol. 316: Symposium on Non-Well-Posed Problems and Logarithmic Convexity. Edited by R. J. Knops. V, 176 pages. 1973. DM 18,–

Vol. 317: Séminaire Bourbaki – vol. 1971/72. Exposés 400–417. IV, 361 pages. 1973. DM 26,–

Vol. 318: Recent Advances in Topological Dynamics. Edited by A. Beck, VIII, 285 pages. 1973. DM 24,–

Vol. 319: Conference on Group Theory. Edited by R. W. Gatterdam and K. W. Weston. V, 188 pages. 1973. DM 18,–

Vol. 320: Modular Functions of One Variable I. Edited by W. Kuyk. V, 195 pages. 1973. DM 18,–

Vol. 321: Séminaire de Probabilités VII. Edité par P. A. Meyer. VI, 322 pages. 1973. DM 26,–

Vol. 322: Nonlinear Problems in the Physical Sciences and Biology. Edited by I. Stakgold, D. D. Joseph and D. H. Sattinger. VIII, 357 pages. 1973. DM 26,–

Vol. 323: J. L. Lions, Perturbations Singulières dans les Problèmes aux Limites et en Contrôle Optimal. XII, 645 pages. 1973. DM 42,–

Vol. 324: K. Kreith, Oscillation Theory. VI, 109 pages. 1973. DM 16,–

Vol. 325: Ch.-Ch. Chou, La Transformation de Fourier Complexe et L'Equation de Convolution. IX, 137 pages. 1973. DM 16,–

Vol. 326: A. Robert, Elliptic Curves. VIII, 264 pages. 1973. DM 22,–

Vol. 327: E. Matlis, 1-Dimensional Cohen-Macaulay Rings. XII, 157 pages. 1973. DM 18,–

Vol. 328: J. R. Büchi and D. Siefkes, The Monadic Second Order Theory of All Countable Ordinals. VI, 217 pages. 1973. DM 20,–

Vol. 329: W. Trebels, Multipliers for (C, α)-Bounded Fourier Expansions in Banach Spaces and Approximation Theory. VII, 103 pages. 1973. DM 16,–

Vol. 330: Proceedings of the Second Japan-USSR Symposium on Probability Theory. Edited by G. Maruyama and Yu. V. Prokhorov. VI, 550 pages. 1973. DM 36,–

Vol. 331: Summer School on Topological Vector Spaces. Edited by L. Waelbroeck. VI, 226 pages. 1973. DM 20,–

Vol. 332: Séminaire Pierre Lelong (Analyse) Année 1971-1972. V, 131 pages. 1973. DM 16,–

Vol. 333: Numerische, insbesondere approximationstheoretische Behandlung von Funktionalgleichungen. Herausgegeben von R. Ansorge und W. Törnig. VI, 296 Seiten. 1973. DM 24,–

Vol. 334: F. Schweiger, The Metrical Theory of Jacobi-Perron Algorithm. V, 111 pages. 1973. DM 16,–

Vol. 335: H. Huck, R. Roitzsch, U. Simon, W. Vortisch, R. Walden, B. Wegner und W. Wendland, Beweismethoden der Differentialgeometrie im Großen. IX, 159 Seiten. 1973. DM 18,–

Vol. 336: L'Analyse Harmonique dans le Domaine Complexe. Edité par E. J. Akutowicz. VIII, 169 pages. 1973. DM 18,–

Vol. 337: Cambridge Summer School in Mathematical Logic. Edited by A. R. D. Mathias and H. Rogers. IX, 660 pages. 1973. DM 42,–

Vol. 338: J. Lindenstrauss and L. Tzafriri, Classical Banach Spaces. IX, 243 pages. 1973. DM 22,–

Vol. 339: G. Kempf, F. Knudsen, D. Mumford and B. Saint-Donat, Toroidal Embeddings I. VIII, 209 pages. 1973. DM 20,–

Vol. 340: Groupes de Monodromie en Géométrie Algébrique. (SGA 7 II). Par P. Deligne et N. Katz. X, 438 pages. 1973. DM 40,–

Vol. 341: Algebraic K-Theory I, Higher K-Theories. Edited by H. Bass. XV, 335 pages. 1973. DM 26,–

Vol. 342: Algebraic K-Theory II, "Classical" Algebraic K-Theory, and Connections with Arithmetic. Edited by H. Bass. XV, 527 pages. 1973. DM 36,–

Vol. 343: Algebraic K-Theory III, Hermitian K-Theory and Geometric Applications. Edited by H. Bass. XV, 572 pages. 1973. DM 38,–

Vol. 344: A. S. Troelstra (Editor), Metamathematical Investigation of Intuitionistic Arithmetic and Analysis. XVII, 485 pages. 1973. DM 34,–

Vol. 345: Proceedings of a Conference on Operator Theory. Edited by P. A. Fillmore. VI, 228 pages. 1973. DM 20,–

Vol. 346: Fučik et al., Spectral Analysis of Nonlinear Operators. II, 287 pages. 1973. DM 26,–

Vol. 347: J. M. Boardman and R. M. Vogt, Homotopy Invariant Algebraic Structures on Topological Spaces. X, 257 pages. 1973. DM 22,–

Vol. 348: A. M. Mathai and R. K. Saxena, Generalized Hypergeometric Functions with Applications in Statistics and Physical Sciences. VII, 314 pages. 1973. DM 26,–

Vol. 349: Modular Functions of One Variable II. Edited by W. Kuyk and P. Deligne. V, 598 pages. 1973. DM 38,–

Vol. 350: Modular Functions of One Variable III. Edited by W. Kuyk and J.-P. Serre. V, 350 pages. 1973. DM 26,–

Vol. 351: H. Tachikawa, Quasi-Frobenius Rings and Generalizations. XI, 172 pages. 1973. DM 18,–

Vol. 352: J. D. Fay, Theta Functions on Riemann Surfaces. V, 137 pages. 1973. DM 16,–

Vol. 353: Proceedings of the Conference on Orders, Group Rings and Related Topics. Organized by J. S. Hsia, M. L. Madan and T. G. Ralley. X, 224 pages. 1973. DM 20,–

Vol. 354: K. J. Devlin, Aspects of Constructibility. XII, 240 pages. 1973. DM 22,–

Vol. 355: M. Sion, A Theory of Semigroup Valued Measures. V, 140 pages. 1973. DM 16,–

Vol. 356: W. L. J. van der Kallen, Infinitesimally Central-Extensions of Chevalley Groups. VII, 147 pages. 1973. DM 16,–

Vol. 357: W. Borho, P. Gabriel und R. Rentschler, Primideale in Einhüllenden auflösbarer Lie-Algebren. V, 182 Seiten. 1973. DM 18,–

Vol. 358: F. L. Williams, Tensor Products of Principal Series Representations. VI, 132 pages. 1973. DM 16,–

Vol. 359: U. Stammbach, Homology in Group Theory. VIII, 183 pages. 1973. DM 18,–

Vol. 360: W. J. Padgett and R. L. Taylor, Laws of Large Numbers for Normed Linear Spaces and Certain Fréchet Spaces. VI, 111 pages. 1973. DM 16,–

Vol. 361: J. W. Schutz, Foundations of Special Relativity: Kinematic Axioms for Minkowski Space Time. XX, 314 pages. 1973. DM 26,–

Vol. 362: Proceedings of the Conference on Numerical Solution of Ordinary Differential Equations. Edited by D. Bettis. VIII, 490 pages. 1974. DM 34,–

Vol. 363: Conference on the Numerical Solution of Differential Equations. Edited by G. A. Watson. IX, 221 pages. 1974. DM 20,–

Vol. 364: Proceedings on Infinite Dimensional Holomorphy. Edited by T. L. Hayden and T. J. Suffridge. VII, 212 pages. 1974. DM 20,–

Vol. 365: R. P. Gilbert, Constructive Methods for Elliptic Equations. VII, 397 pages. 1974. DM 26,–

Vol. 366: R. Steinberg, Conjugacy Classes in Algebraic Groups (Notes by V. V. Deodhar). VI, 159 pages. 1974. DM 18,–

Vol. 367: K. Langmann und W. Lütkebohmert, Cousinverteilungen und Fortsetzungssätze. VI, 151 Seiten. 1974. DM 16,–

Vol. 368: R. J. Milgram, Unstable Homotopy from the Stable Point of View. V, 109 pages. 1974. DM 16,–

Vol. 369: Victoria Symposium on Nonstandard Analysis. Edited by A. Hurd and P. Loeb. XVIII, 339 pages. 1974. DM 26,–

Vol. 370: B. Mazur and W. Messing, Universal Extensions and One Dimensional Crystalline Cohomology. VII, 134 pages. 1974. DM 16,–